Mathematical
Surveys
and
Monographs
Volume 174

Hereditary Noetherian Prime Rings and Idealizers

Lawrence S. Levy

J. Chris Robson

American Mathematical Society
Providence, Rhode Island

2010 *Mathematics Subject Classification.* Primary 16–XX, 18Gxx, 19Axx.

Author photos courtesy of J. Chris Robson

For additional information and updates on this book, visit
www.ams.org/bookpages/surv-174

Library of Congress Cataloging-in-Publication Data

Levy, Lawrence S., 1933–
Hereditary noetherian prime rings and idealizers / Lawrence S. Levy, J. Chris Robson.
p. cm. — (Mathematical surveys and monographs ; v. 174)
Includes bibliographical references and index.
ISBN 978-0-8218-5350-4 (alk. paper)
1. Algebra, Homological. 2. Grothendieck groups. I. Robson, J. C. (James Christopher), 1940– II. Title.

QA169.L445 2011
512′.46—dc22

2011003905

♾ The paper used in this book is acid-free and falls within the guidelines established to ensure permanence and durability.
Visit the AMS home page at http://www.ams.org/

10 9 8 7 6 5 4 3 2 1   16 15 14 13 12 11

Contents

Introduction and Standard Notation

HNP (hereditary Noetherian prime) rings are perhaps the only noncommutative Noetherian rings whose projective modules, both finitely generated and infinitely generated, have nontrivial direct sum behaviour and a structure theorem describing that behaviour. This book provides a full account of that structure and behaviour, as well as much of what is known about nonprojective finitely generated modules, injective modules, and the ring structure that underlies all of this. In doing so, it revisits several areas of theory which are used in the solution and which have previously only been available scattered through a large number of research papers. Happily, but perhaps not unexpectedly, this has allowed substantial simplifications, improved clarity, fresh insights and new results.

History[1]**:** We start by providing a historical perspective. Steinitz and others, by the early 1900s, had extended the fundamental theorem of abelian groups to a structure theorem for finitely generated modules over commutative Dedekind domains. There are many equivalent definitions of Dedekind domains. The most apt for our purposes is that they are commutative hereditary Noetherian integral domains; but they can equally well be defined, for example, as integrally closed Noetherian domains of Krull dimension 1 or as Noetherian domains whose nonzero ideals are all invertible.

Steinitz's theorems include a precise description of all isomorphism classes of finitely generated torsionfree modules and their direct sums. In modern terminology, the description involves just two invariants, both additive in direct sums: the rank (i.e. the number of indecomposable summands, each isomorphic to an ideal) and an element of an abelian group called the ideal class group of the Dedekind domain. Somewhat more recent results are that direct sums of ideals of a Dedekind domain coincide with its projective modules and that all infinitely generated projective modules are free.

In the light of Goldie's Theorem of 1958, it was natural to seek a similar theory for an appropriate class of noncommutative Noetherian rings which included all Dedekind domains. To a greater or lesser extent, this was considered and accomplished during the 1960s and 1970s for two special types of HNP ring: *classical hereditary orders* (HNP rings that are finitely generated modules over their centre, which is a Dedekind domain) and noncommutative *Dedekind prime rings* (Noetherian prime rings whose nonzero one-sided ideals are not only projective but also generators). However, the theory for a general HNP ring was much more elusive. There was spasmodic progress over a long period by several authors with the final complete solution only coming in 1999/2000 in a series of papers by the authors of this book.

[1]Further historical comments and precise references appear in the sections entitled 'Notes on Chapter(s)'

The structure: Let R be an HNP ring, and P_R a nonzero finitely generated projective module. We now give a description of the structure of P and then outline this book's approach to it. We associate two isomorphism invariants with P: the 'Steinitz class' $\mathcal{S}(P)$, and the 'genus' $\Psi(P)$. The Steinitz class $\mathcal{S}(P)$ is an element of an abelian group $G(R)$ called the 'ideal class group' of R (because it becomes that in the commutative case); and the genus is a noncommutative replacement for rank. It is a function from the prime spectrum of R to the non-negative integers which, in the commutative case, is simply a constant function equal to the rank of P.

The structure theory first determines precisely what functions can occur as genera. Then it shows that these invariants $\mathcal{S}(P)$ and $\Psi(P)$ can be independently assigned, are additive in direct sums and, if $\operatorname{udim}(P) > 1$ (i.e. P is not indecomposable), they determine P up to isomorphism.

Outline: Perhaps surprisingly, the book starts with what might appear a diversion: a relatively complete account in Chapters 1,2 of idealizer subrings of an arbitrary ring, S say. However, idealizers play a crucial role in the theory which follows; so it seems best to meet them first. Some explanation is needed to see why they are relevant. In studying the structure of an HNP ring R and its modules, we need to study all rings S between R and its Goldie quotient ring R_{quo}. The main results depend upon the case when S_R is finitely generated; and we show that, in that situation, R can be obtained from S by iterating the process of forming a special type of idealizer subring that we call a 'basic idealizer'. It turns out that the larger ring here is, in a sense, a localization of the smaller one. The extremely tight relationship between these rings R, S and their simple modules, which is made explicit in §13 and §14, ultimately allows a reduction to the special case when R is a Dedekind prime ring, where the module structure is readily determined.

Chapter 4 introduces the notion of a 'tower', described very roughly as follows. A nonzero extension relation between certain pairs of simple modules over the HNP ring R partitions the simple R-modules into finite sets, each such set being either cyclicly or linearly ordered. We call these sets 'cycle towers' and 'faithful towers', respectively. We omit further comment on these here, except to say that they are involved in either the statement or proof of almost every important result from there on. Then Chapter 5, 'Integral overrings', introduces a noncommutative extension of the familiar concept of the same name in commutative rings. After all this preparation, Chapter 6 studies the isomorphism invariants of our projective module P_R and proves Theorem 35.13, the main structure theorem. The remaining HNP parts of this book can be viewed as applications of its first six chapters.

Some appendices provide material required in the earlier work but not about idealizers or HNP rings. In particular, we give a self-contained account of modules over Artinian serial rings, in a form suitable for our work. This is relevant because all proper factor rings of an HNP ring are Artinian serial rings. Another topic, which has more than one application here, concerns a surjective homomorphism $P \twoheadrightarrow U$ of modules over an arbitrary ring S, with P projective and U of finite length, and deals with questions about the lifting to P of decompositions of U.

0.1. Acknowledgments. Writing this book has diverted the authors from other activities. We both thank Ruth and Lyn for their gracious acceptance of this.

0.2. Standard Notation and Terminology. We list here some notation and terminology that is used throughout the book. See also the index sections at the end of the book; and see [**McConnell-Robson 01**], referred to henceforth as [**McR 01**], for other standard ring-theoretic terminology and notation.

By *ring* we mean a ring with an identity element, and a *subring* shares the same identity element.

An *HNP (hereditary, Noetherian, prime) ring* is a right and left Noetherian prime ring in which every right and every left ideal is a projective module. To avoid trivialities, we also assume that the ring is not Artinian. If, further, it is an integral domain, we term it an HNP domain

Throughout, as in the preceding definition, omission of the term 'right' or 'left' denotes that the ring satisfies both right and left conditions.

$M_n(R)$ denotes the ring of $n \times n$ matrices over a ring R.

Module means 'right module' unless otherwise specified; and M_R denotes a right R-module.

Ideal means 'two-sided ideal' unless otherwise specified.

(S, R)-*bimodule* means a bimodule which is a left S-module and a right R-module.

Maps of right modules (i.e. right module homomorphisms) will be viewed and written as acting on the left. In particular, this makes every M_R into an (S, R)-bimodule where $S = \mathrm{End}(M_R)$, the endomorphism ring of M.

Likewise, if N is a left R-module, denoted N_R, then its maps are viewed as acting from the right and then N becomes an (R,S)-bimodule where $S = \mathrm{End}(_RN)$.

$\mathrm{rad}(M)$ and $J(M)$ denote the Jacobson radical of a module M and $\mathrm{top}(M)$ denotes $M/\mathrm{rad}(M)$.

$\mathrm{soc}(M)$ and $\mathrm{soc}_R(M)$ denote the socle of an R-module M.

$M^{(n)}$ denotes the direct sum of n copies of M.

ann, ann_R, rann, lann denote, respectively, the annihilator, the annihilator in R, the right annihilator and the left annihilator.

$\lambda(M_R)$ denotes the length of a module M_R of finite length.

$E(M)$ denotes the injective envelope of a module M.

fd = flat dimension and pd = projective dimension.

$\subset$ means strict containment.

$\hookrightarrow$ and $\twoheadrightarrow$ denote maps that are, respectively, (1,1) and onto.

Part 1

Idealizer Rings

CHAPTER 1

Basic Idealizers

This chapter introduces the idealizer subring $\mathbb{I}_S(A)$ of a right ideal A in a ring S. Its main aim is to investigate, in §4 and §5, the 'basic idealizer' case — when A is not two-sided and $S/A \cong U^{(n)}$ for some n and some simple module U_S — this being the case that underpins much that follows. The preceding sections lead up to this by considering the relationship between S and $\mathbb{I}_S(A)$ under less stringent restrictions on A.

1. Idealizers and Endomorphisms

In this preliminary section we introduce the notion of an idealizer and link this with endomorphisms.

1.1. DEFINITION. Let A be a right ideal of a ring S. The subring $\mathbb{I}(A)$ or, if the ring concerned needs emphasis, $\mathbb{I}_S(A)$ defined by

$$\mathbb{I}(A) = \{x \in S \mid xA \subseteq A\}$$

is called the *idealizer* of A in S. There is a corresponding notion for any left ideal B of S with its *(left) idealizer* being denoted by $\mathbb{I}(B)$, $\mathbb{I}_S(B)$ or $\mathbb{I}({}_SB)$. For example, if

$$S = M_2(\mathbf{Z}), \quad A = \begin{pmatrix} \mathbf{Z} & \mathbf{Z} \\ 0 & 0 \end{pmatrix} \quad \text{and} \quad B = \begin{pmatrix} 0 & \mathbf{Z} \\ 0 & \mathbf{Z} \end{pmatrix},$$

then A is a right ideal of S, B a left ideal and $\mathbb{I}(A) = \begin{pmatrix} \mathbf{Z} & \mathbf{Z} \\ 0 & \mathbf{Z} \end{pmatrix} = \mathbb{I}(B)$.

1.2. REMARK. We note two immediate consequences of the definition:

(i) $\mathbb{I}(A)$ is a ring, namely the largest subring of S in which A is an ideal, and the factor ring $\mathbb{I}(A)/A$ is sometimes called the *eigenring* of A;

(ii) if B is an ideal of S with $B \subseteq A$ then $\mathbb{I}_{S/B}(A/B) = (\mathbb{I}_S(A))/B$.

One can identify $\operatorname{End}(S_S)$ with S acting on itself via left multiplication; and then $\mathbb{I}(A) = \{\lambda \in \operatorname{End}(S_S) \mid \lambda(A) \subseteq A\}$. This leads to the following result.

1.3. LEMMA. *Let A, B be right ideals of a ring S.*

(i) *$\mathbb{I}(A)/A \cong \operatorname{End}_S(S/A)$ acting via left multiplication.*

(ii) *$\{s \in S \mid sA \subseteq B\}/B \cong \operatorname{Hom}_S(S/A, S/B)$ acting via left multiplication.*

PROOF. First we consider (ii). To simplify notation, let $\operatorname{Hom}_S(S/A, S/B) = H$ and $\{s \in S \mid sA \subseteq B\} = C$; so $B \subseteq C$. Given any element $c \in C$, let λ_c denote the endomorphism of S given by left multiplication by c. Now λ_c maps A into B; so λ_c restricts to an element of H. This restriction is the zero homomorphism precisely when $cS \subseteq B$; thus we have obtained an injective map from C/B to H. We wish to show it is surjective. So, let $h \in H$ and suppose that $h(1 + A) = x + B$. Then

$h(s + A) = xs + B$ for all $s \in S$ and in particular, since $h(a + A) = h(0) = 0$ for each $a \in A$, then $xA \subseteq B$. Thus $x \in C$ and h is the restriction of λ_x. Thus the map from C/B to H is indeed surjective.

We note that (i) is the special case of (ii) when $A = B$. It can be checked that the isomorphism given in (i) is a ring isomorphism. □

1.4. COROLLARY. *Let A be a right ideal of any ring S, and let C/A be a fully invariant submodule of $(S/A)_S$. Then $\mathbb{I}_S(A) \subseteq \mathbb{I}_S(C)$.*

PROOF. For any $x \in \mathbb{I}_S(A)$, left multiplication by x is an endomorphism of $(S/A)_S$. Full invariance of C/A therefore implies that $x(C/A) \subseteq C/A$. Since $xA \subseteq A$ we also have $xC \subseteq C$, as desired. □

The existence of a link between idealizers and endomorphisms leads to the Dual Basis Lemma playing an important role in the basic theory of idealizer rings. We therefore include here a brief account of this.

1.5. DEFINITIONS (Duals and related products). Let R be any ring and M a right R-module. The *dual* of M is denoted by M^* , or M_R^* if the ring needs emphasis, and is defined to be $\operatorname{Hom}(M, R)$. Recall from 0.2 that M is an (S, R)-bimodule where $S = \operatorname{End} M_R$. Similarly, M^* may be viewed as an (R, S)-bimodule: for $f \in M^*$, $r \in R$, and $\phi \in S$, rf is the map $m \to r{\cdot}f(m)$ in M^*, and $f\phi$ is the map $m \to f(\phi(m))$ in M^*. Verification of 'bimodule associativity', that is $(rf)\phi = r(f\phi)$, is straightforward.

We often make use of two related 'products'. One is $M \otimes_R M^* \to MM^* \subseteq S$: for $m_1 \in M$ and $f \in M^*$, m_1f is the R-endomorphism given by $m \to m_1{\cdot}f(m)$. The other is the more obvious $M^* \otimes_S M \to M^*M \subseteq R$, defined by $fm_1 = f(m_1)$. As usual, the notation MM^* and M^*M denotes the additive groups generated by the 'monomials' that define them. We note that these products are bimodule maps. Therefore MM^* and M^*M are ideals of the rings S and R respectively. The overall facts about the two rings and the two modules concerned, together with their products, can be summed up by saying that the set of formal 2×2 matrices $\begin{pmatrix} R & M^* \\ M & S \end{pmatrix}$ forms a ring via the given products. This ring is sometimes called the *ring of the Morita context.* (See [**McR 01**, 1.1.6, 1.9.1] for more details.)

1.6. LEMMA (Dual Basis Lemma).

(i) *A module M_R over any ring R is finitely generated and projective if and only if $MM^* = \operatorname{End}(M_R)$ (equivalently, $1_{\operatorname{End}(M_R)} \in MM^*$).*

(ii) *If M_R is projective with a set of n generators, then the same is true of ${}_RM^*$; and $M \cong M^{**}$ via the canonical identification which maps $m \in M$ to the map $M^* \to R$ given by $f \to f(m)$.*

PROOF. (i) Suppose first that $1 \in MM^*$; so we have an expression $1 = \sum_{i=1}^n m_if_i$ with each $m_i \in M$ and $f_i \in M^*$. Let e_i be the element of $R^{(n)}$ whose i^{th} coordinate is 1 and whose other coordinates are zero. Now consider the map $\beta{:}R^{(n)} \twoheadrightarrow M$ defined by $\beta(e_i) = m_i$. Bearing in mind the equation $m = 1m = \sum_i m_if_i(m)$, we define $\alpha{:}M \to R^{(n)}$ by $\alpha(m) = \sum_{i=1}^n e_if_i(m)$ and note that $\beta\alpha = 1_M$. Thus M, being isomorphic to a direct summand of $R^{(n)}$, is projective and is generated by the elements $\{m_i \mid i = 1, \dots, n\}$.

Conversely, suppose that M is finitely generated and projective, generated by the elements $m_1, \dots, m_n$. The map $g : R^{(n)} \twoheadrightarrow M$, defined by $g(e_i) = m_i$ with the

e_i as above, is a surjection which is split by some injective map $f : M \hookrightarrow R^{(n)}$. Let f_i: $M \to R$ be the ith coordinate map of f. Then, for each $m \in M$, we have $f(m) = \sum_i e_i f_i(m)$ and therefore

$$m = gf(m) = \sum_i g(e_i)f_i(m) = \sum_i m_i f_i(m).$$

Therefore we have the desired relation $1 = \sum_i m_i f_i$.

(ii) The canonical right R-module homomorphism $\theta : M \to M^{**}$ sends each $m \in M$ to the homomorphism $\theta(m)$: $M^* \to {}_RR$ given by $(f)\theta(m) = f(m)$, (with the map of the left module M^* written on the right). We will show θ is an isomorphism. We know, from (i), that there is an expression $1 = \sum_i m_i f_i \in \operatorname{End}(M_R)$, since M_R is finitely generated and projective. Therefore $m = \sum_i m_i f_i(m)$ for every $m \in M$. If $\theta(m) = 0$, then $(f_i)\theta(m) = f_i(m) = 0$ for every i, and therefore $m = \sum_i m_i f_i(m) = 0$. Thus θ is (1,1).

Next, take any left R-module homomorphism α: $M^* \to {}_RR$. For every $g \in M^*$ we have $g = \sum_i g(m_i)f_i$. Applying α yields

$$(g)\alpha = \sum_i g(m_i)(f_i)\alpha.$$

However, if we let $m = \sum_i m_i(f_i)\alpha$ then

$$(g)\theta(m) = g(m) = \sum_i g(m_i)(f_i)\alpha = (g)\alpha.$$

Thus θ is onto and so is an isomorphism.

Finally, the fact that the left R-module ${}_RM^* = (M_R)^*$ is projective and generated by $f_1, \dots, f_n$ follows from the left-handed version of (i) applied to M^* and the isomorphism $(M^*)^* \cong M$. □

Indeed, for finitely generated projective right R-modules M, the double duality functor $M \twoheadrightarrow M^{**}$ is equivalent to the identity functor. However, this fact will not be needed here.

1.7. Corollary. *For every finitely generated projective M_R, the 'trace' ideal $\operatorname{tr}_R(M) = M^*M$ of M is idempotent.*

Proof. The associativity of the Morita ring, mentioned in 1.5, provides the equations $(M^*M)(M^*M) = M^*(MM^*)M = M^* \operatorname{End}(M_R)M = M^*M$. □

1.8. Remark. Suppose that R is a field and the elements m_i in the Dual Basis Lemma form a basis of M. Then the 'dual base' $\{m_i\}$ and $\{f_i\}$ above form a pair of dual bases in the sense of linear algebra.

To see this start with the relation $m = \sum_i m_i f_i(m)$, which holds for every $m \in M$. Taking $m = m_j$ for some j yields $m_j = \sum_i m_i f_i(m_j)$. Then linear independence of basis elements of vector spaces yields $f_i(m_j) = \delta_{ij}$, the Kronecker delta, for every i, j.

2. Subidealizers of Generative Right Ideals

The rich theory of idealizers which fills this chapter is built upon the case when the right ideal has certain special properties. This section introduces one of the relevant properties — being generative — and delineates the consequences for the idealizer ring and, usefully, for a more general class of rings. We now give the appropriate definitions.

2.1. DEFINITION. Let A be a right ideal of a ring S. Any subring T such that $\mathbb{I}_S(A) \supseteq T \supset A$ is called a *subidealizer* of A. Hence $1_S \in T$ and A is an ideal of T.

2.2. DEFINITION. A right ideal A of a ring S is said to be *generative* provided that $SA = S$. For example:

(i) if S is a simple ring then every nonzero right ideal is generative;

(ii) if $S = M_n(D)$ for some ring D and $a \in S$ is a matrix which has an entry which is a unit of D then aS is a generative right ideal of S — thus if $S = M_2(\mathbf{Z})$ then $e_{11}S = \begin{pmatrix} \mathbf{Z} & \mathbf{Z} \\ 0 & 0 \end{pmatrix}$ is a generative right ideal.

We should note that any generative right ideal A is automatically a generator right module; i.e. the ring is a sum of homomorphic images of A_S. However, a right ideal which is a generator right module is not necessarily generative: any proper nonzero ideal in any commutative principal ideal domain is a counterexample.

2.3. PROPOSITION. *Let A be a generative right ideal of a ring S and let R be a subidealizer of A. Then:*

(i) *S_R and ${}_RA$ are finitely generated projective;*

(ii) *$S \otimes_R S \cong S \otimes_R A \cong S$ via multiplication;*

(iii) *$(S/R) \otimes_R S = 0 = (R/A) \otimes_R A$ and $(S/R) \otimes_R A \cong S/A \cong (R/A) \otimes_R S$.*

PROOF. (i) First consider S_R. By the Dual Basis Lemma [1.6] it suffices to show that $1 \in SS^*$ where 1 denotes the identity endomorphism of S_R. The hypothesis $SA = S$ yields an expression $\sum_{i=1}^n s_i a_i = 1$. Since $A \subseteq R$, left multiplication by any element of A is a map in $S^* = \operatorname{Hom}(S_R, R_R)$. Moreover, left multiplication by each product $s_i a_i$ is an endomorphism of S_R. Therefore the expression $\sum_{i=1}^n s_i a_i = 1$ shows that S_R is finitely generated projective.

Likewise right multiplication by any $s \in S$ gives a map in $\operatorname{Hom}({}_RA, {}_RR)$. Therefore the expression $\sum_{i=1}^n s_i a_i = 1$ shows that $1 \in A^*A$ and hence ${}_RA$ is finitely generated projective.

(ii) We note first that, since S_R is projective and thus flat, the embeddings ${}_RA \subseteq {}_RR \subseteq {}_RS$ yield embeddings $S \otimes_R A \subseteq S \otimes_R R \subseteq S \otimes_R S$. Then, viewing all terms involved as subsets of $S \otimes_R S$, we see that

$$S \otimes_R S = SA \otimes_R S = S \otimes_R AS = S \otimes_R A = S \otimes_R AR = SA \otimes_R R = S \otimes_R R.$$

The multiplication map $S \otimes_R S \to S$ when restricted to $S \otimes_R R$ is the canonical isomorphism $S \otimes_R R \cong S$. This, by the equalities above, yields the isomorphisms $S \otimes_R S \cong S \cong S \otimes_R A$.

(iii) The short exact sequence of right R-modules

$$0 \to R \to S \to S/R \to 0$$

when tensored on the right by ${}_RS$, gives the exact sequence

$$R \otimes S \to S \otimes S \to (S/R) \otimes S \to 0.$$

But (ii) shows that $S \otimes S \cong S$ via multiplication, and the image in S of $R \otimes S$ is $RS = S$. Thus $(S/R) \otimes S = 0$.

On the other hand if we had tensored the given short exact sequence by A we would have obtained the exact sequence

$$R \otimes A \to S \otimes A \to (S/R) \otimes A \to 0.$$

In this case, $S \otimes A \cong S$ via multiplication; and the image in S of $R \otimes A$ is $RA = A$. Thus $(S/R) \otimes A \cong S/A$.

The remaining facts are proved similarly using the short exact sequence

$$0 \to A \to R \to R/A \to 0$$

and tensoring it on the right by S or A. $\square$

2.4. REMARK. If R is a subring of a ring S, then S is sometimes called a *left localization* of R if $S \otimes_R S \cong S$ and S_R is flat. If, further, S_R is finitely generated projective, S is a *finite left localization* of R. This applies, of course, to the rings R and S in 2.3 above.

2.5. PROPOSITION. *Let R be a subring of a ring S and suppose that $S \otimes_R S \cong S$ via multiplication. Let M and N be right S-modules and L a left S-module. Then:*

(i) *$M \otimes_R S \cong M$ via multiplication;*
(ii) *if M_R is projective then M_S is projective;*
(iii) *$\mathrm{Hom}_R(M, N) = \mathrm{Hom}_S(M, N)$;*
(iv) *if M_R is injective then M_S is injective;*
(v) *$M \otimes_R L \cong M \otimes_S L$.*

PROOF. (i) $M \otimes_R S \cong M \otimes_S S \otimes_R S \cong M \otimes_S S \cong M$ via multiplication.

(ii) If M_R is a direct summand of a free right R-module, then tensoring over R by S shows that M_S is a direct summand of a free right S-module.

(iii) Let $\phi\colon M \to N$ be an R-homomorphism. It induces the S-homomorphism $\phi' = \phi \otimes 1$ from $M \otimes_R S$ to $N \otimes_R S$. Then after identifying $M \otimes_R S$ and $N \otimes_R S$ with M and N respectively, using (i), we get $\phi' = \phi$. In other words, ϕ is an S-homomorphism.

(iv) Suppose first that M_R is injective. Let I be a right ideal of S and let $\alpha : I \to M$ be an S-homomorphism. To demonstrate that M_S is injective, we need only show that α can be lifted to a homomorphism $S \to M$. Of course α is also an R-module homomorphism which, since M_R is injective, lifts to an R-homomorphism $S \to M$; and this, by (iii), is also an S-homomorphism.

(v) $M \otimes_R L \cong (M \otimes_S S) \otimes_R (S \otimes_S L) \cong M \otimes_S S \otimes_S L \cong M \otimes_S L$. $\square$

Next we apply this to a subidealizer of a generative right ideal of a ring where more can be shown.

2.6. PROPOSITION. *Let A be a generative right ideal of a ring S and let R be a subidealizer of A. Then:*

(i) *all the statements in 2.5 hold;*
(ii) *for each nonzero ideal B of S, $B \cap R \neq 0$;*
(iii) *if M is a right S-module then $M \otimes_R A \cong M$ via multiplication; and M_R is projective if and only if M_S is projective;*
(iv) *if $S \supseteq {}_RX \supseteq A$ then $S \otimes_R X \cong SX = S$; $({}_RX)^* \cong \{s \in S \mid Xs \subseteq R\}$ via right multiplication; and, viewing this isomorphism as an identification, $S \supseteq X^* \supseteq A$;*
(v) *$({}_RA)^* = S$ and $(S_R)^* = A$, in each case acting via multiplication;*

(vi) *if M is a right S-module and L a left S-module then* $\operatorname{Tor}_n^R(M,L) \cong \operatorname{Tor}_n^S(M,L)$ *for all n.*

PROOF. Note first that 2.3 shows not only that $S \otimes_R S \cong S$ via multiplication but also that S_R is projective and hence flat. These facts will be used throughout this proof without further comment.

(i) Clear.

(ii) Note that $AB \subseteq B$ and $SA = S$. Thus $0 \neq B = SB = SAB \subseteq SB = B$ and so $0 \neq AB \subseteq B \cap R$.

(iii) The first claim is clear since $M \otimes_R A \cong M \otimes_S S \otimes_R A \cong M \otimes_S S \cong M$ via multiplication. For the second claim, note that since S_R is projective, any free right S-module will be projective over R and so too is a direct summand such as M_S. The converse is covered directly by 2.5(ii).

(iv) Since S_R is flat, $S \otimes_R X \subseteq S \otimes_R SX$. However

$$S \otimes_R SX \cong S \otimes_R S \otimes_S SX \cong S \otimes_S SX \cong SX$$

via multiplication. Since $S \otimes_R X \twoheadrightarrow SX$ under multiplication, we deduce that this epimorphism is in fact an isomorphism. Finally, $S \supseteq SX \supseteq SA = S$ and so $SX = S$.

Next we turn to $({}_RX)^*$. If $Xs = 0$ for some $s \in S$ then $0 = SXs \supseteq SAs = Ss$ and so $s = 0$. Thus we need only show that each $\phi \in \operatorname{Hom}({}_RX, R)$ is given by right multiplication by some element of S. However, given ϕ, then $1 \otimes \phi : S \otimes_R X \to S \otimes_R R$. Now we have just seen that $S \otimes_R X \cong SX = S$. Also $S \otimes_R R \cong SR = S$. Thus $1 \otimes \phi : S \to S$ and so is given by right multiplication, as required.

(v) It is immediate from (iv) that $({}_RA)^* = S$, acting via right multiplication. Consequently $(S_R)^* = A$ acting via left multiplication.

(vi) We start with any short exact sequence $0 \to K_S \to P_S \to M_S \to 0$ with P_S projective. Since P_S is flat, the long exact Tor sequence (described in 53.17) demonstrates that $\operatorname{Tor}_1^S(M,L) \cong \ker(K \otimes_S L \to P \otimes_S L)$ and likewise, since (iii) shows that P_R is projective and so flat, $\operatorname{Tor}_1^R(M,L) \cong \ker(K \otimes_R L \to P \otimes_R L)$. Hence, using the isomorphisms given by 2.5(v) above, $\operatorname{Tor}_1^S(M,L) \cong \operatorname{Tor}_1^R(M,L)$. Similarly $\operatorname{Tor}_{k+1}^S(M,L) \cong \operatorname{Tor}_k^S(K,L)$ and $\operatorname{Tor}_{k+1}^R(M,L) \cong \operatorname{Tor}_k^R(K,L)$. Induction on k allows us to suppose that $\operatorname{Tor}_k^S(K,L) \cong \operatorname{Tor}_k^R(K,L)$ and the result follows. □

Further results along the lines of 2.5 and 2.6, but under stronger hypotheses, will be found in 4.13.

3. Idealizers of Isomaximal and Semimaximal Right Ideals

In this section we introduce the second property of the right ideal A which will be heavily involved throughout the remainder of the chapter.

3.1. DEFINITIONS. Let A be a right ideal of a ring S. If the right S-module S/A is semisimple (necessarily of finite length) we say A is *semimaximal*. If, further, $S/A \cong U^{(n)}$ for some simple module U_S and some $n \geq 1$ (and so S/A is semisimple isotypic), we say A is *isomaximal of type U*.

We follow the convention that the zero module is semisimple, being the empty direct sum of simple modules. Thus S itself is semimaximal.

For example, if D is a division ring and $S = M_n(D)$, every right ideal is isomaximal and, if nonzero, is also generative. If $S = M_2(\mathbf{Z})$ and

$$A = \begin{pmatrix} \mathbf{Z} & \mathbf{Z} \\ 2\mathbf{Z} & 2\mathbf{Z} \end{pmatrix}, \quad B = \begin{pmatrix} \mathbf{Z} & \mathbf{Z} \\ 6\mathbf{Z} & 6\mathbf{Z} \end{pmatrix}, \qquad C = M_2(2\mathbf{Z}) \quad \text{and} \quad D = M_2(4\mathbf{Z}),$$

then A is a generative isomaximal right ideal, B is a generative semimaximal right ideal, C is an isomaximal right ideal which is not generative and D is a right ideal which is neither semimaximal nor generative.

There is a useful criterion for being semisimple isotypic.

3.2. LEMMA. *If a nonzero module M has finite length and no proper, fully invariant submodules, then M is semisimple isotypic.*

PROOF. Since $\operatorname{soc}(M)$ is fully invariant and nonzero, then $M = \operatorname{soc}(M)$ and so M is semisimple. If M had more than one isotypic component, then each of these would be a proper, fully invariant submodule. Hence the result. $\square$

The next result shows that, in studying the nature of the idealizer of an isomaximal right ideal A, there is no loss in assuming that A is generative, as defined in 2.2.

3.3. LEMMA. *Let A be an isomaximal right ideal of a ring S. If A is not generative, then A is an ideal and hence $\mathbb{I}_S(A) = S$.*

PROOF. Let $S/A = U^{(n)}$, with U simple. Then either $UA = 0$ or $UA = U$. If $UA = 0$, then $(S/A)A = 0$ and therefore $SA \subseteq A$; that is, A is an ideal. If $UA = U$ then $(S/A)A = S/A$, which implies $SA = S$. $\square$

For future use, we note the following related fact.

3.4. LEMMA. *Let A be an isomaximal right ideal of a ring S. Then A is generative if and only if there is an isomaximal right ideal, A' say, with $A' \subset A$.*

PROOF. Let $B = \operatorname{ann}(S/A) = \operatorname{ann}(U)$.

Suppose that A is generative. Now $B = \cap C_i$ where the intersection is of all maximal right ideals C_i with $S/C_i \cong U$. Since A is generative, A is not an ideal and so $A \supset B$. Thus there is a maximal right ideal C such that $S/C \cong U$ and $C \not\supseteq A$. We let $A' = A \cap C$.

Conversely, if A is not generative then, by 3.3, A is an ideal and so $A = B$. Evidently $B \subseteq A'$ for every isomaximal right ideal A' of type U. $\square$

The next two results start the process of reducing questions about idealizers of semimaximal right ideals to the generative isomaximal case.

3.5. LEMMA. *Let S be a ring and $A \neq S$ a semimaximal right ideal. Then:*

(i) *$\mathbb{I}_S(A)/A$ is a semisimple Artinian ring and is simple if A is isomaximal;*

(ii) *there is a finite set of isomaximal right ideals A_i of distinct types such that $A = \cap A_i$;*

(iii) *A is generative if and only if each A_i is generative.*

PROOF. (i) $\mathbb{I}_S(A)/A \cong \operatorname{End}(S/A)$ by 1.3 and hence is semisimple Artinian and is simple if A is isomaximal.

(ii) Let U_i be one of the simple factor modules of S/A and

$$A_i = \cap\{M_S \mid S \supset M \supseteq A, \ \ S/M \cong U_i\}.$$

Then A_i is isomaximal of type U_i and $A = \cap A_i$.

(iii) Evidently, if $SA = S$ then $SA_i = S$. Now suppose that each A_i is generative and yet A is not. We aim at a contradiction. Since $SA \neq S$ then $A \subseteq SA \subseteq M$ for some maximal ideal M of S. Since $(S/A)_S$ has finite length and M is a maximal ideal, S/M is a simple Artinian ring. Thus $(S/M)_S$ is semisimple isotypic. But $M \supseteq A$, so the simple type of $(S/M)_S$ must be that of some A_i. To ease notation, suppose that $i = 1$. Then $M = \operatorname{ann} U_1 \subseteq A_1$ and also $M \neq A_1$ since A_1 is generative. Thus A_1/M is nonzero isotypic of type U_1. This contradicts the fact that A_1/A is semisimple with composition factors coming from $U_2, \ldots, U_k$ and so completes the proof. □

3.6. PROPOSITION. *Let S be a ring and let $\{A_i \mid i = 1, \ldots, k\}$ be a finite set of isomaximal right ideals of distinct simple type. Let $A = \cap_1^k A_i$ and let A' be the (possibly empty) intersection of all those A_i which are generative. Then A' is a generative semimaximal right ideal and $\mathbb{I}_S(A) = \cap_i \mathbb{I}_S(A_i) = \mathbb{I}_S(A')$.*

PROOF. First we show that $\mathbb{I}_S(A) = \cap \mathbb{I}_S(A_i)$. Let $x \in \mathbb{I}_S(A)$. From 1.3 we know that left multiplication by x induces an endomorphism of S/A. Evidently A_i/A is an invariant submodule of S/A; so 1.4 shows that $\mathbb{I}_S(A) \subseteq \mathbb{I}_S(A_i)$ and hence $\mathbb{I}_S(A) \subseteq \cap \mathbb{I}_S(A_i)$. On the other hand, if $x \in \mathbb{I}_S(A_i)$ for each i then $xA_i \subseteq A_i$ and so $x(\cap_i A_i) \subseteq \cap_i A_i$.

Next note, by 3.3, that $\mathbb{I}_S(A_i) = S$ whenever A_i is not generative. Hence $\mathbb{I}_S(A) = \cap\{\mathbb{I}_S(A_i) \mid A_i \text{ generative}\}$ and the latter term equals $\mathbb{I}_S(A')$ by the first paragraph of this proof. □

Since our interest in idealizers lies in the rings they provide, the preceding proposition suggests that we should concentrate on idealizers of generative isomaximal right ideals. That is the focus of the next few sections. In fact we show, in 8.10, that every semimaximal idealizer can also be obtained by forming a succession of idealizers of generative isomaximal right ideals.

4. Basic Idealizers

We now turn to the study of the idealizer R of an isomaximal generative right ideal of a ring S. The section establishes, in 4.4 and 4.8, a remarkably tight connection between the simple modules of R and S, and then explores some of its consequences.

We start with some useful terminology.

4.1. DEFINITION. We say that $R = \mathbb{I}_S(A)$ *is a basic idealizer of type U* if A is a generative isomaximal right ideal of S of type U. Thus $S/A \cong U^{(n)}$ for some $n \geq 1$.

4.2. LEMMA. *Let $R = \mathbb{I}_S(A)$ be a basic idealizer of type U.*

(i) *R/A is a simple Artinian ring. Indeed if $S/A \cong U^{(n)}$ then $R/A \cong M_n(\operatorname{End}(U))$.*

(ii) *A is a maximal ideal of R and is idempotent.*

PROOF. (i) This follows from 1.3.

(ii) Note that, since A is generative, $SA = S$; and so $A^2 = ASA = AS = A$. □

Next we demonstrate some symmetry in certain special cases. Note that, if S is simple Artinian, then $\mathbb{I}(A)$ is basic for every right ideal $A \neq 0, S$.

4.3. LEMMA. *Let S be a ring and $R = \mathbb{I}_S(A)$ be a basic (right) idealizer.*

(i) *Suppose that S is a simple Artinian ring and $A = eS$ with $e = e^2$. Then $R = \mathbb{I}_S(eS) = \mathbb{I}_S(S(1-e))$ and so is a basic left idealizer from S.*

(ii) *More generally, suppose that $S/\operatorname{ann}(S/A)_S$ is simple Artinian. Then R is a basic left idealizer from S.*

PROOF. (i) Let $e' = 1 - e$. From $S = (eSe) \oplus (eSe') \oplus (e'Se) \oplus (e'Se')$, one easily calculates that $\mathbb{I}_S(eS) = eS + Se' = \mathbb{I}_S(Se')$. Also, since $eS \neq 0, S$ then the same is true of Se'.

(ii) Let $C = \operatorname{ann}(S/A)_S$. Evidently $C \subseteq R$ and (as was noted in 1.2) one can check that $R/C = \mathbb{I}_{S/C}(A/C)$. This, by (i), is a basic (left) idealizer for a generative isomaximal left ideal of S/C. This lifts to a left ideal A' with $C \subset A' \subset S$; and then $R = \mathbb{I}(A')$. $\square$

The next result describes all simple R-modules; and their isomorphism types are dealt with by 4.8.

4.4. THEOREM. *Let $R = \mathbb{I}_S(A)$ be a basic idealizer of type U, with $S/A \cong U^{(n)}$.*

(i) *If X is a simple S-module not isomorphic to U, then X_R is simple.*

(ii) *U_R has a unique composition series of length 2; and if its top and bottom R-composition factors are named V and W respectively, then $V \ncong W$ and*

$$(S/R)_R \cong V^{(n)} \qquad \text{and} \qquad (R/A)_R \cong W^{(n)}. \tag{4.4.1}$$

(iii) *Every simple right R-module is of the form V, W or X as described in (i) and (ii).*

PROOF. We start with some discussion related to both (i) and (ii). First note that 4.2(i) shows that R/A, viewed as a right R-module, is a direct sum of n copies of a simple right R-module, W say; and W is distinguished amongst simple right R-modules by having A as its annihilator.

Next, given any simple right S-module, we write it in the form S/B for some maximal right ideal B. Let C/B be a proper R-submodule of S/B. Since $A \subseteq R$,

$$(CA + B)/B = (C/B)A \subseteq C/B \subset S/B.$$

Thus $(CA + B)/B$ is a proper S-submodule of the simple S-module S/B; hence $(CA + B)/B = 0$ and so $CA \subseteq B$. Thus $C \subseteq D = \{d \in S \mid dA \subseteq B\}$. Moreover, 1.3 shows that, for each $c \in C$, left multiplication by c induces an element of $\operatorname{Hom}_S(S/A, S/B)$.

(i) Suppose now that $S/B = X \ncong U$ and so $\operatorname{Hom}_S(S/A, S/B) = 0$. Then left multiplication by c must induce the zero map; that is, $cS \subseteq B$ for each $c \in C$. Thus $C \subseteq B$ and so $C/B = 0$. Thus S/B has no nonzero proper R-submodules; so it is a simple R-module.

(ii) Suppose next that $S/B \cong U$ and that C/B is a proper R-submodule. The remarks above tell us that $C \subseteq D = \{d \in S \mid dA \subseteq B\}$; and $D \neq S$ since $SA = S \nsubseteq B$. We deduce that D/B is the unique maximal R-submodule of $S/B \cong U$ and so $(S/D)_R$ is simple, say $(S/D)_R \cong V$. Note that $VA = V$ since $SA = S$; so $V \ncong W$.

Next, choose a set of n maximal right ideals B_i with $\cap B_i = A$, and so each $S/B_i \cong U$. For each i, let D_i/B_i be the maximal R-submodule provided by the preceding paragraph. Then

$$\cap D_i = \{d \in S \mid dA \subseteq \cap B_i\} = \{d \in S \mid dA \subseteq A\} = R$$

and so $\oplus_{i=1}^n D_i/B_i \cong R/A \cong W^{(n)}$. We see from this that $D_i/B_i \cong W$. Thus we have shown that U has a unique composition series of length 2, and that the two composition factors are nonisomorphic. Since $S/A \cong U^{(n)}$ and $R/A \cong W^{(n)}$ we deduce that $S/R \cong V^{(n)}$.

(iii) Let Y_R be simple. It is enough to show that Y is an R-composition factor of some S-module of finite length, since, using (i) and (ii), every S-composition series can be refined to an R-composition series whose composition factors are as described.

Now $Y \cong R/E$ for some maximal right ideal E of R; and we can assume, in the notation of (ii), that $Y \not\cong V$. We know from (ii) that $(S/R)_R$ has finite length, and so too, of course, has R/E. Therefore $(S/E)_R$ has finite length, and has a composition factor isomorphic to Y.

Next consider the R-module ES/E which is a submodule of S/E and so has finite length. Now $ES/E \cong E \otimes_R (S/R)$ as right R-modules. However, $E \otimes_R (S/R)$ is a sum of right R-submodules of the form $e \otimes (S/R)$, where $e \in E$. Each of these is a homomorphic image of (S/R); i.e. by (ii), of $V^{(n)}$. Hence all composition factors of ES/E are isomorphic to V which, by hypothesis, is not isomorphic to Y.

Finally, we note that $S/ES \cong (S/E)/(ES/E)$. Since S/E has an R-composition factor isomorphic to Y and ES/E does not, S/ES is the desired S-module of finite length that has an R-composition factor isomorphic to Y. □

4.5. NOTATION. To avoid repetition, when $R = \mathbb{I}_S(A)$ is a basic idealizer of type U and so U_R is uniserial of length 2 with composition factors V, W, we will simply say that *$R = \mathbb{I}_S(A)$ is a basic idealizer of type $U = [VW]$* and say that R *slices* U into $[VW]$.

The next few results concern the 'new' simple modules V and W.

4.6. PROPOSITION. *Let $R = \mathbb{I}_S(A)$ be a basic idealizer of type $U = [VW]$. Then, viewing W as an R-submodule of U_S,*

$$R = \{s \in S \mid Ws \subseteq W\}, \quad A = \operatorname{ann}_S(W) \quad \textit{and} \quad W = \operatorname{ann}_U(A).$$

PROOF. It is enough to prove these assertions with W replaced by $W^{(n)}$, regarded as an R-submodule of $U_S^{(n)}$, with $n \geq 1$ chosen so that $S/A \cong U^{(n)}$. However, there is a unique R-submodule of S/A that corresponds to $W^{(n)}$ under any isomorphism $(S/A)_R \cong U^{(n)}$, namely R/A, since $V \not\cong W$ [4.4(ii)]. So we may replace $W^{(n)}$ by R/A. However, if $(R/A)s \subseteq R/A$ then $s \in R$, as desired. The second assertion is proved similarly. For the third assertion, suppose that $(s + A)A = 0$ in S/A. Then $sA \subseteq A$ and therefore $s \in \mathbb{I}_S(A) = R$. Thus $s + A \in R/A$, as desired. □

Keep the notation in 4.6; and recall [2.6] that, for any S-module X we have $X \otimes_R S \cong X \otimes_R A \cong X$. Thus we already know the effect of such tensoring by S and A on all simple R-modules other than V and W.

4.7. PROPOSITION. *Let $R = \mathbb{I}_S(A)$ be a basic idealizer of type $U = [VW]$.*

(i) *$VA = V$ and $WA = 0$.*
(ii) *$V \otimes_R S = 0$ and $V \otimes_R A \cong U$.*
(iii) *$W \otimes_R S \cong U$ and $W \otimes_R A = 0$.*
(iv) *$\operatorname{pd}(V_R) \leq 1$ and $\operatorname{pd}(W_R) \leq \operatorname{pd}(U_S)$.*

PROOF. (i) 4.6 shows that $WA = 0$. To see that $VA = V$, note that $S/R \cong V^{(n)}$, by 4.4, and $SA = S$; so $V^{(n)}A = V^{(n)}$.

(ii) 2.3 shows that $(S/R) \otimes S = 0$ and hence $V \otimes S = 0$. The same result also shows that $(S/R) \otimes A \cong S/A$. But $S/A \cong U^{(n)}$ and so $V^{(n)} \otimes A \cong U^{(n)}$, by (4.4.1). Hence the result holds.

(iii) This is proved in a similar fashion.

(iv) Since S_R is projective, by 2.3, $(S/R)_R \cong V^{(n)}$ has projective dimension at most 1; so the same is true of V_R. Next, since any S-projective resolution of U_S is also an R-projective resolution, we see that $\mathrm{pd}(U_R) \le \mathrm{pd}(U_S)$. The existence of the short exact sequence $0 \to W \to U \to V \to 0$ now implies the result, using, e.g., [**McR 01**, 7.1.6]. □

4.8. COROLLARY. *Let $R = \mathbb{I}_S(A)$, a basic idealizer of type $U = [VW]$. Then:*

(i) *V is the unique simple R-module with the property that $V \otimes_R S = 0$;*
(ii) *W is the unique simple R-module with the property that $W \otimes_R A = 0$;*
(iii) *the distinct simple R-isomorphism classes are precisely the simple S-isomorphism classes but with that of U replaced by those of V and W.*

PROOF. Let $X \ncong U$ be a simple S-module. Recall [4.4] that every simple R-module is either isomorphic to such an X or to V or W. Recall, from 2.6(i),(iii), that $X \otimes_R S \cong X \otimes_R A \cong X$. However, 4.7 asserts that $V \otimes_R S = 0$ and $W \otimes_R S \cong U$, proving (i), and that $V \otimes A \cong U$ and $W \otimes_R A = 0$, proving (ii).

By (i) and (ii), we know that no two of V, W and X are isomorphic. Therefore, to prove (iii), it is enough to prove that if Y_S is simple and $Y_S \ncong X_S$ then $Y_R \ncong X_R$; and this holds since $\mathrm{Hom}_R(X,Y) = \mathrm{Hom}_S(X,Y) = 0$ [2.6(i) and 2.5(iii)]. □

4.9. LEMMA. *Let $R = \mathbb{I}_S(A)$ be a basic idealizer of type $U = [VW]$ and let W' be an R- submodule of $U^{(a)}$, for some a, with $W' \cong W$. Then W' is the R-socle of some S-submodule U' of $U^{(a)}$ with $U' \cong U$.*

PROOF. The nonzero S-submodule $U' = W'S$ of $U^{(a)}$ generated by W' is a homomorphic image of the simple S-module $W \otimes_R S \cong U$. Hence $U' \cong U$, as desired. □

4.10. LEMMA. *Let $R = \mathbb{I}_S(A)$ be a basic idealizer of type $U = [VW]$. If M is an R-submodule of some S-module, there exists a commutative diagram with exact rows*

$$\begin{array}{ccccc} M/MA & \hookrightarrow & MS/MA & \twoheadrightarrow & MS/M \\ \downarrow(\cong) & & \downarrow(\cong) & & \downarrow(\cong) \\ W^{(a)} & \hookrightarrow & U^{(a)} & \twoheadrightarrow & V^{(a)} \end{array} \tag{4.10.1}$$

where the vertical maps are isomorphisms and a is a cardinal number; and a is finite if M_R is finitely generated.

PROOF. We have a short exact sequence $MR/MA \hookrightarrow MS/MA \twoheadrightarrow MS/MR$ of R-modules, where we have written MR in place of M for emphasis. Note that $(MS/MA)_S \cong M \otimes_R (S/A)$. This is a sum of S-submodules of the form $m \otimes (S/A)$, with m ranging over the members of a generating set for M_R; and each of these is a homomorphic image of $U^{(n)}$ where $S/A \cong U^{(n)}$. Thus $(MS/MA)_S \cong U^{(a)}$ as S-modules, and hence as R-modules, for some a which is finite if M_R is finitely generated. Similarly, with the help of (4.4.1), we get $MR/MA \cong W^{(c)}$ and

$MS/MR \cong V^{(b)}$ as R-modules for suitable cardinal numbers b, c. Substituting into the short exact sequence above yields a short exact sequence of R-modules:

$$0 \to W^{(c)} \to U^{(a)} \to V^{(b)} \to 0.$$

Since the map $U^{(a)} \to V^{(b)}$ is a surjection and its kernel contains no copies of the top composition R-factor V of U, we have $b = a$. Similarly, since the image of this map contains no copies of W, the kernel must be the entire R-socle $W^{(a)}$ of $U^{(a)}$. Thus $c = a$. The rest now follows. □

We note one consequence.

4.11. PROPOSITION. *Let $R = \mathbb{I}_S(A)$ be a basic idealizer of type $U = [VW]$. If $S_R \supseteq X_R \supseteq R_R$ then X_R is finitely generated projective.*

PROOF. We know from 4.7 that V has projective dimension at most 1. Since S_R is projective [2.3], the short exact sequence

$$0 \to X \to S \to V^{(k)} \to 0$$

shows that X_R is projective. Moreover, $X/R \subseteq S/R$ which has finite length [4.4] and so X_R is finitely generated. □

4.12. THEOREM. *If $R = \mathbb{I}_S(A)$ is a basic idealizer then ${}_RS$ is flat.*

PROOF. We have seen, in 2.6(vi) , that if M is a right S-module and N a left S-module then $\mathrm{Tor}_n^R(M, N) \cong \mathrm{Tor}_n^S(M, N)$.

Next we show that $\mathrm{Tor}_1^R(S/R, S) = 0$. We consider the short exact sequence

$$0 \to R \to S \to S/R \to 0.$$

When tensored over R with S this gives us [see 53.17.1] the long exact sequence

$$0 \to \mathrm{Tor}_1^R(S/R, S) \to R \otimes_R S \to S \otimes_R S \to (S/R) \otimes_R S \to 0$$

using the fact that $\mathrm{Tor}_1^R(S, S) = 0$ since S_R is flat. The isomorphism of $R \otimes_R S$ and $S \otimes_R S$ under the given map implies that $\mathrm{Tor}_1^R(S/R, S) = 0$, as desired.

Next, let B be any right ideal of R. By 4.10, BS/B is a direct sum of copies of V and hence is a direct summand of a direct sum of copies of S/R. Hence, from above, we deduce that $\mathrm{Tor}_1^R(BS/B, S) = 0$. But we know, as noted above, that $\mathrm{Tor}_1^R(S/BS, S) = \mathrm{Tor}_1^S(S/BS, S)$; and the latter term is zero since ${}_SS$ is flat. We deduce that $\mathrm{Tor}_1^R(S/B, S) = 0$. Finally, the short exact sequence

$$0 \to B \to S \to S/B \to 0$$

yields the long exact sequence

$$\ldots \to \mathrm{Tor}_1^R(S/B, S) \to B \otimes_R S \to S \otimes_R S \to (S/B) \otimes_R S \to 0$$

and so the map $B \otimes_R S \to S \otimes_R S$ is an embedding. The same is therefore true of the map $B \otimes_R S \to R \otimes_R S$. By [**Anderson and Fuller 92**, 19.17], this is equivalent to ${}_RS$ being flat. □

This result, combined with 2.3(ii), shows that a basic idealizer satisfies the hypotheses of the next proposition which gives further results along the lines of 2.6.

4.13. PROPOSITION. *Let R be a subring of a ring S such that $S \otimes_R S \cong S$ via multiplication and ${}_RS$ is flat — as is true when R is a basic idealizer from S.*

(i) *If M and N are R-submodules of right S-modules then $M \otimes_R S \cong MS$ via multiplication and* $\operatorname{Hom}_R(M, N) \subseteq \operatorname{Hom}_S(MS, NS)$.
(ii) *If M is a right S-module then M_R is injective if and only if M_S is injective.*
(iii) *If M and N are right S-modules then* $\operatorname{Ext}^1_R(M, N) = \operatorname{Ext}^1_S(M, N)$, *when viewed as equivalence classes of short exact sequences; (and if R is a basic idealizer from S, then* $\operatorname{Ext}^n_R(M, N) \cong \operatorname{Ext}^n_S(M, N)$ *for each n, via the forgetful functor from S to R.)*
(iv) *$I = (I \cap R)S$ for every right ideal I of S.*

PROOF. (i) Note first that, since $_RS$ is flat, then $M \otimes_R S \subseteq MS \otimes_R S$. However

$$MS \otimes_R S \cong MS \otimes_S S \otimes_R S \cong MS \otimes_S S \cong MS$$

via multiplication. Since $M \otimes_R S \twoheadrightarrow MS$ under multiplication, we deduce that this epimorphism is in fact an isomorphism.

Next, we note that every $\phi \in \operatorname{Hom}_R(M, N)$ induces $\phi \otimes 1\colon M \otimes_R S \to N \otimes_R S$ which, by the preceding paragraph, we may consider to be an S-homomorphism $MS \to NS$.

(ii) The case when M_R is injective is covered by 2.5(iv). So we suppose that M_S is injective. Let I be a right ideal of R and let $\alpha : I \to M$ be an R-homomorphism. We need only show that α can be lifted to a homomorphism $R \to M$. We form tensor products by S over R. Since $_RS$ is flat, we get $I \otimes_R S$ embedded in $R \otimes_R S$ and also $\alpha \otimes 1 : I \otimes_R S \to M \otimes_R S$. Using (i), $M \otimes_R S \cong MS = M$, $R \otimes_R S \cong S$ and $I \otimes_R S \cong IS$, all via multiplication. Since M_S is injective, the map $\alpha \otimes 1$ from IS to M lifts to a map from S to M. Restricted to R, this is the required lifting of α.

(iii) Let $\mathcal{E} : 0 \to N \to X \to M \to 0$ be a short exact sequence of R-modules. Since $_RS$ is flat, tensoring by S gives a short exact sequence of right S-modules

$$\mathcal{E}' : 0 \to N \otimes_R S \to X \otimes_R S \to M \otimes_R S \to 0.$$

The multiplication maps $N \otimes_R S \to N$ and $M \otimes_R S \to M$ are isomorphisms; so we deduce that $X \cong X \otimes_R S$ and that $\mathcal{E}$ was already a short exact sequence of S-modules. We know, from 2.5(iii), that $\operatorname{Hom}_R = \operatorname{Hom}_S$ for right S-modules. The process of forming Ext from short exact sequences, which is described in 53.11, is now readily seen to be identical whether one considers R-modules or S-modules. Hence $\operatorname{Ext}^1_R(M, N) = \operatorname{Ext}^1_S(M, N)$.

Suppose that R is a basic idealizer from S. This, by 2.6(iii), gives the additional property that projective S-modules are also projective over R. We now proceed to prove, by induction on n, that $\operatorname{Ext}^n_R(M, N) \cong \operatorname{Ext}^n_S(M, N)$, the cases $n = 0, 1$ being already known. Consider the start of an S-projective resolution of M_S, say $0 \to K \to P \to M \to 0$. If we apply $\operatorname{Hom}_S(-, N)$ to this, we get $\operatorname{Ext}^{n+1}_S(M, N) \cong \operatorname{Ext}^n_S(K, N)$ for $n \geq 1$, using 53.8. As noted above, P is also projective over R and so, if we apply $\operatorname{Hom}_R(-, N)$ instead, we get $\operatorname{Ext}^{n+1}_R(M, N) \cong \operatorname{Ext}^n_R(K, N)$. The inductive hypothesis applied to K_S tells us that $\operatorname{Ext}^n_R(K, N) \cong \operatorname{Ext}^n_S(K, N)$. Hence $\operatorname{Ext}^{n+1}_R(M, N) \cong \operatorname{Ext}^{n+1}_S(M, N)$ as required.

(iv) Note that $I/(I \cap R) \cong (I + R)/R \subseteq S/R$. Hence $(I/(I \cap R)) \otimes_R S = 0$ and so $I = IS = (I \cap R)S$. □

Next we obtain some results directly involving the right ideal A.

4.14. PROPOSITION. *Let $R = \mathbb{I}(A)$ be a basic idealizer from S (or, more generally, let A be a generative right ideal of S and R be a subidealizer of A such that*

$_RS$ is flat). If $S \supseteq Y_R \supseteq R$ then $(Y_R)^* \cong \{s \in S \mid sY \subseteq R\}$ via left multiplication and, if we view this isomorphism as an identification, then $R \supseteq Y^* \supseteq A$.

PROOF. For the nontrivial part of this, take $\phi \in Y^* = \operatorname{Hom}_R(Y, R)$. Since Y is an R-submodule of S_S, 2.3(ii)(a) shows that ϕ extends to an element of $\operatorname{Hom}_S(YS, RS) = \operatorname{Hom}_S(S_S, S_S)$, and hence ϕ equals left multiplication by some element of S, as desired. The inclusion $Y^* \subseteq R$ holds because $1 \in R \subseteq Y$. □

Using this, we now demonstrate that in a basic idealizer situation, S is a minimal extension of R in the following sense.

4.15. PROPOSITION. *Let $R = \mathbb{I}_S(A)$ be a basic idealizer of type U. Then there are no rings strictly between R and S.*

PROOF. Let T be a ring with $R \subseteq T \subseteq S$. From 4.11 we see that T_R is finitely generated projective and 4.14 shows that $T^* \cong \{r \in R \mid rT \subseteq R\}$. Now $R \supseteq T^* \supseteq A$ and, moreover, T^* is an ideal of R. Since A is a maximal ideal of R, we deduce that $T^* = A$ or $T^* = R$. Since T_R is projective, $T^{**} = T$. However, $A^* = S$ and $R^* = R$. □

Next we give a left-handed version of 4.11.

4.16. COROLLARY. *Let $R = \mathbb{I}_S(A)$ be a basic idealizer. If $A \subseteq Y \subseteq S$ with Y a left R-submodule of S, then $_RY$ is flat; and if $A \subseteq Y \subseteq R$ then $_RY$ is finitely generated projective.*

PROOF. We know that $_RA$ is projective and so $\operatorname{pd}(_R(R/A)) \leq 1$. It follows that, if W' is the simple left R-module annihilated by A, then $\operatorname{pd}(_RW') \leq 1$ and so $\operatorname{fd}(_RW') \leq 1$. The same is then true of S/Y since this is a direct sum of copies of W'. Then the short exact sequence $0 \to Y \to S \to S/Y \to 0$ shows that $_RY$ is flat.

Finally, suppose that $Y \subseteq R$. Note that each of R/Y and Y/A is a finite direct sum of copies of W'. So $_RY$ is projective and, since $_RA$ is finitely generated, $_RY$ is finitely generated. □

4.17. COROLLARY. *Let $R = \mathbb{I}_S(A)$ be a basic idealizer. Let M, N be right S-modules and K an R-submodule of M such that $M/K \cong N$ as R-modules. Then K is an S-submodule of M.*

PROOF. We know from 2.3 and 2.5 that $M \otimes_R S \cong M$ via multiplication, and likewise for N. The flatness of $_RS$ given by 4.12 shows that tensoring the exact sequence $0 \to K \to M \to N \to 0$ by S yields the exact sequence $0 \to K \otimes_R S \to M \to N \to 0$. Identifying $K \otimes S$ with its image in M therefore yields $K = KS$, as desired. □

The final two results of this section show, respectively, how module-theoretic and ring-theoretic properties pass up or down in the basic idealizer situation.

4.18. COROLLARY. *Let $R = \mathbb{I}_S(A)$ be a basic idealizer, and let M be an S-module.*

(i) *M_S has finite length if and only if M_R has finite length.*
(ii) *M_S is Noetherian if and only if M_R is Noetherian.*
(iii) *If M_S is Noetherian then M_S is uniserial if and only if M_R is uniserial.*

PROOF. (i) 4.4 shows that an S-composition series for M can be refined to an R-composition series of no more than double the length; and the rest is clear.

(ii) Evidently, if M_R is Noetherian then so too is M_S. Conversely, suppose that M_S is Noetherian. Let N be any R-submodule of M. Then $NS \supseteq N \supseteq NA$. Each of NS and NA are finitely generated over S and hence over R. Now, by 4.10, $NS/N \cong V^{(a)}$, $N/NA \cong W^{(a)}$ and $NS/NA \cong U^{(a)}$ for some a. Since $(NS/NA)_S$ is Noetherian, a must be finite; and then $(N/NA)_R$ has finite length and N_R is finitely generated.

(iii) Say M_R is uniserial and N_1 and N_2 are two S-submodules of M_S. They are also R-submodules and so must be comparable. Hence M_S is uniserial.

Conversely, say M_S is uniserial. Choose any $m \in M$. Now mS/mA is an S-homomorphic image of S/A which, in turn, is isomorphic to $U^{(n)}$, and yet mS/mA is uniserial. Hence either $mS/mA = 0$ or else $mS/mA \cong U$, which, as an R-module, is uniserial of length 2. Thus each cyclic R-submodule of M is either an S-submodule or is the unique R-submodule lying between two consecutive S-submodules. Hence, any two cyclic R-submodules are comparable and their sum is the larger. Consequently, each finitely generated R-submodule is cyclic and any two are comparable. □

Note that right serial rings are defined at the start of §50.

4.19. THEOREM. *Let $R = \mathbb{I}_S(A)$ be a basic idealizer of type $U = [VW]$.*

(i) *S is right Artinian if and only if R is right Artinian.*
(ii) *S is right Noetherian if and only if R is right Noetherian.*
(iii) *S is right hereditary if and only if R is right hereditary.*
(iv) *S is right Artinian, right serial if and only if R is right Artinian, right serial.*

PROOF. We review some facts that this proof uses repeatedly, in addition to those in 4.18. Since A is generative, S_R is finitely generated, and S-modules are projective if and only they are projective as R-modules [2.6(iii)]. In particular, S_R is projective.

(i) Suppose that S is right Artinian, so S_S has finite length. Then S_R has finite length, and so its submodule R_R has finite length. Conversely, if R is right Artinian, then the finitely generated right R-module S has finite length. Hence the same is true of S_S.

(ii) If R is right Noetherian then so is the finitely generated right R-module S_R. Therefore S_S is Noetherian too. Conversely, if S is right Noetherian then 4.18(ii) shows that S_R is Noetherian and so too is its R-submodule R_R.

(iii) Suppose first that R is right hereditary. Let C be a right ideal of S. Since R is right hereditary, the submodule C_R of the projective R-module S_R is again projective, and therefore, as noted above, C_S is projective.

Conversely, suppose that S is right hereditary, let B be a right ideal of R, and consider the following short exact sequence of R-modules.

$$0 \to B \to BS \to BS/B \to 0$$

Since S is right hereditary, $(BS)_S$ is projective, and hence $(BS)_R$ is projective. Thus the short exact sequence is the start of a projective resolution of BS/B. However, $BS/B \cong V^{(a)}$ for some a [4.10] and $\text{pd}(V) \leq 1$ [4.7]; so $\text{pd}(BS/B) \leq 1$. Hence B_R is projective.

(iv) The Artinian property is dealt with by (i). First suppose that R is right serial and let B be an indecomposable direct summand of S_S. Say $B_R = C \oplus D$. Since $SA = S$, we have $B = BA = CA \oplus DA$. Since CA and DA are S-submodules of B, one of them must be zero; say $CA = 0$. But then $B = DA \subseteq D$, and hence $C = 0$. Thus we see that B_R is indecomposable.

Since B_S is finitely generated projective, so is B_R. But every finitely generated indecomposable projective right module over the right Artinian right serial ring R is uniserial (by the Krull-Schmidt theorem, applied to free R-modules). Therefore B_R is uniserial, and hence B_S is too. Thus S is a right serial ring.

Conversely, suppose that S is right serial, and let $J = J(S)$. Since A is isomaximal and generative we have $A \supset J$. Therefore A/J has an idempotent generator in S/J; and this lifts to an idempotent, e say, in S such that $A = eS + J$. Let $1 = e_1 + e_2 + \cdots + e_n$ be a decomposition of 1 into a sum of orthogonal primitive idempotents of S such that $e = e_1 + e_2 + \cdots + e_m$ for some $m < n$. We now show that every $e_i \in R$.

If $1 \leq i \leq m$, we have $e_i = ee_i \in eS \subseteq A \subseteq R$. If $m+1 \leq i \leq n$, we have $e_i e \in J \subseteq A$; and hence $e_i A = e_i(eS + J) \subseteq J \subseteq A$. Therefore $e_i \in \mathbb{I}_S(A) = R$.

Thus we see that $R = e_1 R \oplus \ldots \oplus e_n R$. Since each $e_i S$ is uniserial over S it is also uniserial over R; hence so is its R-submodule $e_i R$. □

4.20. REMARK. We will see, in 14.6, that S is an HNP ring if and only if R is an HNP ring.

5. Extensions of Simple Modules in Idealizers

In this section, given a basic idealizer R from a ring S, we will describe all endomorphisms of simple R-modules and all extensions of simple R-modules by simple R-modules. The description will be in terms of information about simple modules over S.

5.1. NOTATION. Let $R = \mathbb{I}_S(A)$, a basic idealizer of type $U = [VW]$. The embedding of W in U gives a short exact sequence $W \overset{\alpha}{\hookrightarrow} U \overset{\beta}{\twoheadrightarrow} V$. We will use this notation throughout this section.

General facts about Ext from §53 will be used in this section, often without further comment.

5.2. PROPOSITION. *Let $R = \mathbb{I}_S(A)$ be a basic idealizer.*

(i) *Let H be a right R-module such that* $\mathrm{Hom}_R(H, N) = 0$ *for all N_S; then* $\mathrm{Ext}^n_R(H, N) = 0$ *for all N_S.*

(ii) *Let H be a right R-module such that* $\mathrm{Hom}_R(M, H) = 0$ *for all M_S; then* $\mathrm{Ext}^n_R(M, H) = 0$ *for all M_S.*

PROOF. (i) We proceed by induction. We consider the start $0 \to N \to I \to K \to 0$ of an S-injective resolution of N_S and apply $\mathrm{Hom}_R(H, -)$. All the Hom terms are zero; and so too are all the $\mathrm{Ext}^n_R(H, I)$, since, by 4.13(ii), I_R is injective. Hence $\mathrm{Ext}^{n+1}_R(H, N) \cong \mathrm{Ext}^n_R(H, K)$ for all $n \geq 0$. However, the latter is zero by the induction hypothesis applied to K.

(ii) This is proved similarly, using the start of an S-projective resolution of M. □

Recall from 4.8(iii) that the simple R-modules other than V and W are precisely the simple S-modules other than U. We have already seen in 2.6(i) and 4.13(iii) that, for pairs of S-modules, their homomorphism and extension groups over R are the same as those over S. This describes all the endomorphism rings and extension groups of simple R-modules which do not involve V or W. So that leaves only the cases which involve one or more of V and W.

Before starting the theory, consider any S-module extension $U \hookrightarrow M \twoheadrightarrow Y$ of U by an S-module Y. We may view M as an R-module and, as such, it has a submodule isomorphic to W. Then $V = U/W \hookrightarrow M/W \twoheadrightarrow Y$ is an R-module extension of V by Y. Analogously, any extension of Y_S by U_S leads to an extension of Y_R by W_R. The first theorem of this section will show that there are no nonsplit extensions involving Y and one or other of V and W other than those just described; in particular, there are none of Y by V or of W by Y.

Note, in relation to the statements of (iii) and (iv) below, that $\mathrm{End}_S(Y) = \mathrm{End}_R(Y)$, and $(\alpha, 1), (1, \beta)$ are as described in §53.

5.3. THEOREM. *Let $R = \mathbb{I}_S(A)$, a basic idealizer of type $U = [VW]$, and let α, β be as in 5.1. Let Y be any S-module. Then, for all $n \geq 1$:*

(i) $\mathrm{Hom}_R(V, Y) = 0$ *and* $\mathrm{Ext}^n_R(V, Y) = 0$;
(ii) $\mathrm{Hom}_R(Y, W) = 0$ *and* $\mathrm{Ext}^n_R(Y, W) = 0$;
(iii) $\mathrm{Hom}_R(Y, V) \cong \mathrm{Hom}_S(Y, U)$ *as* $\mathrm{End}_S(Y)$*-modules via* β, *and* $\mathrm{Ext}^1_R(Y, V) \cong \mathrm{Ext}^1_S(Y, U)$ *as* $\mathrm{End}_S(Y)$*-modules via* $(1, \beta)^{-1}$;
(iv) $\mathrm{Hom}_R(W, Y) \cong \mathrm{Hom}_S(U, Y)$ *as* $\mathrm{End}_S(Y)$*-modules via* α, *and* $\mathrm{Ext}^1_R(W, Y) \cong \mathrm{Ext}^1_S(U, Y)$ *as* $\mathrm{End}_S(Y)$*-modules via* $(\alpha, 1)^{-1}$.

PROOF. (i) Let $\theta \in \mathrm{Hom}(V, Y)$ and consider the commutative square

$$\begin{array}{ccc} V & \xrightarrow{\theta} & Y \\ \downarrow & & \downarrow f \\ V \otimes_R S & \xrightarrow{\theta \otimes 1} & Y \otimes_R S \end{array} \tag{5.3.1}$$

in which the vertical maps are induced by the embedding of R in S. Recall that f is an isomorphism and that $V \otimes_R S = 0$. So the combined map from V to $Y \otimes_R S$ is zero. Thus $\theta = 0$ and $\mathrm{Hom}(V, Y) = 0$. We can now apply 5.2(ii) to deal with Ext^n.

(ii) Let θ: $Y \to W$. Tensor with A to get a commutative square:

$$\begin{array}{ccc} Y \otimes_R A & \xrightarrow{\theta \otimes 1} & W \otimes_R A \\ \downarrow g & & \downarrow \\ Y & \xrightarrow{\theta} & W \end{array} \tag{5.3.2}$$

in which the vertical maps are multiplication. Now $W \otimes A = 0$ and g is an isomorphism. So the combined map from $Y \otimes A$ to W is zero. Hence $\theta = 0$ and $\mathrm{Hom}(Y, W) = 0$. 5.2(iii) now completes the proof.

(iii) Start with the short exact sequence $W \overset{\alpha}{\hookrightarrow} U \overset{\beta}{\twoheadrightarrow} V$ and apply $\mathrm{Hom}_R(Y, -)$, getting the long exact sequence

$$0 \to \mathrm{Hom}_R(Y, W) \to \mathrm{Hom}_R(Y, U) \xrightarrow{\beta} \mathrm{Hom}_R(Y, V) \to$$
$$\mathrm{Ext}^1_R(Y, W) \to \mathrm{Ext}^1_R(Y, U) \xrightarrow{(1,\beta)} \mathrm{Ext}^1_R(Y, V) \to 0 \to \dots$$

in which (ii) tells us that $\mathrm{Hom}_R(Y,W)=0$ and $\mathrm{Ext}^n_R(Y,W)=0$. This immediately gives $\mathrm{Hom}_R(Y,U)\cong\mathrm{Hom}_R(Y,V)$ and $\mathrm{Ext}^1_R(Y,U)\cong\mathrm{Ext}^1_R(Y,V)$. However, 2.6(i) and 4.13(iii) show that $\mathrm{Hom}_R(Y,U)\cong\mathrm{Hom}_S(Y,U)$ and $\mathrm{Ext}^1_R(Y,U)\cong\mathrm{Ext}^1_S(Y,U)$.

(iv) Applying $\mathrm{Hom}_R(-,Y)$ to the short exact sequence $W\overset{\alpha}{\to}U\overset{\beta}{\to}V$ gives the long exact sequence:

$$0\to\mathrm{Hom}_R(V,Y)\to\mathrm{Hom}_R(U,Y)\xrightarrow{\alpha}\mathrm{Hom}_R(W,Y)\to$$
$$\mathrm{Ext}^1_R(V,Y)\to\mathrm{Ext}^1_R(U,Y)\xrightarrow{(\alpha,1)}\mathrm{Ext}^1_R(W,Y)\to\mathrm{Ext}^2_R(V,Y)\to\dots.$$

We know from (i) that $\mathrm{Hom}_R(V,Y)=0$ and $\mathrm{Ext}^n_R(V,Y)=0$. The results follow as in (iii). □

Since all simple R-modules other than V and W are also S-modules, we now know all extension groups of one simple R-module by another except those involving two from V and W. We also know the endomorphism rings of all simple R-modules other than those of V and W. The remainder of this section will fill these remaining gaps.

We start with two elementary results which are applicable over any ring.

5.4. LEMMA. *Let X,Y be semisimple modules over any ring and let $M\cong X\oplus Y$. Then every short exact sequence $X\hookrightarrow M\twoheadrightarrow Y$ is split.*

PROOF. It suffices to show that the image of X under the given monomorphism $X\hookrightarrow M$ is a direct summand of M, and this holds because M is semisimple. □

We repeatedly use the following consequence of 5.4 without explicit mention.

5.5. COROLLARY. *If X and Y are simple modules over any ring and $X\hookrightarrow M\twoheadrightarrow Y$ is a nonsplit short exact sequence, then M is uniserial.*

PROOF. If M is not uniserial then $M\cong X\oplus Y$ and the lemma applies. □

Now the endomorphism rings of V and W can be described.

5.6. THEOREM. *Let $R=\mathbb{I}_S(A)$, a basic idealizer of type $U=[VW]$, and let α,β be as in 5.1. Then $\mathrm{End}_R(U)\cong\mathrm{End}_R(W)$ via the injection $W\overset{\alpha}{\hookrightarrow}U$ and $\mathrm{End}_R(U)\cong\mathrm{End}_R(V)$ via the surjection $U\overset{\beta}{\twoheadrightarrow}V$.*

PROOF. Since W is the only proper submodule of U_R, restriction and corestriction induce ring homomorphisms $\mathrm{End}_R(U)\to\mathrm{End}_R(W)$ and $\mathrm{End}_R(U)\to\mathrm{End}_R(V)$. Both of these maps are monomorphisms because $V\not\cong W$. We now show these maps are surjections.

We know that there is an isomorphism $\theta:U\to U\otimes A$ given, in reverse, by multiplication. Thus each $u\in U$ can be written $u=\Sigma u_ia_i$ for some $u_i\in U$, $a_i\in A$ and $\theta(u)=\Sigma u_i\otimes a_i$. There is a sequence of mappings

$$U\overset{\theta}{\to}U\otimes_R A\overset{\beta\otimes 1}{\twoheadrightarrow}V\otimes_R A\overset{\mu}{\twoheadrightarrow}V$$

in which μ is the multiplication map. This sequence maps u to $\Sigma(\beta(u_i)a_i=\beta(\Sigma(u_ia_i))=\beta(u)$ and so the sequence forms a factorization of β. Let $\sigma=(\beta\otimes 1)\theta$. We see, using 4.7, that σ is an isomorphism.

Choose any $f \in \mathrm{End}(V)$. We form the diagram

$$\begin{array}{ccccc} U & \stackrel{\sigma}{\to} & V \otimes A & \stackrel{\mu}{\twoheadrightarrow} & V \\ \downarrow g & & \downarrow f \otimes 1 & & \downarrow f \\ U & \stackrel{\sigma}{\to} & V \otimes A & \stackrel{\mu}{\twoheadrightarrow} & V \end{array}$$

noting that, since σ is an isomorphism, there exists g such that the left-hand square commutes. This g lifts f, as desired.

On the other hand, let $f \in \mathrm{End}(W)$. Since ${}_RS$ is flat and $U = WS = W \otimes_R S$, $f \otimes 1$ is an element of $\mathrm{End}(U)$ that lifts f. □

Next we turn to the Ext groups involving two of V, W. We already know that U is a nonsplit extension of W by V; and, in the way described in the paragraph preceding 5.3, extensions of U by U lead to extensions of V by W. The final theorem of this section will show that there are no extensions involving two of V and W other than those just outlined.

5.7. THEOREM. *Let $R = \mathbb{I}_S(A)$, a basic idealizer of type $U = [VW]$, and let α, β be as in 5.1. Then:*

(i) $\mathrm{Ext}^1_R(V, V) = 0$;
(ii) $\mathrm{Ext}^1_R(W, W) = 0$;
(iii) $\mathrm{Ext}^1_R(W, V) \cong \mathrm{Ext}^1_S(U, U)$ *via* $(\alpha, \beta)^{-1}$;
(iv) $\mathrm{Ext}^1_R(V, W)$ *is a one-dimensional vector space over each of the division rings* $\mathrm{End}_R(V)$ *and* $\mathrm{End}_R(W)$. *Hence it is one-dimensional on each side over* $\mathrm{End}_S(U)$.

PROOF. (i) We assume that there is a nonsplit R-module extension of the form

$$0 \to V \to M \to V \to 0 \tag{5.7.1}$$

and note, by 5.5, that M is uniserial. We seek a contradiction.

Note that $MA = M$ since $VA = V$. Consequently, there is an R-module surjection $M' = M \otimes_R A \twoheadrightarrow M$. Tensoring (5.7.1) by the flat left R-module A and remembering that $V \otimes_R A \cong U$, we get the short exact sequence of S-modules

$$0 \to U \to M' = M \otimes_R A \to U \to 0. \tag{5.7.2}$$

Suppose (5.7.2) were nonsplit (as a sequence of S-modules). Then, by 5.5, M'_S would be uniserial of length 2, and so, by 4.4 and 4.18, M'_R would be uniserial of length 4 with R-composition factors V, W, V, W from top to bottom. Since the sequence V, V is not an initial segment of V, W, V, W there could be no surjection $M' \twoheadrightarrow M$. Hence we may suppose that (5.7.2) is split. Then $M' \cong U^{(2)}$; but the only homomorphic image of this having two composition factors, each isomorphic to V, is the direct sum. This contradicts M being uniserial and so completes the proof of (i).

(ii) Recall that $WA = 0$ by 4.7 and $A^2 = A$ by 4.2. Hence, given any short exact sequence of the form $W \hookrightarrow M \twoheadrightarrow W$, we see that $MA = 0$ and so M is a module over the simple Artinian ring R/A. Therefore the sequence splits, as required.

(iii) Applying $\mathrm{Hom}_R(-, V)$ to the short exact sequence $W \stackrel{\alpha}{\to} U \stackrel{\beta}{\to} V$ gives the long exact sequence:

$$\ldots \to \mathrm{Ext}^1_R(V, V) \to \mathrm{Ext}^1_R(U, V) \xrightarrow{(\alpha,1)} \mathrm{Ext}^1_R(W, V) \to \mathrm{Ext}^2_R(V, V) \to \ldots .$$

Since $\operatorname{pd} V \leq 1$, by 4.7, we have $\operatorname{Ext}^2_R(V,V) = 0$ [53.10]; and, by (i), $\operatorname{Ext}^1_R(V,V) = 0$. Therefore $(\alpha, 1)$ is an isomorphism. However, 5.3(iii) shows that $\operatorname{Ext}^1_S(U,U) \cong \operatorname{Ext}^1_R(U,V)$ via $(1,\beta)$ and so $\operatorname{Ext}^1_S(U,U) \cong \operatorname{Ext}^1_R(W,V)$ via (α,β).

(iv) We apply $\operatorname{Hom}(-,W)$ and $\operatorname{Hom}(V,-)$ to the short exact sequence $W \hookrightarrow U \twoheadrightarrow V$, getting the following two long exact sequences:

$$\ldots \to \operatorname{Hom}(U,W) \to \operatorname{Hom}(W,W) \overset{\partial}{\to} \operatorname{Ext}^1_R(V,W) \to \operatorname{Ext}^1_R(U,W) \to \ldots \text{ ; and}$$

$$\ldots \to \operatorname{Hom}(V,U) \to \operatorname{Hom}(V,V) \overset{\partial}{\to} \operatorname{Ext}^1_R(V,W) \to \operatorname{Ext}^1_R(V,U) \to \ldots.$$

The Ext-groups at the ends of the displayed sequences are zero, by 5.3, and the Hom-groups at the beginnings are obviously zero. Therefore the connecting homomorphism ∂ in each sequence is an isomorphism. Moreover, [53.6] these maps are homomorphisms over $\operatorname{End}(W) = \operatorname{End}(V) = \operatorname{End}(U)$; so $\operatorname{Ext}^1_R(V,W)$ is indeed one dimensional over these division rings. □

5.8. COROLLARY. *Let $R = \mathbb{I}_S(A)$, a basic idealizer of type $U = [VW]$. Then, up to isomorphism:*

(i) *U_R is the unique nonsplit extension of W by V;*
(ii) *V is the unique simple R-module X such that $\operatorname{Ext}^1_R(X,W) \neq 0$;*
(iii) *W is the unique simple R-module X such that $\operatorname{Ext}^1_R(V,X) \neq 0$.*

PROOF. (i) 5.7(iv) shows that $\operatorname{Ext}^1_R(V,W)$ is one-dimensional over both division rings $\operatorname{End}(V)$ and $\operatorname{End}(W)$. It is an elementary consequence (see 53.16) that the nonsplit extension is unique.

(ii) (iii) Combine 5.7 and 5.3. □

CHAPTER 2

Iterated and Multichain Idealizers

Having dealt with basic idealizers we now turn to the effect of iterating the process. The main results in this chapter demonstrate that this effect can equally be obtained in a single step — by simultaneously idealizing a carefully chosen collection of generative isomaximal right ideals — and then use this fact.

6. Iterated Basic Idealizers

Essentially all of the idealizer-related results in this book, except for the introductory material in §1, involve basic idealizers or iterated basic idealizers. The main results of the present section show the way many of the results about basic idealizers extend to the iterated case.

6.1. Definition. We say that a ring R is an *iterated basic idealizer* from S or, more precisely, an *n-step iterated basic idealizer* from a ring S if there is a chain of rings

$$S = S_0 \supset S_1 \supset \ldots \supset S_n = R \tag{6.1.1}$$

with $n \geq 0$, in which each S_{i+1} is a basic idealizer from S_i. We call any chain of the form (6.1.1) a *basic idealizer path of length n* from S to R.

Although basic idealizer paths are not unique [see e.g. 7.5], their length n is unique, as we show in 6.6.

6.2. Proposition. *If R is an iterated basic idealizer from S then the following hold.*

(i) *S_R is finitely generated projective and $(S/R)_R$ has finite length.*
(ii) *$S \otimes_R S \cong S$ via multiplication and ${}_RS$ is flat.*
(iii) *If M is a right S-module then $M \otimes_R S \cong M$ via multiplication.*
(iv) *If M is a right S-module then M_S is projective (resp. injective; resp. has finite length; resp. is uniserial of finite length) if and only if the same is true of M_R.*
(v) *S is right Artinian (resp. right Noetherian, resp. right hereditary, resp. right Artinian and right serial) if and only if the same is true of R.*
(vi) *Each simple R-composition factor of $(S/R)_R$ has projective dimension at most 1.*
(vii) *If $S_R \supseteq X_R \supseteq R_R$ then X_R is finitely generated projective, its dual, $X^* = \{r \in R \mid rX \subseteq R\}$ via the canonical multiplication map, and ${}_R(X^*)$ is finitely generated projective.*
(viii) *If S is right Artinian then R is also an iterated basic left idealizer from S.*

PROOF. The general attack here is as follows. Let T denote the ring immediately preceding R in the iteration. Induction on the number of idealizing steps allows us to assume that T has the properties claimed. Then the corresponding property of a basic idealizer carries the property from T to R. The properties listed in (i) to (v) and in (viii) are readily obtained in this way. We will, therefore, just indicate for each the source for the basic case.

(i) 2.3 and 4.18.

(ii) 2.3 and 4.12.

(iii) 2.5(i).

(iv) 2.6(iii), 4.13(ii), 4.18(i) and 4.18(iii), respectively.

(v) 4.19.

(viii) 4.3.

We will give a little more detail for the other parts.

(vi) Say $R = \mathbb{I}_T(A)$ with A isomaximal generative of type U; so $T/A \cong U^{(n)}$ for some n. We make the inductive hypothesis that if Y is any simple T-composition factor of S/T then $\mathrm{pd}(Y_T) \leq 1$. Since any T-projective resolution of Y is an R-projective resolution, then $\mathrm{pd}(Y_R) \leq 1$ also. Now consider U itself. We know that U_R has composition length 2 with factors V and W as usual; and $T/R_R \cong V^{(n)}$. So we know from 4.7 that $\mathrm{pd}(V_R) \leq 1$. If U is not a T-composition factor of S/T, then all composition factors of S/R have been dealt with. On the other hand, if U is a simple T-composition factor of S/T then $\mathrm{pd}(W_R) \leq 1$ by 4.7 again. So the result is proved.

(vii) These facts follow from (vi) as in the proof of 4.11. □

6.3. COROLLARY (Down-Up). *Let R be an iterated basic idealizer from a ring S, and let A' be a right ideal of S. Then $(A' \cap R)S = A'$.*

PROOF. $S \otimes_R S \cong S$ via multiplication, and ${}_RS$ is flat [6.2]; and 4.13 shows that this implies precisely this assertion. □

We make extensive use of the following details about simple R-modules versus simple S-modules.

6.4. THEOREM. *Let R be an n-step iterated basic idealizer from S.*

(i) *Each simple S-module is a uniserial R-module of length at most $n+1$ whose R-composition factors are mutually nonisomorphic and, except possibly for the top one, unfaithful.*

(ii) *All but at most n (isomorphism classes of) simple S-modules are also R-simple*

(iii) *Suppose that the simple S-modules which are not R-simple are $U_1, \dots, U_t$. Then $\sum_{i=1}^{t} \lambda(U_i)_R = n+1$.*

(iv) *Each simple R-module arises as an R-composition factor of some unique simple S-module.*

6.5. NOTATION. In the light of (iv) above, given any iterated basic idealizer R from a ring S, and given a simple R-module V, we let $V^\wedge$ denote the unique (isomorphism class of) simple S-module of which it is an R-composition factor. It is, of course, possible that V is a simple S-module, in which case $V^\wedge = V$.

PROOF. We proceed by induction on n, with the case $n = 1$ being covered by §3. Let T be the last ring obtained before R in the basic idealizer path from S to

R. Then T is an $(n-1)$-step iterated basic idealizer from S and so we may assume all these results for T. We need to extend them to R, which is a basic idealizer from T. Let X_T be the simple T-module sliced by the final idealizing step, say into $[VW]$. By (iv), applied to T and S, there is a unique simple S-module $X^\wedge$ of which X is a T-composition factor.

(i) Let U_S be some simple S-module. Then U_T is uniserial of length at most n and its composition factors are mutually nonisomorphic. So X can occur at most once in its T-composition factors and does so only if $U \cong X^\wedge$. 6.2(iv) tells us that U_R is uniserial. Moreover, X is the only simple T-module which is not simple over R and it has length 2 over R. Hence U_R has length at most $n+1$. By 4.8(iii) applied to T and R, the (isomorphism types of) simple R-modules are the same as those of T, but with X replaced by V and W. Therefore the R-composition factors of U are mutually nonisomorphic.

It remains to discuss faithfulness. By induction, all except possibly the top composition factor of U_T are unfaithful. Let X' be any composition factor of U_T other than its top factor. Then $X'A' = 0$ with A' a nonzero ideal of T. Let $A = A' \cap R$. Then $AS = A' \neq 0$ by 6.3, and $X'A = 0$, showing that X'_R and its composition factors are unfaithful. The top composition factor of U_T remains R-simple (in which case there is nothing more to prove) unless it is X. However, T slices X in the form $X_R = [VW]$ with W being R-unfaithful.

(ii) All but at most $n-1$ simple S-modules are T-simple and at most one extra simple S-module is sliced in moving from T to R. So induction gives this.

(iii) By induction, X is either a simple S-module which has not been sliced in forming T (and so $X = X^\wedge$) or else X is a T-composition factor of $X^\wedge$ which is one of the simple S-modules which have been sliced. We consider these two cases separately.

In the former case, one more simple S-module, $X^\wedge$, has been sliced, and so is no longer simple over R but rather has length 2. All other simple S-modules have the same length over R as over T. So (iii) follows in this case.

In the latter case, X is the unique T-composition factor of $X^\wedge$ which is sliced. Now $X^\wedge$ is a uniserial T-module of finite length; so, as in the proof of (i), $X^\wedge$ is a uniserial R-module of length 1 more. Once again (iii) follows.

(iv) By our induction hypothesis, the (isomorphism types of) simple T-modules are those of S except that those simple S-modules which are not T-simple are replaced by their T-composition factors; and X occurs in just one simple S-module, $X^\wedge$. As noted in the proof of (i), the simple isomorphism types of R are those of T except that X is replaced by V, W which thus occur uniquely as R-composition factors of $X^\wedge$. Hence the simple R-modules are precisely those of S but with $U_1, \ldots, U_t$ replaced by their R-composition factors. Hence (iv) is proved. □

6.6. COROLLARY. *Let R be an n-step iterated basic idealizer from S. Then every basic idealizer path from S to R has length n.*

PROOF. 6.4 (iii) describes n in terms independent of the path followed. □

6.7. DEFINITION. Let R be an iterated basic idealizer from a ring S. If U_S is the only simple S-module which fails to be R-simple, we say R (or the path from S to R) has *simple type* U.

6.8. COROLLARY. *Let R be an n-step iterated basic idealizer from S and let U be a simple S-module.*

(i) *If the idealizer path from S to R has simple type U, then $\lambda(U_R) = n+1$.*
(ii) *The idealizer path from S to R has simple type U if and only if, at each idealizing step, the simple module being sliced is one of the simple composition factors, over that ring, of U.*

PROOF. (i) Clear from 6.4(iii).

(ii) The path will have simple type U if and only if no simple S-module other than U gets sliced. However, at each ring T along the path, the simple T-modules are precisely the simple S-modules, but with those previously sliced replaced by their T-composition factors. □

6.9. NOTATION. Let R be an iterated basic idealizer from a ring S and let U be a simple S-module. We know from 6.4 that U_R is uniserial of finite length, say with composition factors $V_1, \ldots, V_t$ from top to bottom, and that U is the unique simple S-module having these R-composition factors. It will sometimes be convenient to write $U = [V_1 \ldots V_t]$. Then $V_1 = U/\operatorname{rad}(U) = \operatorname{top}(U_R)$, the *$R$-top* of U and $V_t = \operatorname{soc}(U_R)$, the *$R$-socle* of U.

Note also that $U = V_i^{\wedge}$ for each i. We prove in 6.15 that this U is the unique uniserial R-module with the listed sequence of composition factors. It is, of course, possible that U is a simple R-module, in which case U is also the R-top and the R-socle of U_S.

6.10. THEOREM. *Let R be an iterated basic idealizer from S and let X and Y be simple R-modules. Then:*
(i) $\operatorname{End}_R(X) \cong \operatorname{End}_S(X^{\wedge})$;
(ii) $\operatorname{Ext}^1_R(X,Y) = 0$ *in all but possibly the following cases:*
  (a) *if $Y_R = \operatorname{top}(Y^{\wedge})$ and $X = \operatorname{soc}_R(X^{\wedge})$ then* $\operatorname{Ext}^1_R(X,Y) \cong \operatorname{Ext}^1_S(X^{\wedge}, Y^{\wedge})$;
  (b) *if $X^{\wedge} = Y^{\wedge} = [V_1 \ldots V_t]$ and $\{X,Y\} = \{V_i, V_{i+1}\}$ for some i, then* $\operatorname{Ext}^1_R(X,Y) \cong \operatorname{End}_S(X^{\wedge})$.

PROOF. Suppose R is an n-step iterated basic idealizer from S. We proceed by induction on n. When $n = 1$, this is true by §4. Let T be the final ring reached before R. So T is $n-1$ steps from S and R is a basic idealizer from T. Thus we may assume this result for simple T-modules and we need to extend it to R.

Suppose that U_T is the simple T-module sliced in forming R and that it has R-composition factors V, W. Note that, by construction, V, W are successive composition factors V_i, V_{i+1} of some simple S-module. We know that the simple R-modules are just those of T, but with U replaced by V and W.

(i) By 2.6(i), the End groups are left unchanged in passing from T to R. So we need only consider the case when $X = V$ or $X = W$. By 5.6, $\operatorname{End}_R(V) \cong \operatorname{End}_R(W) \cong \operatorname{End}_T(U)$; and the induction hypothesis tells us that $\operatorname{End}_T(U) \cong \operatorname{End}_S(U^{\wedge})$. However $U^{\wedge} = V^{\wedge} = W^{\wedge}$ by definition.

(ii) By 2.6 and 4.13(iii), the Ext groups and End groups remain unchanged in passing from T to R. So the only cases we need to consider are those involving one or both of V and W.

We start with the case of $\operatorname{Ext}^1_R(V,Y)$ for some simple R-module Y, which subdivides into three cases. If Y is a simple T-module then, by 5.3(i), $\operatorname{Ext}^1_R(V,Y) = 0$ as predicted by the theorem. ((a) doesn't apply since V evidently is not the R-bottom of a simple S-module and (b) doesn't since $V_{i+1} = W \not\cong Y$.) Suppose next that $Y = V$. It is clear again that neither (a) nor (b) apply; and 5.7(i) tells us that $\mathrm{Ext}^1(V, V) = 0$. Finally, suppose that $Y = W$. By 5.7(iv), $\mathrm{Ext}^1_R(V, W) \cong \mathrm{End}_T(U)$. However, (i) shows that $\mathrm{End}_T(U) \cong \mathrm{End}_S(U^\wedge) = \mathrm{End}_S(V^\wedge)$. This again fits the theorem since (b) applies.

The other remaining cases are all readily dealt with in an analogous way. □

6.11. LEMMA. *Let R be any subring of S with ${}_RS$ flat and let M_R be of finite length. Then*

(i) *$M \otimes_R S = 0$ if and only if $Y \otimes_R S = 0$ for each composition factor Y of M;*
(ii) *the natural map $\tau : M \to M \otimes_R S$ is an injection if and only if $X \otimes_R S \neq 0$ for all simple submodules X of M.*

PROOF. (i) Straightforward.

(ii) Let $K = \ker(\tau)$ and suppose $K \neq 0$ and X is a simple submodule of K. The image of $X \otimes_R S$ in $M \otimes_R S$ is zero and so, by flatness, $X \otimes_R S = 0$. The converse is clear. □

6.12. COROLLARY. *Let R be an iterated basic idealizer from a ring S and V a simple R-module. Then $V \otimes_R S = 0$ unless $V = \mathrm{soc}(V^\wedge)$, in which case $V \otimes_R S \cong V^\wedge$.*

PROOF. Let $U = V^\wedge$ and $W = \mathrm{soc}(U_R)$. Simplicity of U_S implies that $WS = U$. Since U is an S-module, $U \otimes_R S \cong U$ [6.2(iii)]. Therefore tensoring the inclusion map $W \hookrightarrow U$ with the flat module ${}_RS$ shows that the surjection $W \otimes_R S \twoheadrightarrow WS = U$ is an isomorphism, proving the second assertion. Tensoring the short exact sequence $W \hookrightarrow U \twoheadrightarrow X = U/W$ with ${}_RS$ now shows that $X \otimes_R S = 0$. Hence, by 6.11 above, $V \otimes_R S = 0$ for each composition factor V of U/W. □

As we saw in 6.3 (Down-Up), if R is an iterated basic idealizer from a ring S, and A a right ideal of S then $(A \cap R)S = A$. The next proposition discusses a circumstance under which a converse holds.

6.13. PROPOSITION (Up-Down). *Let R be an iterated basic idealizer from a ring S, and B a right ideal of R such that R/B has finite length. Then $BS \cap R = B$ if and only if $X = \mathrm{soc}(X^\wedge)$ for every simple R-submodule X of R/B.*

PROOF. Clearly $BS \cap R \supseteq B$; and the natural homomorphism $\alpha : R/B \to S/BS$ is an injection if and only if $BS \cap R = B$. However, $(R/B) \otimes_R S \cong S/BS$ and so α may be identified with the map $\tau : R/B \to (R/B) \otimes_R S$. By 6.11, this is an injection if and only if, for all simple submodules X of R/B, we have $X \otimes_R S \neq 0$; i.e. (using 6.12) $X = \mathrm{soc}(X^\wedge)$. □

For ease of use later, we note a special case.

6.14. COROLLARY. *Let R be a basic idealizer of type $U = [VW]$. Let B be a right ideal of R such that R/B has finite length. Then $BS \cap R = B$ if and only if no simple R-submodule of R/B is isomorphic to V.*

PROOF. Clear from 6.13. □

We recall, see 6.9, that if U_S is simple then $U_R = [V_1 \dots V_n]$ is uniserial. The next result shows some additional uniqueness.

6.15. THEOREM. *Let R be an iterated basic idealizer from a ring S, and U_S a simple module. If M_R is a uniserial module with the same sequence of composition factors (from top to socle) as U_R, then $M \cong U_R$.*

PROOF. Let $W = \text{soc}(M_R)$. By hypothesis, $W \cong \text{soc}(U_R)$. Therefore, by 6.12, $W \otimes_R S \cong U$ and $V \otimes_R S = 0$ for all composition factors V of U/W. These are also, of course, all the composition factors of M/W. Hence, by 6.11, $M/W \otimes_R S = 0$. Tensoring the short exact sequence $W \hookrightarrow M \twoheadrightarrow M/W$ by the flat module ${}_RS$ yields $U \cong M \otimes_R S$. Let ν be the natural map $M \to M \otimes_R S$. If $\ker(\nu) \neq 0$ then $W \subseteq \ker(\nu)$, since M is uniserial. Hence the image of $W \otimes_R S$ in $M \otimes_R S$ would be zero. Since $W \otimes_R S \cong U \neq 0$, this would contradict the flatness of ${}_RS$. Hence we see that $\ker(\nu) = 0$. Thus U_R has a submodule isomorphic to M. Since U_R and M have the same length, $M \cong U_R$. □

7. 2-step Iterated Basic Idealizers

In order to better understand n-step iterated basic idealizers, it proves helpful to consider, in some detail, the behaviour of 2-step iterated basic idealizers, say $S \supset T \supset R$, with $T = \mathbb{I}_S(A)$, a basic idealizer of type $U = [VW]$. In order to do that, we first describe a one-one correspondence between the generative isomaximal right ideals of T and a certain subset of those of S.

Recall, from 4.8, that the simple T-modules are precisely those of S except that U is replaced by V and W.

7.1. DEFINITION. Let $T = \mathbb{I}_S(A)$, a basic idealizer of type U, with $U_T = [VW]$ and let U' be any simple S-module with $U' \not\cong U$ (and so U'_T is also simple). For generative isomaximal right ideals B' of T, let $(B')^{\uparrow S}$ (*B' up to S*) be as follows:

(β_1) $(B')^{\uparrow S} = B'S$ if B' has type U'_T;
(β_2) $(B')^{\uparrow S} = B'S$ if B' has type W;
(β_3) $(B')^{\uparrow S} = B'A$ if B' has type V.

For generative isomaximal right ideals A' of S, let:

(α_1) $(A')_{\downarrow T} = A' \cap T$ if A' has type U';
(α_2) $(A')_{\downarrow T} = A' \cap T$ if $A' \supset A$ (and hence has type U);
(α_3) $(A')_{\downarrow T} = B'$ where $B'/A' = \text{soc}_T(T/A')$, if $A' \subset A$ (and hence has type U).

Otherwise $(A')_{\downarrow T}$ (*A' down to T*) is not defined. By 'the set β_i', we mean the set of all B' of the type described in (β_i); and we write $B' \in \beta_i$. Similarly, we define 'the set α_i'.

7.2. THEOREM. *Let $T = \mathbb{I}_S(A)$, a basic idealizer of type U, with $U_T = [VW]$. Then the functions $B' \to (B')^{\uparrow S}$ and $A' \to (A')_{\downarrow T}$ are mutually inverse, inclusion-preserving bijections between the sets β_i and α_i for $i = 1, 2, 3$.*

In particular:

(i) *if U'_S is simple with $U' \not\cong U$, then $A' \in \alpha_1$ has type U'_S if and only if $(A')_{\downarrow T}$ has type U'_T;*
(ii) *$(A')_{\downarrow T}$ has type W or V if and only if $A' \supset A$ or $A' \subset A$, respectively.*

Remark. The function $A' \to (A')_{\downarrow T}$ is not inclusion-preserving on the set $\bigcup_i \alpha_i$. For example, take $A' \supset A \supset A''$. Then $(A')_{\downarrow T} \in \beta_2$ and $(A'')_{\downarrow T} \in \beta_3$. Since these have different types, namely W and V, they cannot be related by inclusion.

PROOF. The proof of the theorem comes in three separate parts dealing with $i = 1, 2, 3$. Frequent use will be made of the facts that, since A is isomaximal of type U, say $S/A \cong U^{(n)}$, then $S/T \cong V^{(n)}$ and $T/A \cong W^{(n)}$ [4.4]; and that $(A' \cap T)S = A'$ [6.3].

Case i = 1. Here, $(A')_{\downarrow T} = A' \cap T$ and $(B')^{\uparrow S} = B'S$.

First, let A' be a generative isomaximal right ideal of S of type $U' \not\cong U$. Since the simple T-modules are precisely those of S but with U replaced by V and W, U' is T-simple and is not isomorphic to V or W. Now $(S/T)_T \cong V^{(n)}$; and, by hypothesis, S/A' is a direct sum of copies of U'. Hence $T + A' = S$. Therefore $T/(A' \cap T) \cong (S/A')_T$. Thus $A' \cap T$ is indeed isomaximal of type U'.

We claim that $A' \cap T$ is generative. Otherwise the isomaximal $A' \cap T$ is an ideal of T [3.3]; and since it has type U' we have $U'(A' \cap T) = 0$. Therefore we reach the contradiction $0 = U'(A' \cap T)S = U'A' = (U'S)A' = U'(SA') = U'$, the final equality since A' is generative in S. Thus $B' = A' \cap T \in \beta_1$; and $(B')^{\uparrow S} = (A' \cap T)S = A'$, as desired.

Conversely, let B' be a generative isomaximal right ideal of T of type U'. Since U' is an S-module, $U' \otimes_T S \cong U'$ [2.6(i)] and so $(T/B') \otimes_T S \cong (U')^{(m)}$ for some m. Thus $S/B'S \cong U'^{(m)}$ and hence $B'S$ is isomaximal of type U'. Now, $TB' = T$ since B' is generative in T; hence $SB'S = STB'S = S$ and so $B'S$ is generative in S. Thus $A' = B'S \in \alpha_1$. Finally, each simple submodule X of T/B is isomorphic to U', and hence not to V. So, by 6.14, $B'S \cap T = B'$; that is, $(A')_{\downarrow T} = B'$.

The 'inclusion-preserving' statement is obvious in this case.

Case i = 2. Here, $(A')_{\downarrow T} = A' \cap T$ and $(B')^{\uparrow S} = B'S$.

Choose $A' \in \alpha_2$ and let $B' = A' \cap T$. Since $A' \supset A$ we have $B' \supseteq A \cap T = A$. Moreover, the inclusion is proper since $A = B'$ implies $A = B'S = (A' \cap T)S = A'$, contradicting $A' \in \alpha_2$. We know that $T/A \cong W^{(n)}$ and so $T/B' \cong W^{(m)}$ for some $m < n$. Hence B' is isomaximal of type W. Since $B' \supset A$ which is isomaximal, 3.4 asserts that B' is generative. Thus $B' = (A')_{\downarrow T} \in \beta_2$. Moreover, $A' = (B')^{\uparrow S}$ since $B'S = A'$.

Conversely, consider $B' \in \beta_2$ and let $A' = B'S$. Since $WA = 0$ [4.7] and B' has type W, we have $A \subseteq B'$. However, $A \neq B'$ since A is an ideal of T and B' is generative. Therefore $A \subset B' \subseteq B'S = A'$. Since A is generative isomaximal in S, so too is A'. Thus $A' \in \alpha_2$. To check $(A')_{\downarrow T} = B'$ (i.e. $B'S \cap T = B'$), apply 6.14, noting that T/B' contains no copies of V.

The 'inclusion-preserving' statement is again obvious in this case.

Case i = 3. Here $(A')_{\downarrow T} = B'$ where $B'/A' = \mathrm{soc}_T(T/A')$ and $(B')^{\uparrow S} = B'A$.

First, let $A' \in \alpha_3$ be given, and define B' by the socle formula above. Since A' is isomaximal of type U and $S \supset T \supset A \supset A'$, the semisimplicity of $(S/A')_S$ ensures that A/A' is an S-direct summand of S/A' and so $A/A' \cong U^{(b)}$ for some $b > 0$. Further, A/A' is also a T-direct summand of S/A' and hence also of T/A'. Therefore

$$T/A' \cong A/A' \oplus T/A \cong U^{(b)} \oplus W^{(n)}. \tag{7.2.1}$$

However, $U_T = [VW]$ which is uniserial of length 2. So (7.2.1) implies that $B'/A' = \mathrm{soc}_T(T/A') \cong W^{(b)} \oplus W^{(n)}$ and hence $T/B' \cong V^{(b)}$. Thus B' is isomaximal of type V, as desired.

We claim that B' is generative in T. If not, then, by 3.3, B' is an ideal of T; and so, since B' has type V, we have $VB' = 0$. However $(S/T)_T \cong V^{(n)}$ and so $SB' \subseteq T$. But then $SA' \subseteq SB' \subseteq T$, contrary to the hypothesis that A' is generative in S. Hence $B' \in \beta_3$, as desired.

Finally, since $B'/A' \cong W^{(b+n)}$ and since $WA = 0$ [4.7] then $B'A \subseteq A'$; and, since A is generative in S, $B'A \supseteq A'A = A'SA = A'S = A'$. So $B'A = A'$; that is, $(B')^{\uparrow S} = A'$.

Conversely, let $B' \in \alpha_3$ and $A' = (B')^{\uparrow S} = B'A$. We need to establish the various properties of A'. Since A is generative in S and B' is generative in T we have $SA' = SB'A = STB'A = SA = S$. Thus A' is generative in S.

The rest of this proof uses the following facts, from 4.7: $V \otimes_T S = 0$; $W \otimes_T S \cong U$; and $V \otimes_T A \cong U$.

Since A is an ideal of T, $B'A \subseteq A$ and so $0 \neq S/B'A$. Now B' is isomaximal of type V, so $T/B' \cong V^{(p)}$ for some $p > 0$. Hence $S/B'S \cong (T/B') \otimes_T S \cong V^{(p)} \otimes_T S = 0$. Therefore $S = B'S$ and so

$$0 \neq S/B'A = B'S/B'A. \tag{7.2.2}$$

Since B' is a T-submodule of an S-module, we have $B' \otimes_T S \cong B'S$ via multiplication [4.13(i)]; and so $S/B'A = B'S/B'A \cong (B'/B'A) \otimes_T S$. Since $T/A \cong W^{(n)}$ and $B'/B'A$ is a right T/A-module, $B'/B'A \cong W^{(c)}$ for some c. Therefore $S/B'A \cong W^{(c)} \otimes_T S \cong U^{(c)}$ and $c \neq 0$ by (7.2.2). So $A' = B'A$ has type U in S. As already observed, $B'A \subseteq A$ and $T/B' \cong V^{(p)}$ with $p > 0$. Hence

$$A/B'A \cong (T/B') \otimes_T A \cong V^{(p)} \otimes_T A \cong U^{(p)} \neq 0$$

and so $A' = B'A \subset A$. Thus $A' \in \alpha_3$.

Next, we show that $(A')_{\downarrow T} = B'$; that is, $B'/A' = \mathrm{soc}_T(T/A')$. Since we now know that $A' = B'A$ is generative and isomaximal of type U in S and $A' \subset A$, we can apply (7.2.1), getting $T/B'A = T/A' \cong U^{(b)} \oplus W^{(n)}$. From the preceding paragraph, $B'/B'A \cong W^{(c)}$ and $T/B' \cong V^{(p)}$ with c, p nonzero. Hence $B'/B'A = \mathrm{soc}(T/B'A)$ as required.

Having checked that the two functions yield mutually inverse bijections $\alpha_3 \leftrightarrow \beta_3$, it remains to prove that these bijections are inclusion-preserving. It is clear that the function $B' \to B'A$ is inclusion-preserving. Conversely, suppose $A', A'' \in \alpha_3$ with $A'' \subseteq A'$ and let $B' = (A')_{\downarrow T}$, $B'' = (A'')_{\downarrow T}$. We have just seen that $B'/A' = B'/B'A \cong W^{(c)}$ and $T/B' \cong V^{(p)}$. Thus

$$B' = \cap \{M_T \mid T \supset M \supseteq A' \quad \text{and} \quad T/M \cong V\}.$$

This formula, applied to B'', demonstrates that $B'' \subseteq B'$. □

We now apply 7.2 to show that every 2-step iterated basic idealizer is the intersection of two (one-step) basic idealizers.

7.3. THEOREM. *Let $S \supset T = \mathbb{I}_S(A) \supset R = \mathbb{I}_T(B')$ be a 2-step iterated basic idealizer path, and $A' = (B')^{\uparrow S}$. Then $T' = \mathbb{I}_S(A')$ is a basic idealizer and $R = T \cap T'$.*

PROOF. Let A have type U_S, where $U_T = [VW]$; and let X_T be the simple T-module sliced by R. We argue together the first two cases listed in 7.1; and then deal with the third case.

Cases (i), (ii). Suppose either $X = U'_T$ for some simple $U'_S \not\cong U$ or $X = W$. In each of these cases $A' = B'S$ and $B' = A' \cap T$. To see that $R = T \cap T'$, consider

$x \in T$. We need to show that $x \in \mathbb{I}_T(B') \Leftrightarrow x \in T'$. However,

$$xB' \subseteq B' \Rightarrow xB'S \subseteq B'S \Leftrightarrow x \in T' \Rightarrow x(A' \cap T) \subseteq A' \cap T \Leftrightarrow xB' \subseteq B'.$$

Case (iii). Here $X = V$, $A' = B'A$, and $B'/(A' \cap T) = \mathrm{soc}(T/(A' \cap T)$. To see that $R = \mathbb{I}_T(B')$, consider $x \in T$. Then

$$xB' \subseteq B' \Rightarrow xB'A \subseteq B'A \Leftrightarrow xA' \subseteq A' \Leftrightarrow x(A' \cap T) \subseteq A' \cap T.$$

However, $B'/(A' \cap T)$ is an invariant submodule of $T/(A' \cap T)$, being its socle. So $x(A' \cap T) \subseteq A' \cap T \Rightarrow xB' \subseteq B'$. □

Remark: We note that, in the result above, either the simple types of A and A' differ or else A and A' are distinct and comparable by inclusion. This condition on A, A' is imposed in the following result. We shall see in 8.18 that if this condition fails, then $T \cap T'$ cannot be a 2-step iterated basic idealizer from S.

7.4. COROLLARY. *Let $T = \mathbb{I}_S(A)$ and $T' = \mathbb{I}_S(A')$, basic idealizers of respective types U, U'. Suppose that either $U \ncong U'$, or else A and A' are comparable by inclusion and $A \neq A'$. Then $T \cap T'$ is a 2-step iterated basic idealizer from S, namely $T \cap T' = \mathbb{I}_T(B')$, where $B' = (A')_{\downarrow T}$.*

PROOF. B' is generative isomaximal in T by 7.2, so we can apply 7.3. □

7.5. COROLLARY (Change of path). *Let $S \supset T \supset R$ be a 2-step iterated basic idealizer path, where T slices U_S into $U_T = [VW]$.*

(i) *If R slices U'_T for some simple $U'_S \ncong U$, then there is another 2-step iterated basic idealizer path $S \supset T' \supset R$ where T' slices U'_S and R slices U_T.*
(ii) *If R slices W, then there is another 2-step iterated basic idealizer path $S \supset T' \supset R$ where T' slices U and R slices V.*
(iii) *If R slices V, then there is another 2-step iterated basic idealizer path $S \supset T' \supset R$ where T' slices U and R slices W.*

PROOF. In more detail, let the given path be $S \supset T = \mathbb{I}_S(A) \supset R = \mathbb{I}_T(B')$. Let $T' = \mathbb{I}_S(A')$ where $A' = (B')^{\uparrow S}$. Then $R = T \cap T'$, by 7.3. Since $T' = \mathbb{I}_S(A')$ is a basic idealizer, we can set $B = A_{\downarrow T'}$, and so $B^{\uparrow S} = A$. Then $S \supset T' = \mathbb{I}_S(A') \supset R' = \mathbb{I}_{T'}(B)$ is a 2-step iterated basic idealizer path, and a second application of 7.3 (with the roles of A and A' interchanged) yields $R' = T' \cap T = R$. In each case, 7.1 now shows that A' and B have the required types. □

7.6. REMARK. Let $S \supset T \supset R$ be a 2-step iterated basic idealizer in which T slices U_S and R slices X_T. Let $S \supset T = \mathbb{I}_S(A) \supset R = \mathbb{I}_T(B')$ be a 2-step iterated basic idealizer path which slices U_S and then X. If $X \cong W$ then $U_R = [V_1V_2V_3]$ where $V \cong V_1$ and $W \cong [V_2V_3]$. If $X \cong V$ then $U_R \cong [V_1V_2V_3]$ where $V \cong [V_1V_2]$ and $W \cong [V_3]$.

8. Multichain Idealizers

Let R be an iterated basic idealizer from a ring S. There is an equivalent description of such rings R — intrinsically in terms of S — as the simultaneous idealizer of certain families of right ideals of S called 'multichains'. This will be used explicitly in §25. It also gives insight into several properties of iterated basic idealizers that are not otherwise apparent. For example, 8.16 shows that every decreasing chain of rings from S to R can be refined to a basic idealizer path; and 8.17 describes explicitly all basic idealizer paths from S to R.

8.1. DEFINITIONS. Let S be a ring. A chain $\mathcal{A} = \{A_1 \subset A_2 \subset \ldots \subset A_n\}$ of generative isomaximal right ideals of S is called a *basic chain* and n is its *length*. The subring $\mathbb{I}_S(\mathcal{A}) = \cap\{\mathbb{I}_S(A_i) \mid A_i \in \mathcal{A}\}$ is called a *basic chain idealizer* from S. If A_1, and hence each A_i, has type U, we say $\mathcal{A}$ and $\mathbb{I}_S(\mathcal{A})$ have *type* U .

More generally, a *multichain* $\mathcal{A}$ *of type* $\{U_1, \ldots, U_t\}$ is the union of a finite set of basic chains $\mathcal{A}_1, \ldots, \mathcal{A}_t$ of distinct respective types $U_1, \ldots, U_t$. Each $\mathcal{A}_i$ we term a *component basic chain* in $\mathcal{A}$. We define $\mathbb{I}_S(\mathcal{A}) = \cap\{\mathbb{I}_S(A_i) \mid A_i \in \mathcal{A}\}$ and call this a *multichain idealizer of type* $\{U_1, \ldots, U_t\}$ *from* S . The *number of members of* $\mathcal{A}$ is the number of right ideals $A_i \in \mathcal{A}$, denoted by $|\mathcal{A}|$.

We adopt the convention that a basic chain or multichain may be empty, in which case its type is empty and its idealizer is $\mathbb{I}_S(\varnothing) = S$.

Remark. The requirement that distinct component basic chains must have distinct types will be explained by 8.18.

First, we aim to provide an inductive process for dealing with multichain idealizers.

8.2. NOTATION. Let $T = \mathbb{I}_S(A)$, a basic idealizer, and let $\mathcal{C}$ be a set of right ideals of S such that the function $\downarrow T$ (defined in 7.1) can be applied to each member of $\mathcal{C}$. Then we let $\mathcal{C}_{\downarrow T}$ denote the resulting set of right ideals of T. Similarly we define $\mathcal{D}^{\uparrow S}$ for appropriate sets of right ideals $\mathcal{D}$ of T.

This function behaves well on appropriate multichains.

8.3. LEMMA. *Let* $T = \mathbb{I}_S(A)$, *a basic idealizer,* $\mathcal{A}$ *a multichain in* S *containing* A, *and* $\mathcal{B}$ *a multichain in* T. *Then:*

(i) $(\mathcal{A} - \{A\})_{\downarrow T}$ *is a multichain in* T, *with component basic chains* $(\mathcal{C} - \{A\})_{\downarrow T}$ *where* $\mathcal{C}$ *ranges over the component basic chains of* $\mathcal{A} - \{A\}$;

(ii) $\mathcal{B}^{\uparrow S} \cup \{A\}$ *is a multichain in* S, *and* $A \notin \mathcal{B}^{\uparrow S}$;

(iii) $(\mathcal{A} - \{A\})_{\downarrow T} = \mathcal{B}$ *if and only if* $\mathcal{B}^{\uparrow S} = \mathcal{A} - \{A\}$.

PROOF. (i) Evidently, $\mathcal{A} - \{A\}$ is a multichain and, by 7.1, the function $\downarrow T$ applies to each of its members. By 7.2, the functions $\downarrow T$ and $\uparrow S$ are inclusion-preserving bijections between each pair of sets $\alpha_i \leftrightarrow \beta_i$ in 7.1. It follows that, if $\mathcal{A}_i$ is a component basic chain of $\mathcal{A}$, then $\mathcal{B}_i = (\mathcal{A}_i)_{\downarrow T}$ is a basic chain in T.

Clearly $(\mathcal{A} - \{A\})_{\downarrow T} = \cup_i \mathcal{B}_i$. To see that this is a multichain and each $\mathcal{B}_i$ is a component basic chain of it, we need to show that $\mathcal{B}_i$ and $\mathcal{B}_j$ have distinct types when $i \neq j$; equivalently, the type of any $A' \in \mathcal{A} - \{A\}$ is determined by that of $B' = (A')_{\downarrow T}$. If $B' \in \beta_2$ or β_3, this is clear from the bijections in the previous paragraph. The remaining case is $B' \in \beta_1$, in which case $A' \in \alpha_1$ and has some type $U' \not\cong U$. In this situation, 7.2(i) states that A' has type $(U')_S$ if and only if $(A')_{\downarrow T}$ has type U'_R, completing the proof of (i).

(ii)(iii) By 7.1, the function $\uparrow S$ applies to each member of $\mathcal{B}$. The bijections $\alpha_i \leftrightarrow \beta_i$ show that $A \notin \mathcal{B}^{\uparrow S}$, proving part of (ii). These bijections also prove (iii), although we do not yet know that $\mathcal{B}^{\uparrow S}$ is a multichain in S. We omit the details, which are similar to the proof, in (i), that $(\mathcal{A} - \{A\})_{\downarrow T}$ is a multichain in T.

To complete the proof of (ii) we need to show that $\mathcal{B}^{\uparrow S} \cup \{A\}$ is a multichain. Since A has type U, it suffices to show that A is comparable, by inclusion, with all elements of type U in the multichain $(B')^{\uparrow S}$ (if any exist). The correspondences in 7.2 and 7.1 show that $(B')^{\uparrow S}$ has type U if and only if B' has type W or V. In the former case we have $A \subset (B')^{\uparrow S}$, and in the latter case we have $(B')^{\uparrow S} \subset A$. □

The next result provides the inductive tool.

8.4. THEOREM. *Let $\mathcal{A}$ be a multichain in S, $A \in \mathcal{A}$ and $T = \mathbb{I}_S(A)$. Then $\mathcal{B} = (\mathcal{A} - \{A\})_{\downarrow T}$ is a multichain in T and $\mathbb{I}_S(\mathcal{A}) = \mathbb{I}_T(\mathcal{B})$.*

PROOF. $\mathcal{B}$ is a multichain by 8.3. By definition:

$$\mathbb{I}_S(\mathcal{A}) = \cap\{\mathbb{I}_S(A_i) \mid A_i \in \mathcal{A}\} = \cap\{T \cap \mathbb{I}_S(A_i) \mid A_i \in \mathcal{A} - \{A\}\}.$$

We know from 7.4 that each $T \cap \mathbb{I}_S(A_i) = \mathbb{I}_T(B_i)$ where $B_i = (A_i)_{\downarrow T}$. Thus $\mathbb{I}_S(\mathcal{A}) = \cap\{B_i \mid B_i \in \mathcal{B}\} = \mathbb{I}_T(\mathcal{B})$. □

Next we consider the relationship between the types of the two multichains involved in the inductive step.

8.5. LEMMA. *Let $T = \mathbb{I}_S(A)$, a basic idealizer of type $U = [VW]$. Let $\mathcal{A}$ be a multichain in S containing A, and $\mathcal{B} = (\mathcal{A} - \{A\})_{\downarrow T}$. Suppose $\mathcal{A}$ has type $\{U_1, \dots, U_t\}$, $\mathcal{A}_1$ is the component basic chain in $\mathcal{A}$ of type $U = U_1$ and $\mathcal{U}' = \{U_2, \dots, U_t\}$. Then:*

$\mathcal{B}$ has type $\mathcal{U}'$ if $\mathcal{A}_1 = \{A\}$;
$\mathcal{B}$ has type $\mathcal{U}' \cup \{V\}$ if $\mathcal{A}_1 \neq \{A\}$ and A is maximal in $\mathcal{A}_1$;
$\mathcal{B}$ has type $\mathcal{U}' \cup \{W\}$ if $\mathcal{A}_1 \neq \{A\}$ and A is minimal in $\mathcal{A}_1$;
$\mathcal{B}$ has type $\mathcal{U}' \cup \{V\} \cup \{W\}$ otherwise.

PROOF. Take $A' \in \mathcal{A} - \{A\}$ and apply 7.2. If A' has type U then $(A')_{\downarrow T}$ has type W or V according as $A' \supset A$ or $A' \subset A$ respectively. Otherwise $(A')_{\downarrow T}$ has one of the types U_i with $i \neq 1$. □

8.6. THEOREM. *Let R be a subring of S. The following are equivalent:*

(i) *R is an n-step iterated basic idealizer from S;*
(ii) *R is an n-member multichain idealizer from S.*

PROOF. (i) ⇒ (ii) Let T be the first ring after S in the idealizing path; say $T = \mathbb{I}_S(A)$. By induction we may assume that R is an $(n-1)$-member multichain idealizer from T; say $R = \mathbb{I}_T(\mathcal{B})$. Let $\mathcal{A} = (\mathcal{B})^{\uparrow S} \cup \{A\}$. Then, by 8.3, $\mathcal{A}$ is a multichain in S and $(\mathcal{A} - \{A\})_{\downarrow T} = \mathcal{B}$. Then 8.4 shows that $\mathbb{I}_S(\mathcal{A}) = \mathbb{I}_T(\mathcal{B})$.

(ii) ⇒(i): Choose $A \in \mathcal{A}$, let $T = \mathbb{I}_S(A)$, and let $\mathcal{B} = (\mathcal{A} - \{A\})_{\downarrow T}$. Then $R = \mathbb{I}_S(\mathcal{A}) = \mathbb{I}_T(\mathcal{B})$ [8.4]. The latter is the idealizer of a multichain having one fewer member than $\mathcal{A}$. By induction, R is an iterated basic idealizer from T and hence also from S. □

8.7. REMARK. The proof that (ii) ⇒(i) above makes clear that in obtaining R as an iterated basic idealizer we may choose any member of the multichain and take its idealizer as the first step.

8.8. THEOREM. *Let $R = \mathbb{I}_S(\mathcal{A})$, a multichain idealizer of type $\{U_1, \dots, U_t\}$. Then the simple S-modules that do not remain R-simple are $U_1, \dots, U_t$.*

PROOF. Choose some $A \in \mathcal{A}$ of type $U = U_1$, and let $U_T = [VW]$. Then $R = \mathbb{I}_S(\mathcal{A}) = \mathbb{I}_T(\mathcal{B})$ where $\mathcal{B} = (\mathcal{A} - \{A\})_{\downarrow T}$, a multichain in T whose type is $\mathcal{U}' = \{U_2, \dots, U_t\}$ together with some or none of V, W [8.4, 8.5]. Since the multichain $\mathcal{B}$ has one fewer member than $\mathcal{A}$, we may apply induction to conclude that the simple T-modules that do not remain R-simple are $U_2, \ldots, U_t$ together with some or none of V, W.

Since U is non-simple over T, it remains non-simple over the smaller ring R. All remaining simple S-modules $U' \not\cong U$ remain T-simple [4.8]; and, by the previous paragraph, the U' that do not remain R-simple are $U_2, \ldots, U_t$. □

The next result is our first instance where a property of an iterated basic idealizer is much better displayed by viewing it as a multichain idealizer.

8.9. THEOREM. *Let $R = \mathbb{I}_S(\mathcal{A})$, a multichain idealizer of type $\{U_1, \ldots, U_t\}$. Then each U_i is uniserial of length $\lambda(U_i)_R = |\mathcal{A}_i| + 1$ where $\mathcal{A}_i$ is the component basic chain of $\mathcal{A}$ having type U_i.*

PROOF. R is an iterated basic idealizer from S by 8.6. Hence, by 6.4, each U_i is uniserial over R. We proceed via induction on $|\mathcal{A}|$. Choose any $A \in \mathcal{A}$, say of type $U = U_1$, and let $T = \mathbb{I}(A)$. By 4.4, U_T is uniserial of length 2, say $U_T = [VW]$. Further, 8.3 tells us: that $R = \mathbb{I}_T(\mathcal{B})$ where $\mathcal{B}$ is a multichain with $|\mathcal{B}| = |\mathcal{A}| - 1$; that the simple types of $\mathcal{B}$ are $\{U_2, \ldots, U_t\}$ together with 0, 1 or 2 from $\{V, W\}$; that, for $i > 1$, the component basic chain in $\mathcal{B}$ of type U_i has length $|\mathcal{A}_i|$; and that the sum of the lengths (possibly zero) of the component basic chains in $\mathcal{B}$ of type V and W is $|\mathcal{A}_1| - 1$. The result now follows immediately by induction. □

There is an immediate consequence of 8.6 for semimaximal idealizers. Recall that every generative semimaximal right ideal is an intersection of generative isomaximal right ideals of distinct types [3.5].

8.10. THEOREM. *Let $A = \cap_{i=1}^{n} A_i$, an intersection of generative isomaximal right ideals A_i of distinct types. Then the ring $R = \mathbb{I}_S(A)$ is an n-step basic idealizer from S.*

PROOF. The set $\{A_1, \ldots, A_n\}$ is a multichain each of whose component basic chains is a singleton. So 8.6 applies. □

8.11. THEOREM. *Let $R = \mathbb{I}_S(\mathcal{A})$, a multichain idealizer, and let U_S be one of the simple types of $\mathcal{A}$. Let the unique composition series of U_R be $U \supset X_1 \supset \ldots \supset X_m \supset 0$.*

(i) *If $A \in \mathcal{A}$ has type U, then, for some i, $\operatorname{ann}_S(X_i) = A$ and $\operatorname{ann}_U(A) = X_i$.*
(ii) *For each i, there is an $A \in \mathcal{A}$ such that $\operatorname{ann}_S(X_i) = A$.*

PROOF. Recall that a ring R is an n-step iterated basic idealizer from S if and only if R is an n-member multichain idealizer from S [8.6].

(i) In expressing R as an iterated basic idealizer from S, we can choose the first idealizer step to be $T = \mathbb{I}_S(A)$, as noted in 8.7. Let $U_T = [VW]$. Then 4.6 tells us that $A = \operatorname{ann}_S(W)$ and $W = \operatorname{ann}_U(A)$. We know that U_R is uniserial; and W, being a T-submodule is, a fortiori, an R-submodule too. Thus $W = X_i$ for some i and the result follows.

(ii) Let the type of $\mathbb{I}_S(\mathcal{A})$ be $\{U_1, \ldots, U_t\}$. In forming R from S as an n-step iterated basic idealizer, 6.4 demonstrates that each step adds one to the sum of the composition series lengths of $U_1, \ldots, U_t$ (which remain uniserial throughout the idealization process). Thus the total number of proper nonzero R-submodules of $U_1, \ldots, U_t$ is n, which is the number of members of $\mathcal{A}$. Since, by (i), each A

has a corresponding member X_i of the R-composition series of some U_j, this has exhausted the supply of X_i. Thus (ii) holds. □

8.12. COROLLARY. *Given two multichain idealizers, if* $\mathbb{I}_S(\mathcal{A}) = \mathbb{I}_S(\mathcal{A}')$ *then* $\mathcal{A} = \mathcal{A}'$.

PROOF. This holds because 8.11 gives a description of the members of $\mathcal{A}$ purely in terms of the R-composition series of those simple S-modules that do not remain R-simple. □

8.13. LEMMA. *Let* $R = \mathbb{I}_S(\mathcal{A})$, *a multichain idealizer. Then the set of maximal subrings* T *relative to* $S \supset T \supseteq R$ *coincides with the set of rings* $\mathbb{I}_S(A')$ *with* $A' \in \mathcal{A}$.

PROOF. We already know, from 4.15, that $\mathbb{I}_S(A')$ is a maximal subring of S for each $A' \in \mathcal{A}$; and, of course, $\mathbb{I}_S(A') \supseteq \mathbb{I}_S(\mathcal{A})$.

Conversely, let T be as described and let I be the intersection of the members of $\mathcal{A}$. For each $A \in \mathcal{A}$, we have $IA \subseteq I \subseteq A$, and hence $I \subset \mathbb{I}_S(\mathcal{A}) = R$. Since each $A \in \mathcal{A}$ is generative and isomaximal, I is generative and semimaximal in S [3.5].

Next, let $B = \{b \in S \mid bS \subseteq T\} \subseteq T$. Then B is a right ideal of S and $TB \subseteq B$; hence $T \subseteq \mathbb{I}_S(B) \subseteq S$. Now $I \subseteq B$ because $IS = I \subseteq R \subseteq T$. Hence B (like I) is generative in S and yet $B \neq S$ since $1 \notin B$. So B is not an ideal of S and hence $\mathbb{I}_S(B) \subset S$. Since $T \subseteq \mathbb{I}_S(B)$, maximality of T shows that $T = \mathbb{I}_S(B)$.

Since $I \subseteq B$, B is generative and semimaximal in S. So $B = \cap_i B_i$ with the B_i generative isomaximal right ideals of S, of distinct types [3.5]. Therefore $T = \mathbb{I}_S(B) = \cap_i \mathbb{I}_S(B_i) \subset S$ [3.6]; and hence maximality of T implies that $T = \mathbb{I}_S(B_i)$ for some i. Moreover, since $I \subseteq B \subseteq B_i$, the type of the isomaximal B_i is one of the simple types of I, and hence some simple U_S in the type of $\mathcal{A}$.

Let $U_T = [VW]$. Since $T = \mathbb{I}_S(B_i)$, we have $\text{ann}_S(W) = B_i$, by 4.6. Since $0 \neq W \subset U$, the module W_R is a term (other than 0 and U) in the unique composition series of the uniserial U_R; and so 8.11 implies that $\text{ann}_S(W) = A'$ for some $A' \in \mathcal{A}$. Therefore $B_i = A' \in \mathcal{A}$. □

8.14. THEOREM. *Let* $R = \mathbb{I}_S(\mathcal{A})$ *be a multichain idealizer from* S *and* R' *a ring such that* $S \supseteq R' \supseteq R$. *Then* R' *is a multichain idealizer from* S *and* R *is a multichain idealizer from* R'.

In more detail:

(i) *the function* $\mathcal{A}' \to R' = \mathbb{I}_S(\mathcal{A}')$ *is an inclusion-reversing bijection between the collections of subsets* $\mathcal{A}' \subseteq \mathcal{A}$ *and subrings* R' *as above;*

(ii) *if* $R' = \mathbb{I}_S(\mathcal{A}')$ *then* $R = \mathbb{I}_{R'}(\mathcal{C})$ *for some multichain* $\mathcal{C}$ *in* R' *with* $|\mathcal{A}| = |\mathcal{A}'| + |\mathcal{C}|$;

(iii) *if* R'' *is a ring with* $S \supseteq R' \supseteq R'' \supseteq R$, *then* R'' *is a multichain idealizer from* R'.

PROOF. (i),(ii): If $\mathcal{A}'$ is a subset of $\mathcal{A}$, it is immediate that $S \supseteq \mathbb{I}_S(\mathcal{A}') \supseteq R$. On the other hand, suppose that $S \supset R' \supseteq R$. Let T be a subring of S maximal with respect to $S \supset T \supseteq R'$. By 8.13, $T = \mathbb{I}_S(A)$ for some $A \in \mathcal{A}$; and by 8.4, $R = \mathbb{I}_T(\mathcal{B})$ where $\mathcal{B} = (\mathcal{A} - \{A\})_{\downarrow T}$ and thus $|\mathcal{B}| = |\mathcal{A}| - 1$. By induction on $|\mathcal{A}|$, we may assume that $R' = \mathbb{I}_T(\mathcal{B}')$ for some $\mathcal{B}' \subseteq \mathcal{B}$, that $R = \mathbb{I}_{R'}(\mathcal{C})$ is a multichain idealizer from R' and that $|\mathcal{B}'| + |\mathcal{C}| = |\mathcal{B}|$. Let $\mathcal{A}' = (\mathcal{B}')^{\uparrow S} \cup \{A\}$. Then $\mathcal{A}'$ is a multichain in S and $(\mathcal{A}' - \{A\})_{\downarrow T} = \mathcal{B}'$, by 8.3; and so 8.4 shows that $\mathbb{I}_S(\mathcal{A}') = \mathbb{I}_T(\mathcal{B}') = R'$. Note that $|\mathcal{A}'| = |\mathcal{B}'| + 1$; hence (ii) follows. Uniqueness of $\mathcal{A}'$ is proved in 8.12; and it is evident that the bijection is inclusion-reversing. So (i) holds too.

(iii) By applying (i),(ii) to the rings $S \supseteq R'' \supseteq R$, we see that R'' is a multichain idealizer from S. Now apply (i),(ii) to the rings $S \supseteq R' \supseteq R''$. □

8.15. COROLLARY. *Let $R = \mathbb{I}_S(\mathcal{A})$, an n-member multichain idealizer. Then the number of intermediate rings T such that $S \supseteq T \supseteq R$ is 2^n.*

PROOF. By 8.14, the number of such T is the number of subsets of the n-element set $\mathcal{A}$. □

8.16. COROLLARY. *Let R be an iterated basic idealizer from a ring S. Then every chain of rings $S = T_0 \supset T_1 \supset \ldots \supset T_m = R$ can be refined to a basic idealizer path from S to R.*

PROOF. By the equivalence of iterated and multichain idealizers [8.6] we know that R is a multichain idealizer from S. Therefore, by repeated use of [8.14(iii)], every T_i is a multichain idealizer from T_{i-1}, and hence an iterated basic idealizer from T_{i-1}. □

We can now use multichain idealizers to construct all basic idealizer paths intrinsically, in terms of S.

8.17. THEOREM. *Let $R = \mathbb{I}_S(\mathcal{A})$, an n-member multichain idealizer.*

(i) *There is a (1-1)-correspondence between basic idealizer paths from S to R and chains of subsets of $\mathcal{A}$ of the form $\mathcal{A}_1 \subset \ldots \subset \mathcal{A}_n = \mathcal{A}$ such that each $\mathcal{A}_i$ consists of i members. The corresponding idealizer path is $S \supset T_1 \supset \ldots \supset T_n = R$ where each $T_i = \mathbb{I}_S(\mathcal{A}_i)$.*

(ii) *The number of distinct basic idealizer paths from S to R is $n!$.*

PROOF. (i) Let T_i be as described, and let $T_0 = S$. The (1-1) correspondence given by 8.14 shows that each T_i is a maximal subring of T_{i-1}; and 8.16 shows that this inclusion chain from S to R can be refined to a basic idealizer path. So this inclusion chain is already a basic idealizer path.

Conversely, let $T' \supset T''$ be consecutive terms in some basic idealizer path from S to R. Then T'' is a maximal subring of T' by 4.15. The (1-1) correspondence given by 8.14 ensures that there exist $\mathcal{A}' \subset \mathcal{A}'' \subseteq \mathcal{A}$ such that $T' = \mathbb{I}_S(\mathcal{A}')$ and $T'' = \mathbb{I}_S(\mathcal{A}'')$; and the maximality of T'' tells us that $|\mathcal{A}'| = |\mathcal{A}''| - 1$. The result now follows easily.

(ii) Clear from (i). □

Recall that the definition of a multichain in 8.1 requires that the component basic chains have distinct types. The next result shows that, without this restriction, the subrings obtained need not be iterated basic idealizers. It also shows that the hypothesis about comparability by inclusion cannot be dropped from 7.4.

8.18. COROLLARY. *Let A and A' be generative isomaximal right ideals of S, both of type U, but not comparable by inclusion. Then $\mathbb{I}_S(A) \cap \mathbb{I}_S(A')$ is not a multichain idealizer from S nor does it contain one. In particular, $\mathbb{I}_S(A) \cap \mathbb{I}_S(A')$ is not an iterated basic idealizer from S.*

PROOF. Suppose that the ring $T = \mathbb{I}_S(A) \cap \mathbb{I}_S(A')$ satisfies $S \supset T \supseteq R = \mathbb{I}_S(\mathcal{A})$ for some multichain $\mathcal{A}$. Since $S \supset \mathbb{I}_S(A) \supseteq R$, 8.14 implies that $\mathbb{I}_S(A) = \mathbb{I}_S(\mathcal{A}_1)$ for some multichain $\mathcal{A}_1 \subseteq \mathcal{A}$. But then the uniqueness result 8.12 shows that $\{A\} = \mathcal{A}_1$, and hence $A \in \mathcal{A}$. Similarly, $A' \in \mathcal{A}$. Since $\mathcal{A}$ can contain only one component basic chain of type U, we reach the contradiction that A and A' are comparable by inclusion.

Since T is not a multichain idealizer from S, it is not an iterated basic idealizer from S [8.6]. $\square$

The following result will be needed later, in 25.26.

8.19. COROLLARY. *Let $R = \mathbb{I}_S(\mathcal{A})$, a basic chain idealizer of type U. Let T be an intermediate ring, $S \supseteq T \supset R$. Then $\lambda(U_T) < \lambda(U_R)$.*

PROOF. Since $T \supset R$, the (1,1)-correspondence between subsets of $\mathcal{A}$ and rings between S and R [8.14] shows that $T = \mathbb{I}_S(\mathcal{A}')$ for some subset $\mathcal{A}' \subset \mathcal{A}$. Hence 8.9 gives $\lambda(U_R) = |\mathcal{A}| + 1 > |\mathcal{A}'| + 1 = \lambda(U_T)$. $\square$

9. Triangular Matrix Rings

The main result of this section [9.1] provides a definitive example, of use later. It shows that every basic chain idealizer from a simple Artinian ring S (or, equivalently, every iterated basic idealizer from S [8.6]) is isomorphic to a full block upper triangular matrix ring over the division ring associated with S. The rest of this section demonstrates some consequences for basic chain idealizers over arbitrary rings and gives idealizer-theoretic proofs of several properties of block upper triangular matrix rings.

9.1. THEOREM. *Let $S = M_s(D)$ for some division ring D and some s; and let R be the full $(n+1)$-block upper triangular matrix ring over D with block sizes, say, $s_1, \ldots, s_{n+1}$ (counting from the top).*

(i) *$R = \mathbb{I}_S(\mathcal{A})$ where $\mathcal{A} = \{A_1 \subset \ldots \subset A_n\}$, the n-term basic chain such that each right ideal A_k consists of the top $\sum_{i=1}^k s_i$ rows of S followed by rows of zeros.*

(ii) *Let $\mathcal{A}' = \{A'_1 \subset \ldots \subset A'_n\}$, any basic chain with each $\lambda(A'_k) = \lambda(A_k)$. Then $\mathbb{I}_S(\mathcal{A}') \cong \mathbb{I}_S(\mathcal{A})$ via an inner automorphism of S.*

PROOF. (i) Each $A_k = f_k S$ where f_k is the diagonal matrix with entry 1 in the first $\sum_{i=1}^k s_i$ diagonal positions and zeros elsewhere. We saw in 4.3 that $\mathbb{I}_S(eS) = eS + S(1-e)$ for any idempotent $e \in S$. Applied to the idempotent f_k we see that $\mathbb{I}_S(f_k S)$ is upper triangular with two diagonal blocks, the upper of size $\sum_{i=1}^k s_i$ and the lower of size $s - \sum_{i=1}^k s_i$. It is easily checked that the intersection $\mathbb{I}_S(\mathcal{A})$, of these subrings as k varies, is the block upper triangular ring R.

(ii) Let U_S be the (unique) simple S-module and $k \in \{1, \ldots, n+1\}$. Since S_S is semisimple, there exists B'_k such that each $A'_k = A'_{k-1} \oplus B'_k$, where $A_0 = A'_0 = 0$ and $A_{n+1} = A'_{n+1} = S$. Hence $B'_k \cong U^{(s_k)}$ and $A'_k = \oplus_{i=1}^k B'_i$; in particular, $S = A'_{n+1} = \oplus_{k=1}^{n+1} B'_k$.

Since each $\lambda(A'_k) = \lambda(A_k)$, we similarly find right ideals $B_k \cong U^{(s_k)} \cong B'_k$ such that each $A_k = \oplus_{i=1}^k B_i$; in particular, $S = A_{n+1} = \oplus_{k=1}^{n+1} B_k$. The direct sum ϕ of isomorphisms $B'_k \cong B_k$ is an automorphism of S_S that carries each A'_k onto A_k. Let $u = \phi(1)$, which must be a unit of S; and each $uA'_k = A_k$. Then the inner automorphism $s \to usu^{-1}$ of the ring S carries every right ideal A'_k onto $A_k u^{-1} = A_k$, and hence carries $\mathbb{I}_S(\mathcal{A}')$ onto $\mathbb{I}_S(\mathcal{A})$. $\square$

Some further details about these rings, which will be useful later, need the following definition.

9.2. DEFINITION. Suppose that T is a ring and W a simple T-module such that $T/\operatorname{ann}(W) \cong M_k(D)$ for some division ring D — or, equivalently, $T/\operatorname{ann}(W) \cong W^{(k)}$. We say T *has finite rank* k *at* W and write $\rho(T, W) = k$. Of course, every Artinian ring has finite rank at every simple module.

9.3. LEMMA. *Let $S = M_s(D)$ for some division ring D and some s. Let R be a full block upper triangular ring of S, with diagonal blocks D_i of size s_i and simple D_i-module W_i $(i = 1, \ldots, n+1)$.*

(i) *The top row $U = e_{11}S$ of R is a simple S-module, and a uniserial R-module of length $n+1$.*

(ii) *Each term Y_i $(i \neq 0)$ of the unique composition series*

$$U = Y_0 \supset Y_1 \supset \ldots \supset Y_i \supset \ldots \supset Y_{n+1} = 0 \tag{9.3.1}$$

is formed by replacing the first $s_1 + \cdots + s_i$ entries of U by zero.

(iii) *Each $Y_i/Y_{i+1} \cong W_i$ and $\rho(R, W_i) = s_i$; and every simple R-module is isomorphic to a unique W_i.*

PROOF. The $(n+1)$-block upper triangular matrix ring R is an n-member basic chain idealizer from S [9.1], and hence an n-step iterated basic idealizer from S [8.6]. Recall that 6.4 describes all simple R-modules in terms of composition factors of simple S-modules. In this case, U is the only (isomorphism type) of simple S-module; and then 6.4 tells us that R has precisely $n+1$ types of simple module and that each occurs just once in the unique composition series of U_R.

Let J be the set of strictly block upper triangular matrices in R. Then J is an ideal and $R/J \cong \oplus_i D_i$. Therefore each induced projection map $R \twoheadrightarrow D_i$ is a ring homomorphism, showing that each W_i is a simple R-module. Moreover the projection maps $R \twoheadrightarrow D_i$ have distinct kernels, K_i say — namely the set of matrices in R formed by replacing D_i by zeros. This shows that the $n+1$ simple R-modules W_i are mutually non-isomorphic, and hence constitute all of the non-isomorphic simple R-modules.

Note next that $Y_iK_i = Y_{i+1}$. So an easy induction tells us that each Y_i is a right ideal of R; and we also see that Y_i/Y_{i+1} is a D_i-module. So the set of Y_i form the unique composition series of U_R and $Y_i/Y_{i+1} \cong W_i$. Finally, it is clear that $\rho(R, W_i) = s_i$ since $R/\operatorname{ann}(W_i) = R/K_i \cong D_i$. □

9.4. PROPOSITION. *Let $R = \mathbb{I}_S(\mathcal{A})$ where $\mathcal{A} = \{A_1 \subset A_2 \subset \ldots \subset A_n\}$, a basic chain of type U, and S has finite rank s at U. Let each $A_i/A_{i-1} \cong U^{(s_i)}$, taking $A_0 = \operatorname{ann}_S(U)$, and $S = A_{n+1}$.*

(i) *U_R is uniserial of length $n+1$, with composition factors (say) W_1, W_2, $\ldots$, W_{n+1} from top to bottom.*

(ii) *Each $\rho(R, W_i) = s_i$ and $\sum_i s_i = s$.*

PROOF. Note that the ideal A_0 of S is contained in every A_i and hence in R. Let $\bar{A}_i = A_i/A_0$, $\bar{S} = S/A$, $\bar{R} = R/A_0$. Then U is a simple module over the simple Artinian ring $\bar{S}$, $\rho(\bar{S}, U) = s$, and $\bar{\mathcal{A}} = \{\bar{A}_i\}$ is a basic $\bar{S}$-chain of type U. Moreover, it is easily verified that $\bar{R} = \mathbb{I}_{\bar{S}}(\bar{\mathcal{A}})$. Therefore we may assume for the rest of the proof that $A_0 = 0$ and so that S is a simple Artinian ring, say with $S = M_s(D)$ for some division ring D. Hence R is the $(n+1)$-block upper triangular matrix ring with block sizes s_i described in 9.1(i) and 9.3. So (i) and (ii) now follow immediately from 9.3. □

This has consequences even when S fails to have finite rank at U, since all except the first step in any basic chain idealizer of type U can be viewed as taking place in the situation described above.

9.5. PROPOSITION. *Let S be a ring, $\mathcal{A} = \{A_1 \subset A_2 \subset \ldots \subset A_n\}$ a basic chain of type U and $R = \mathbb{I}_S(\mathcal{A})$. Let $T = \mathbb{I}_S(A_1)$ and $U_T = [VW]$.*

(i) *T has finite rank at W.*

(ii) *R is a basic chain idealizer of type W from T using the basic chain $A_2 \cap T \subset A_3 \cap T \subset \ldots \subset A_n \cap T$.*

(iii) *U_R has a unique composition series of length $n+1$, say with composition factors, from top to bottom, $V = W_1, W_2, \ldots, W_{n+1}$. Then, for each $i \geq 2$, W_i has finite rank, namely $\rho(R, W_i) = s_i$ where $A_i/A_{i-1} \cong U^{(s_i)}$ (taking $A_{n+1} = S$).*

PROOF. (i) Of course A_1 is an ideal of the basic idealizer T. Elementary results about basic idealizers tell us that A_1 is the annihilator of W [4.6] and T/A_1 is a simple Artinian ring [4.2].

(ii) Since each right ideal A_i contains A_1, 7.1 provides the notation $A_i \cap T = (A_i)_{\downarrow T}$; and then 8.2 describes the basic chain in (ii) as $(\mathcal{A} - \{A\})_{\downarrow T}$. However, 8.4 states that $R = \mathbb{I}_T(\mathcal{A} - \{A\})_{\downarrow T}$.

(iii) This follows from 9.4, provided we can show that $(A_{i+1} \cap T)/(A_i \cap T) \cong W^{(s_i)}$. However, 7.2 establishes a (1,1)-correspondence between right ideals B' of T with $B' \supset A_1$ and right ideals A' of S with $A' \supset A_1$. Since $A_i/A_{i-1} \cong U^{(s_i)}$, it follows that $(A_i \cap T)/(A_{i-1} \cap T) \cong W^{(s_i)}$. □

We now apply this to obtain a result, 9.8, of use later (in 25.19).

9.6. DEFINITION. Suppose that there are rings $R \subseteq R'$ and finite length modules $U_R \subseteq V_{R'}$. We will say that an R-composition series $0 \subset U_1 \subset \ldots \subset U_n = U$ *fully embeds* in an R'-composition series $0 \subset V_1 \subset \ldots \subset V_n = V$ if $U_i = V_i \cap U$ for each i. Note the immediate consequence for composition factors that U_{i+1}/U_i embeds in V_{i+1}/V_i for each i.

9.7. LEMMA. *Let $S \subseteq S'$ be rings and A be an isomaximal generative right ideal in S of simple type U, and suppose that $U \otimes_S S' = U'_{S'}$ is simple.*

(a) *Suppose that $S/A \cong U^{(t)}$ and let $A' = AS'$. Then:*
   (i) *A' is generative, $A' \cap S = A$ and $S'/A' \cong U'^{(t)}$;*
   (ii) *U embeds in U'.*

(b) *Let $T = \mathbb{I}(A)$ and $T' = \mathbb{I}(A')$. Then:*
   (i) *U_T and $U'_{T'}$ are uniserial of length 2, say with socles W and W' respectively;*
   (ii) *U_T fully embeds in $U'_{T'}$;*
   (iii) *$W \otimes_T T' \cong W'$.*

PROOF. (a)(i) Since $SA = S$ then $S'A' = S'SAS' = S'$; so A' is generative. From the short exact series

$$0 \to A \to S \to U^{(t)} \to 0$$

we get the exact sequence

$$A \otimes_S S' \to S \otimes_S S' \to U^{(t)} \otimes_S S' \to 0.$$

Now $S \otimes_S S' \cong S'$ under multiplication; so we deduce that $S'/A' \cong U'^{(t)}$. Note that S'/A' has the same length as S/A. Let $A' \cap S = B$. Obviously $B \supseteq A$; so the hypotheses on A are all true of B. Hence, if S/B has length b then the same is true of S'/BS'. However, $A \subseteq B \subseteq A' = AS$ and so $BS' = AS'$. This is a contradiction unless $b = t$; in which case $A = A' \cap S$.

(ii) Let M be a maximal right ideal containing A; so $S/M \cong U$. By (i), applied to M rather than A, we see that $MS' \cap S = M$. Thus $U \cong S/M \hookrightarrow S'/MS' \cong U'$.

(b) We make use of some properties of a basic idealizer.

(i) This is immediate from 4.4.

(ii) By 4.6, $W = \text{ann}_U(A)$ and likewise $W' = \text{ann}_{U'}(A')$. Since $A' = AS$ it is clear that $\text{ann}_U(A) \subseteq \text{ann}_{U'}(A')$; i.e. $W \subseteq W'$. However $WAS' = 0$ since $WA = 0$; so $W \subseteq W' \cap U$. Also $(W' \cap U)A \subseteq W'AS = 0$; and so $W' \cap U \subseteq W$. Thus $W = W' \cap U$ and so U_T fully embeds in U'.

(iii) We know from 4.4 that $T/A \cong W^{(t)}$ Likewise $T'/A' \cong W'^{(t)}$. However $T'/A' = T'/AT' \cong (T/A) \otimes_T T' \cong W^{(t)} \otimes_T T'$. Hence $W \otimes_T T' \cong W'$. □

9.8. PROPOSITION. *Let $S \subseteq S'$ be rings, let $\mathcal{A} = \{A_1 \subset \ldots \subset A_n\}$ be a basic chain in S of simple type U and suppose that $U \otimes_S S' = U'_{S'}$ is simple. Let $\mathcal{A}' = \{A_1S' \subset \ldots \subset A_nS'\}$, $R = \mathbb{I}(\mathcal{A})$ and $R' = \mathbb{I}(\mathcal{A}')$. Then:*

(i) *$\mathcal{A}'$ is a basic chain in S' of type U';*

(ii) *U_R and $U'_{R'}$ are both uniserial of length $n+1$;*

(iii) *U_R is fully embedded in $U'_{R'}$.*

PROOF. (i) This is immediate from 9.7(a)(i).

(ii) This follows from (i) together with 8.9 since each of R and R' is a basic chain idealizer.

(iii) We prove this by induction on n. When $n = 1$, this is shown by 9.7(b)(ii). Now let $A = A_1$ and $A' = AT'$; and let $T = \mathbb{I}_S(A)$ and $T' = \mathbb{I}(A')$. Note that $T \subseteq T'$. By 9.7, each of U_T and $U'_{T'}$ is uniserial of length 2; say with composition factors V, W and V', W' respectively from top to bottom. Further U_T embeds fully in $U'_{T'}$.

We now wish to apply the induction hypothesis to the two rings T and T' and the simple module W_T. First we note that $B_i = A_i \cap T$ is an isomaximal right ideal of type W (since, by 4.4, $T/A \cong W^{(t)}$). Moreover $W \otimes_T T' \cong W'$ which is simple. So the hypotheses on the rings and simple module are as required.

By 9.5, $\mathcal{B} = \{B_2 \subset \ldots \subset B_n\}$ is a basic chain in T and has the same rank sequence as $\{A_2 \subset \ldots \subset A_n\}$; and $\mathbb{I}_T(\mathcal{B}) = \mathbb{I}_S(\mathcal{A}) = R$. Similar facts apply to T', and the basic chain $\{A'_2 \cap T' \subset \ldots \subset A'_n \cap T'\}$ where $A'_i = A_iT'$. We know from (i) that $\{B_2T' \subset \ldots \subset B_nT'\}$ is a basic chain in T', yet again having the same rank sequence. However, it is evident that $B_iT' = (A_i \cap T)T' \subseteq A'_i \cap T'$; and so, since the factors by them share the same multiplicity, they are equal.

We are now in the position to apply the induction hypothesis and see that W_R is fully embedded in $W'_{R'}$. This completes the proof since, by 9.5, the uniserial modules U_R and $U'_{R'}$ have, as their respective unique composition series, V followed by the composition series of W, and V' followed by the composition series of W'. □

We end with a few easy consequences of 9.1, the first of which describes the structure of rings intermediate between a matrix ring over a division ring and its full upper triangular subrings.

9.9. Proposition. *Let $S \supset T \supset R$, where R is an $(n+1)$-block full upper triangular subring of $S = M_s(D)$ for some division ring D. Then each ring T such that $S \supseteq T \supseteq R$ is again a full block upper triangular subring of S; and each of the diagonal blocks of T contains (and has the same size as) a set of consecutive diagonal blocks of R.*

Proof. By 9.1, R is the idealizer of a basic chain $\mathcal{A}$ of right ideals of S. By 8.14, each intermediate ring is the idealizer of a subchain of $\mathcal{A}$. Then 9.1 shows that T is as described. □

Our final consequence is an idealizer-theoretic proof of the following well-known result.

9.10. Corollary. *Every full block upper triangular matrix ring R over a division ring is (right and left) Artinian, hereditary and serial.*

Proof. By 9.1, R is a basic chain idealizer, and hence an iterated basic idealizer, from a simple Artinian, and hence serial, ring S. Hence R is right Artinian, right hereditary and right serial, since these properties are preserved by iterated basic idealizers [6.2].

Since R is an iterated basic idealizer from the right Artinian ring S, R is also an iterated basic left idealizer from S [6.2(viii)]. Hence, as above, R is left hereditary and left serial. □

10. Notes on Chapters 1 and 2

The early history of idealizers is rather thin! The notion of an idealizer subring was introduced by [**Ore 32**] and the isomorphism given in 1.3(i) was noted by [**Fitting 35**]. There appears to have been no further study of them until the 1970s at which point idealizers reappeared, independently, in two places.

On the one hand, [**Jacobinski 71**], considering a classical order in the situation when the base Dedekind domain is complete and local, showed that the order is hereditary precisely when it is the idealizer of a chain of isomaximal right ideals of a maximal order.

On the other hand, [**Hart-Robson 70**] discussed an interesting example which, as became clear in [**Robson 72**], was actually the idealizer $\mathbb{I}(xA_1)$ where A_1 is the first Weyl algebra over a field of characteristic zero (as described in 11.8). The former paper demonstrated that this ring is Noetherian. The latter paper provided, for a general ring S, a substantial theory about the idealizer $\mathbb{I}_S(A)$ of a single semimaximal right ideal A and about subrings obtained by iterating that construction. The main thrust related the properties of S and of any one of these subrings, with special emphasis on their simple modules. This yielded results such as 4.4 and 4.19. It then applied this to the structure theory of HNP rings, and constructed several examples of interest. In particular, it showed that, since A_1 is hereditary, so too is $\mathbb{I}(xA_1)$.

[**Ely 74**] started to bring together these two strands by studying the relationship, for a general ring S, between idealizing a chain of semimaximal right ideals and iterating the process of idealizing a single semimaximal right ideal; and he showed, for HNP rings, that precisely the same subrings are obtained. It follows that Jacobinski's result mentioned above is a consequence of the general theory. [**Robson 75**] extended Ely's result to general rings, and also discussed rings intermediate between the original ring and one of these idealizer subrings.

[**Goodearl 73**] showed that original ring S is always left flat over these idealizer subrings; see 4.12. Combined with earlier results, it follows that S is both a left and a right localization (in the sense of [**Silver 67**], as described in 2.4) of the idealizer ring. Thus we see the unusual feature that we have constructed the 'global' ring from the 'local' ring by idealization. This feature proves a potent weapon when dealing with HNP rings later.

§§1-4 are a carefully refined version of these earlier results about idealizers in a general ring. Our eventual concentration on basic idealizers stems, to some extent, from the treatment of idealizers of HNP rings in [**McR 01**]. The later sections are substantially original. The results on extensions of simple modules in §5 are new, but suggested by some related, partial results for HNP rings in [**Goodearl-Warfield 79**]. The analysis in §§6, 7 and 8 of the interconnections between iterated basic idealizers and multichain idealizers is completely different from the earlier work in [**Ely 74**] and [**Robson 75**]. Thereby, those sections contain new results and insights.

Essentially all of the theory of idealizers presented in these two chapters finds application in later chapters where it is used in studying the structure of HNP rings and their modules and in the construction of salient examples.

Part 2

HNP Rings

CHAPTER 3

Basic Structure

We now turn towards the main theme of this book — hereditary Noetherian prime rings, HNP rings for short. This chapter, after discussing examples of such rings, lays some general foundations for the structure theory of finitely generated projective modules over an HNP ring R. It demonstrates that overrings of R (i.e. rings lying between R and R_{quo}) are all HNP rings and can be viewed as localizations of R at collections of simple modules. The close connection with basic and iterated basic idealizers is then established; and the special relationship between two simple modules V, W such that W is unfaithful and $\text{Ext}(V, W) \neq 0$ is studied, together with some consequences for uniserial R-modules. Finally, this chapter demonstrates a remarkable symmetry between the left and right module structures of R.

Recall, from 0.2, that whenever we discuss an HNP ring R we tacitly assume that $R \neq R_{\text{quo}}$ in order to avoid some trivialities. We note one consequence of this: R has no minimal right ideals [**McR 01**, 3.3.4].

11. Introduction and Examples

Before beginning the work on HNP rings and their modules, it seems sensible to introduce briefly some prototypic examples of HNP rings. So within this section we introduce these examples many of which will be used to provide examples of the behaviour studied. First, however, we mention a decomposition result which in part explains why we restrict our attention to prime rings.

11.1. PROPOSITION. *Let R be a hereditary Noetherian ring. Then R splits as a direct sum of HNP rings and an Artinian hereditary ring.*

PROOF. See [**McR 01**, 5.4.6]. □

11.2. EXAMPLE. The best known examples of HNP rings come from commutative ring theory. A commutative HNP ring is, simply, a Dedekind domain. As specific examples, we can cite any principal ideal domain such as $\mathbb{Z}$ or $F[x]$ for any field F; or the rings of integers of a finite dimensional extension field of $\mathbb{Q}$.

The ideal theory and module theory for Dedekind domains is well-known and described in many texts (e.g. [**Jacobson 80**, Chapter 10]; [**Curtis-Reiner 81**]). Each finitely generated module is the direct sum of a module of finite length and a projective module. The finite length module decomposes as a direct sum of uniserial modules each of which has only one isomorphism class of simple composition factors. The projective module is the direct sum of a collection of ideals, and its isomorphism type is uniquely specified by a pair of independent invariants: the number of nonzero ideals in the direct sum; and the isomorphism type of the product of those ideals.

Further examples follow easily using Morita equivalence.

11.3. DEFINITION. We recall that a right module over any ring is termed a right *progenerator* when it is a finitely generated projective generator. Then two rings S and T are said to be *Morita equivalent* if there is a right S-progenerator M such that $\mathrm{End}(M_S) \cong T$; and this happens if and only if the S- and T-module categories are equivalent. This equivalence is given by mapping a right T-module X to $(X \otimes_T M)_S$. In particular, X_T and its image $(X \otimes_T M)_S$ have isomorphic lattices of submodules. Properties which are preserved under Morita equivalence are said to be *Morita invariant.* For more details see e.g. [**McR 01**, Section 3.5] or [**Anderson and Fuller 92**, §§21, 22].

11.4. LEMMA. *Any ring Morita equivalent to an HNP ring is an HNP ring.*

PROOF. [**McR 01**, 3.5.10]. □

11.5. EXAMPLE. Another well-understood class of examples comes from the study of classical orders in central simple algebras. Let Σ be a simple algebra, finite dimensional over its centre which, in turn, is the field of fractions of a Dedekind domain D. A (classical) *D-order* Λ in Σ is a subring of Σ which contains D, satisfies $D_{\mathrm{quo}}\,\Lambda = \Sigma$, and is finitely generated as a D-module. We then have $\Lambda_{\mathrm{quo}} = \Sigma$.

Each D-order in Σ is contained in a maximal one [**Reiner 75**, (10.4)], called a *maximal D-order* ; and maximal D-orders are HNP rings (see [**Reiner 75**, (21.4)] or [**McR 01**, 5.3.16]).

For an easy noncommutative example, we cite the matrix ring $M_n(D)$ over D, which is an HNP ring [11.4] and is readily seen to be a maximal D-order. The structure of finitely generated modules over a maximal D-order is well-understood. Once again, they decompose as a direct sum of a finite length module and a projective module. Each indecomposable finite length module is uniserial with only one isomorphism type of simple composition factor. The projective summand is readily described as being the direct sum of a free module and a single right ideal.

One fact worth noting here is that infinitely generated projective modules over a maximal D-order are necessarily free (see 46.2, which applies since, as we will see in 23.7, each maximal D-order is a Dedekind prime ring).

11.6. EXAMPLE. Some non-maximal D-orders are also HNP rings and these are called *hereditary D-orders* . The class of hereditary orders is precisely the class of HNP rings which satisfy a polynomial identity (see, e.g. [**McR 01**, 13.9.16]. We can use an idealizer in $M_2(D)$, for any Dedekind domain D, to provide an example. Let M be a maximal ideal of D. Then $S = M_2(D)$ has the generative maximal right ideal $A = \begin{bmatrix} D & D \\ M & M \end{bmatrix}$ with idealizer $R = \mathbb{I}(A) = \begin{bmatrix} D & D \\ M & D \end{bmatrix}$. We know from 4.19 that R is right hereditary and right Noetherian. Symmetry tells us that R is also left hereditary and left Noetherian. Each ideal B of R contains $M_2(M)BM_2(M)$ which is a nonzero ideal of S; hence R is prime.

The structure of finite length modules and of finitely generated projective modules over hereditary D-orders is described in later chapters.

The next few examples are highly noncommutative, being simple hereditary Noetherian domains.

11.7. EXAMPLE. Let F be a field of characteristic zero and let $F(x)$ be the field of rational functions in an indeterminate x. Form the skew polynomial ring $B_1(F) = F(x)[y : \delta]$ with the relation $xy - yx = 1$. (Thus y acts on $F(x)$ as

$\delta = d/dx$). It is easy to check that $B_1(F)$ is a simple principal right and left ideal domain — and so it is hereditary and Noetherian (see e.g. [**McR 01**, 1.3.9]) — and has a division ring of fractions. As will be seen later, it follows almost immediately that its projective modules are all free.

We now cite two results about simple modules [**McConnell-Robson 73**, 4.4, 4.1] which will be of use later: $B_1(F)$ has infinitely many isomorphism types of simple modules (for example, those of the form $B_1(F)/(y+p)B_1(F)$ where p varies through distinct elements of $F[x]$); and the dimension of $\mathrm{Ext}^1(V, W)$ over $\mathrm{End}(V)$ is infinite for all simple $B_1(F)$-modules V, W.

11.8. Example. The next example is a subring of $B_1(F)$ called the first Weyl algebra $A_1(F)$. It is constructed in the same fashion, but starting with the polynomial ring $F[x]$ rather than its field of fractions $F(x)$. So $A_1(F) = F[x][y : \delta]$ with the relation $xy - yx = 1$. [**McR 01**, 7.11.3, 7.11.8] shows that $A_1[F)$ is a simple hereditary Noetherian domain with the same division ring of fractions as $B_1(F)$, but it is not a principal right or left ideal ring. Nevertheless, (see [**McR 01**, 7.11.6]) all its projective modules are free except for those isomorphic to a non-principal right ideal; and there are infinitely many isomorphism types of these right ideals (see e.g. [**Canning-Holland 94**]).

[**McConnell-Robson 73**, §5] contains some results about simple modules which we quote for future reference. First, for all simple $A_1(F)$-modules V, W, the F-dimension of $\mathrm{Ext}^1(V, W)$ over $\mathrm{End}(V)$ is finite. Secondly, whenever $p, q \in F[y]$, the $A_1(F)$-modules $A_1(F)/(x+p)A_1(F)$ and $A_1(F)/(x+q)A_1(F)$ have the following properties: (i) they are simple modules; (ii) they are isomorphic if and only if $p = q$ (so there are infinitely many distinct isomorphism types); (iii) the F-dimension of $\mathrm{Ext}^1\big(A_1(F)/(x+p)A_1(F), A_1(F)/(x+q)A_1(F)\big)$ equals the degree of $p - q$; and (iv) F is the endomorphism ring of each of these modules.

11.9. Example. Now we mention an extreme example, whose behaviour is very different from $A_1(F)$ and $B_1(F)$. It is a simple hereditary Noetherian domain R which has infinitely many isomorphism types of simple modules and each simple module is injective; see [**Cozzens 70**], [**Koifman 70**]. Consequently, there are no non-split extensions of simple modules. (We note that neither $A_1(F)$ nor $B_1(F)$ has any injective simple modules.)

We will see in later sections that, because of the different extension properties of the simple modules over these rings, their iterated basic idealizers provide a rich collection of behaviours which will illuminate many of our results; and later still, in §25, we will construct a general type of idealizer example which exhibits the range of possibilities. Here, however, we describe a straightforward idealizer example to serve as an immediate illustration.

11.10. Example. Let S denote $B_1(F)$ or $A_1(F)$, where F is a field of characteristic zero and let $R = \mathbb{I}(yS)$. Then R is a basic idealizer from S and is a hereditary Noetherian domain.

Proof. We know from above that yS is a maximal right ideal of S; and it is generative because S is simple. Therefore, by definition, R is a basic idealizer from S. Hence, by 4.19, R inherits from S the properties of being a right hereditary and right Noetherian domain.

Note next that $S \cong ySy^{-1} = T$, say. Let $a \in S$; then $ayS \subseteq yS \Leftrightarrow aT \in T \Leftrightarrow a \in T$. Thus $R = \mathbb{I}(yS) = S \cap T$. The symmetric argument, applied to the generative maximal left ideal Ty of T, shows that $\mathbb{I}(Ty)$ is a left Noetherian, left hereditary domain and that $\mathbb{I}(Ty) = T \cap y^{-1}Ty$; i.e. $\mathbb{I}(Ty) = T \cap S = R$. □

We note some details about these idealizer examples for future use.

11.11. LEMMA. *Let R be one of the idealizer rings described in* 11.6 *or* 11.10 *and A the maximal right ideal which was idealized.*

(i) *A is an idempotent maximal ideal of R.*
(ii) *The simple right R-modules coincide with the simple right S-modules except that $(S/A)_R$ is uniserial of length 2 with S/R and R/A as its R-composition factors.*
(iii) *R/A is unfaithful in each case, whereas S/R is unfaithful in the case of* 11.6 *but faithful in the case of* 11.10.
(iv) *${}_RS$ is finitely generated in the case of* 11.6 *but not in the case of* 11.10.

PROOF. (i) (ii) 4.2 and 4.4 give these facts as immediate consequences of R being a basic idealizer.

(iii) It is clear that R/A is unfaithful, with annihilator A. In the case of 11.6 we note that S/A is annihilated by the ideal $B = M_2(M)$ and hence S/R is unfaithful. In the case of 11.10, however, S/yS is faithful over S and hence over R; so S/R must be a faithful R-module.

(iv) In 11.6, ${}_DS$ is finitely generated and hence so too is ${}_RS$. In 11.10, where S is simple and $A = yS$, suppose that ${}_RS$ were finitely generated. We could choose a common denominator c for the generators and then $Sc \subseteq R$. However $yS \subseteq R$ and so, since S is simple, $S = ScyS \subseteq R$, a contradiction. □

12. Preliminary Module Theory

This section contains general module-theoretic results from other sources, and some refinements of these results for HNP ring. We assume familiarity with the basic theory of a semiprime (right and left) Goldie ring R and its ring of quotients, which is a semisimple Artinian ring denoted by R_{quo}. (This theory can be found, for example, in [**McR 01**, Chapters 2, 3]). We begin by reviewing some basic definitions and facts that we use later, often without explicit references. First we recall that an element of R is called *regular* if it is not a zero-divisor in R and that an element of R is regular if and only if it is a unit of R_{quo}.

12.1. LEMMA. *Let R be a semiprime Goldie ring.*

(i) *A right ideal of R is essential if and only if it contains a regular element.*
(ii) *R is prime if and only if R_{quo} is simple Artinian. In that case every nonzero ideal of R contains a regular element.*
(iii) *If M is a finitely generated right R-submodule of R_{quo}, then $M \cong dM \subseteq R$ for some regular element d of R; and if M contains a unit of R_{quo} then the right ideal dM of R is essential.*
(iv) *If T is a subring of R_{quo} and $aRb \subseteq T$, $cTd \subseteq R$ for some units a, b, c, d of R_{quo}, then T is a semiprime Goldie ring and $T_{\text{quo}} = R_{\text{quo}}$. In such a case, if R is prime so too is T.*
(v) *R_{quo} is a maximal essential extension of R_R, and so $R_{\text{quo}} = E(R)$ and is injective.*
(vi) *If S is a ring such that $R \subseteq S \subseteq R_{\text{quo}}$ then S is a semiprime Goldie ring (and is prime if R is prime) and $S_{\text{quo}} = R_{\text{quo}}$.*

(vii) *Let S, R be as in (vi) and M be a right S-module. Then M_S is faithful if and only if M_R is faithful.*

PROOF. (i), (ii), (iv), (vi). See [**McR 01**, 2.3.5, 2.2.1, 3.1.6].

(iii) Choose a finite set of generators of M, and write them with a common left denominator: $d^{-1}r_1, \ldots, d^{-1}r_n$. Then $M \cong dM \subseteq R$.

(v) Evidently R_{quo} is an essential extension of R_R; suppose that $X_R \supseteq R_{\text{quo}}$ is too.

(v) Suppose that $X_R \supseteq R_{\text{quo}}$ is an essential extension over R. Pick any $x \in X$. The set $E = \{r \in R \mid xr \in R\}$ is necessarily an essential right ideal of R by [**McR 01**, 2.2.2(iii)]. So there is a regular element $d \in E$. Say $xd = r$. Then $x = rd^{-1} \in R_{\text{quo}}$ and so $X = R_{\text{quo}}$.

(vii) Suppose M_S is unfaithful. Since $S \subseteq R_{\text{quo}}$ we have $Mrd^{-1} = 0$ for some $r, d \in R$ with $r \neq 0$ and d regular. But then $Mr = 0$, showing that M_R unfaithful. The converse is obvious. □

Two rings R and T related as described in (iv) are termed *equivalent orders in* R_{quo} although we will not be using this terminology here. This, of course, is not the same as saying they are Morita equivalent.

The well-known characterization of right hereditary rings given in (i) below provides a useful source of injective modules over an HNP ring.

12.2. LEMMA.

(i) *A ring is right hereditary if and only if every homomorphic image of every injective right module is again injective.*

(ii) *In particular, if R is an HNP ring, then every homomorphic image of $(R_{\text{quo}})_R$ is injective.*

PROOF. (i) This is an immediate consequence of the facts that right hereditary rings are the rings of global projection dimension ≤ 1, and the right global injective dimension of any ring equals its right global projective dimension [**Rotman 09**, 8.8, 8.14].

(ii) By 12.1(v), $(R_{\text{quo}})_R$ is injective; so this follows from (i). □

Recall (e.g. from [**McR 01**, 2.2.5]) that a *uniform* module is defined to be a nonzero module which is an essential extension of each of its nonzero submodules; and a uniform right ideal is a right ideal which is a uniform module.

12.3. DEFINITION. The *uniform dimension* or *uniform rank* of a module M is the largest integer such that M contains a direct sum of n uniform submodules; or ∞ if no maximum exists. We use the notation $\operatorname{udim}(M)$ or $\operatorname{udim} M_R$ for this. (The infinite case is discussed further in 45.2.)

12.4. LEMMA. *Let R be a semiprime Goldie ring.*

(i) $\operatorname{udim}(R)$ *is finite and equals $\lambda(R_{\text{quo}})_{R_{\text{quo}}}$, the length of R_{quo} over itself; so* $\operatorname{udim}(R_R) = \operatorname{udim}({}_RR)$.

(ii) $\operatorname{udim}(E) = \operatorname{udim}(R)$ *for each essential right ideal E.*

(iii) *A right ideal U of R is uniform if and only if UR_{quo} is a minimal right ideal of R_{quo}.*

(iv) *Any nonzero finitely generated R-submodule U of a minimal right ideal V of R_{quo} is isomorphic to a uniform right ideal of R.*

(v) *If R is a prime Goldie ring then, given any two uniform right ideals of R, each is isomorphic to a submodule of the other. Consequently the direct sum of any* $\mathrm{udim}(R)$ *uniform right ideals is isomorphic to an essential right ideal.*

PROOF. (i)-(iii) See [**McR 01**, 3.4.3, 2.2.10, 3.3.2] respectively.

(iv) By 12.1(iii), there is a regular element $c \in R$ such that $U \cong cU \subseteq R$. Evidently $UR_{\text{quo}} = V$ and $cV \cong V$ since c is a unit of R_{quo}. Now (iii) shows that cU is uniform.

(v) See [**McR 01**, 3.3.4]. □

Recall that an element x of a module M over a semiprime Goldie ring R is *torsion* if $xc = 0$ for some regular element $c \in R$; and M is termed a *torsion module* if each of its elements is torsion. If, on the other hand, M has no nonzero torsion elements, it is *torsion-free.* The collection of all torsion elements in a module M forms a submodule, called the *torsion submodule* and denoted by $\tau(M)$. We note that each submodule of a free R-module is torsion-free.

12.5. LEMMA. *Let R be a semiprime Goldie ring and M a right R-module.*

(i) *$M/\tau(M)$ is torsion-free.*

(ii) *If M is torsion-free with* $\mathrm{udim}(M)$ *finite and $N \subseteq M$ then* $\mathrm{udim}(N) = \mathrm{udim}(M) \Leftrightarrow N$ *is essential in M; and then M/N is a torsion module.*

(iii) *If M is torsion-free and finitely generated then M embeds in $R^{(n)}$ for some n and* $\mathrm{udim}(M_R) = udim((M \otimes R_{\text{quo}})_{R_{\text{quo}}}) = \lambda((M \otimes R_{\text{quo}})_{R_{\text{quo}}})$.

(iv) *If M is nonzero, torsion-free and finitely generated then there is a submodule P of M such that $M/P \cong U$ for some uniform right ideal U of R.*

PROOF. (i) Clear from the definition.

(ii) See [**McR 01**, 2.2.10, 2.2.12].

(iii) See [**McR 01**, 3.4.7, 3.4.3].

(iv) By (iii), we may assume that $M \subseteq R^{(n)}$ for some n and then $M \otimes R_{\text{quo}} \cong MR_{\text{quo}}$. Further, $MR_{\text{quo}} \subseteq {R_{\text{quo}}}^{(n)}$ and $\lambda(MR_{\text{quo}})_{R_{\text{quo}}} = \mathrm{udim}(M_R) = t$, say. Choose a simple R_{quo}-submodule N of MR_{quo}. Since MR_{quo} is semisimple, there exists a map $\theta : MR_{\text{quo}} \twoheadrightarrow N$ with $\ker(\theta) = K$, say; and then $\lambda(K) = t - 1$. Let $P = K \cap M$. Then $P \subset M$ since $\lambda(MR_{\text{quo}}) = t$. Therefore M/P has a nonzero image in N which, by 12.4(iv), is isomorphic to a uniform right ideal of R. □

Applied to HNP rings, this provides the first decomposition result.

12.6. LEMMA. *Let R be an HNP ring. Then each finitely generated module M is the direct sum of a torsion module and a torsion-free module: $M \cong \tau(M) \oplus M/\tau(M)$.*

PROOF. Let $P = M/\tau(M)$. By 12.5 we may assume that $P \subseteq R^{(n)}$ for some n and hence P_R is projective. So the epimorphism $M \twoheadrightarrow P$ splits. □

We now aim towards a description of each of these two summands. The torsion-free case is straightforward.

12.7. LEMMA. *Let R be an HNP ring.*

(i) *If M is a finitely generated R-module then the following are equivalent:*

(a) *M is torsion-free;*

(b) *M is projective;*
(c) *M is isomorphic to a direct sum of uniform right ideals of R.*
(ii) *Each right ideal is isomorphic to a direct sum of uniform right ideals.*

PROOF. (i) (c)$\Rightarrow$(b)$\Rightarrow$(a). This is trivial.

(a)$\Rightarrow$(c). This is trivial if $M = 0$. If $M \neq 0$ then, by 12.5(iv), there is an epimorphism $M \twoheadrightarrow U$ for some uniform right ideal U of R. Since U is projective, this map splits. So $M \cong U \oplus M'$ where M' is the kernel of the map. Note that $\operatorname{udim} M' = \operatorname{udim} M - 1$. By induction on uniform dimension, we may assume the result for M'.

(ii) Clear from (i). □

12.8. LEMMA. *Let R be an HNP ring and P_R, Q_R be finitely generated and projective. If $\operatorname{udim}(P) \leq \operatorname{udim}(Q)$ then P is isomorphic to a submodule of Q.*

PROOF. This holds because P and Q are direct sums of uniform right ideals [12.7] and every uniform right ideal is a submodule of every other [12.4]. □

Before discussing the torsion case, some more generalities are recalled.

12.9. LEMMA. *Let R be semiprime Goldie ring, and M, N right R-submodules of R_{quo}. Suppose that M contains a unit of R_{quo}. Then*

$$\operatorname{Hom}_R(M, N) = \{x \in R_{\text{quo}} \mid xM \subseteq N\}$$

where the identification is via left multiplication. In particular:
(i) $M_R^* = \{x \in R_{\text{quo}} \mid xM \subseteq R\}$*; and*
(ii) $\operatorname{End}(M_R) = O_l(M) = \{x \in R_{\text{quo}} \mid xM \subseteq M\}$ *, the 'left order' of M.*

The corresponding identifications for left modules are $\operatorname{Hom}_R(M, N) = \{x \in R_{\text{quo}} \mid Mx \subseteq N\}$ *and* $\operatorname{End}({}_RM) = O_r(M)$ *, the 'right order' of M.*

PROOF. See [**McR 01**, 3.1.15]. □

12.10. DEFINITION. Let B equal M_R^* or $\operatorname{End}(M_R)$, with M as in 12.9. We refer to the identifications in the lemma as the *standard identifications $B \subseteq R_{\text{quo}}$. We often use these identifications without explicit mention.*

Under these identifications, products like $M_R^* M$ and MM_R^* become simply products inside R_{quo}. The following lemma, which we use very frequently, is an immediate consequence of two earlier results.

12.11. LEMMA. *Let R be a semiprime Goldie ring, and suppose that $M_R \subseteq R_{\text{quo}}$ contains a unit of R_{quo}. Then, in terms of the standard identifications in R_{quo}:*
(i) (Internal Dual Basis Lemma) *M_R is finitely generated projective if and only if $MM_R^* = O_l(M)$.*
(ii) *If M_R is finitely generated projective then the trace ideal* $\operatorname{tr}_R(M) = M_R^*M$ *is a nonzero idempotent ideal of R.*

PROOF. See the Dual Basis Lemma 1.6 and 1.7. □

12.12. LEMMA. *Let R be an HNP ring. The correspondences ${}_RB \to X = ({}_RB)^*$ and $X_R \to (X_R)^*$, when viewed via the standard identifications in R_{quo}, are inverse anti-isomorphisms (i.e. inclusion-reversing) between the lattice of essential left ideals B of R and the lattice of finitely generated submodules X_R such that $R \subseteq X \subset R_{\text{quo}}$. Moreover, every such X_R is finitely generated projective and isomorphic to an essential right ideal and*

(i) $(X/R)_R$ *has finite length;*
(ii) $(X/R)_R$ *is semisimple if and only* ${}_R(R/B)$ *has this property;*
(iii) $(X/R)_R$ *is isotypic semisimple if and only if* ${}_R(R/B)$ *has this property.*

PROOF. Let B be any essential left ideal of R and let $R \subseteq X_R \subset R_{\text{quo}}$, with X finitely generated. B contains a regular element of R, by 12.1, and X obviously does. Therefore we can apply the standard identifications as above, getting $R = R^* \subseteq ({}_RB)^* \subseteq R_{\text{quo}}$ and $(X_R)^* \subseteq R^* = R$.

Since R is an HNP ring, every ${}_RB$ is projective; and every X_R is projective since it is finitely generated and contains a unit of R_{quo} and hence is isomorphic to an essential right ideal of R [12.1]. Therefore 12.11 shows that B^* and X^* are finitely generated modules on the appropriate sides and $B^{**} = B$ and $X^{**} = X$. Moreover, finite generation of X_R shows that $dX \subseteq R$ for some regular $d \in R$ [12.1]. Therefore the left ideal $(X_R)^*$ contains the regular element d, and hence is essential in R.

The claimed lattice anti-isomorphism is now obvious.

(i) The lattice of submodules of X/R is anti-isomorphic to that of R/B which is Noetherian. Therefore X/R is Artinian. However, X is finitely generated and so X/R is Noetherian and hence X/R has finite length.

(ii) This follows from (i) since the property of semisimplicity (every submodule is a direct summand) can be phrased purely in terms of the lattice of submodules, in a way that is preserved by lattice anti-isomorphisms.

(iii) By (ii) we may assume that both modules are semisimple. Note that every homogeneous component of a semisimple module has a unique complement: the sum of the remaining homogeneous components. So a semisimple module is isotypic if and only if no nonzero proper submodule has a unique complement; and the latter statement is preserved by lattice anti-isomorphisms. □

We note an immediate consequence of 12.12.

12.13. LEMMA. *Let R be an HNP ring. Suppose that M_R is a finitely generated module with $R \subseteq M \subseteq R_{\text{quo}}$. Then M is projective and $\lambda(M/R)_R$ is finite. In particular, if $x \in R_{\text{quo}}$ and we let $\bar{x} = x + R \in R_{\text{quo}}/R$, then $(\bar{x}R)_R$ has finite length.*

PROOF. We see from 12.7 that M_R is projective. The rest follows from 12.12. □

12.14. LEMMA. *A right ideal I of an HNP ring R is essential if and only if $(R/I)_R$ has finite length. In particular, if A is a nonzero ideal, then R/A is a (right and left) Artinian ring.*

PROOF. Suppose that I is essential. By the left-right symmetric version of the preceding result, ${}_R(I^*/R)$ has finite length and then so too has R/I.

Conversely, suppose that R/I has finite length. If I is not essential then there is a nonzero right ideal J with $J \cap I = 0$. The canonical map $R \to R/I$ is a monomorphism on J, so J_R has finite length, contrary to the convention that R is not simple Artinian. □

12.15. DEFINITION. A ring R is said to have *restricted minimum condition on the right* if $(R/I)_R$ is Artinian for each essential right ideal I. (Of course, if R is right Noetherian, this says $(R/I)_R$ has finite length.)

12.16. COROLLARY. *Let R be an HNP ring.*

(i) *R has restricted minimum condition (both left and right); so each R/I has finite length.*

(ii) *If M_R is a finitely generated torsion module then M has finite length.*

PROOF. (i) Clear from 12.14.

(ii) For each generator $m \in M$, note that $mR \cong R/\operatorname{rann} m$. By definition $\operatorname{rann} m$ contains a regular element of R and hence is an essential right ideal. The preceding result then demonstrates that mR has finite length. □

12.17. COROLLARY. *Let R be an HNP ring and M, N be finitely generated projective modules with $N \subseteq M$. Then* $\operatorname{udim}(M) = \operatorname{udim}(N) \Leftrightarrow$ *N is an essential submodule of $M \Leftrightarrow M/N$ has finite length.*

PROOF. The equivalence of the first two properties follows directly from 12.5. We now consider the first and last properties.

Suppose that $\operatorname{udim}(M) = \operatorname{udim}(N)$. So if $m \in M$ and A is a nonzero right ideal of R and $mA \neq 0$ then $mA \cap N \neq 0$. Hence $\{r \in R \mid mr \in N\}$ is an essential right ideal for each $m \in M$. Therefore M/N is torsion.

On the other hand, if $\operatorname{udim}(N) < \operatorname{udim}(M)$ then there exists a submodule U of M isomorphic to a uniform right ideal of R and such that $N \cap U = 0$. Hence U embeds in M/N and so M/N does not have finite length. □

Combined with the earlier results, 12.16 gives:

12.18. THEOREM. *Let R be an HNP ring. Then each finitely generated R-module decomposes uniquely as the direct sum of a module of finite length and a projective module; and the projective module decomposes as a direct sum of uniform right ideals.* □

Later sections add substantially to the precision of this result by attacking the question: 'how unique is the direct sum decomposition of the projective summand?' and by partially describing the structure of the torsion summand.

13. Overrings and Simple Modules

In this section[1] we investigate the rings lying between an HNP ring R and its ring of quotients R_{quo}. We will see that each is an HNP ring, is a localization of R (see 13.3 below) and is determined uniquely by a collection of simple R-modules.

13.1. DEFINITION. Any ring S such that $R \subseteq S \subseteq R_{\text{quo}}$ is called an *overring* of R. Note, by 12.1(vi), that $S_{\text{quo}} = R_{\text{quo}}$. If, further, S_R is finitely generated, we call S a *right finite* overring . Note, by 12.12, that in that case S_R is also projective. Note also that a right finite overring need not be also left finite. (See 11.11(iv) for an example).

We have, of course, already come across examples of such overrings. Indeed if A is any nonzero ideal of R then both $O_l(A)$ and $O_r(A)$ are overrings of R; and it is clear that $O_r(A)$ is right finite since, for any regular element $a \in A$, it is isomorphic, as a right R-module, to the right ideal $aO_r(A)$.

In the remainder of this section, we will investigate overrings, leaving till the next section a special study of right finite overrings.

[1]In this section R denotes an HNP ring unless the contrary is specified.

13.2. LEMMA. *Every overring S of R satisfies the following conditions:*

(i) *S is flat as both a left and right R-module.*
(ii) *The canonical multiplication map $\mu : S \otimes_R S \to S$ is a bijection.*

PROOF. (i) Since R is an HNP ring, every finitely generated R-submodule of S_R is projective by 12.12. Thus S is a direct limit of flat R-modules, and hence is flat. Similarly, ${}_RS$ is flat.

(ii) The given map μ is clearly surjective. So consider any element $x = \sum_i s_i \otimes s_i' \in \ker(\mu)$. Then $\mu(x) = \sum_i s_i s_i' = 0$. By (i) applied to S and Q, there are natural monomorphisms $S \otimes_R S \to S \otimes_R Q \to Q \otimes_R Q$. So it is enough to prove that x equals zero, when viewed as an element of $Q \otimes_R Q$. Choose a common left denominator $d \in R$ for the elements s_i. Thus d is regular in R and every $ds_i \in R$. Then $x = \sum_i s_i \otimes s_i' = d^{-1} \otimes_R \sum_i (ds_i)s_i' = d^{-1} \otimes d\mu(x) = 0$ as desired. □

13.3. REMARK. Conditions (i) and (ii) above make S a *localization* of R as defined in 2.4.

13.4. COROLLARY. *Let S be an overring of R.*

(i) *For every right S-module M, the natural map $M \otimes_R S \to M$ is an S-isomorphism.*
(ii) *$I = (I \cap R)S \cong (I \cap R) \otimes_R S$ (via multiplication) for every right ideal I of S.*
(iii) *When restricted to S-modules, $\cong_R = \cong_S$, $\mathrm{Hom}_R = \mathrm{Hom}_S$, $\mathrm{Ext}^1_R = \mathrm{Ext}^1_S$ and $\otimes_R = \otimes_S$. (More precisely, these functors are equivalent when so restricted.)*

PROOF. These are extracts from 2.5(i), (iii) and (v) and 4.13(iii), (iv). □

13.5. THEOREM. *Every overring $S \neq R_{\mathrm{quo}}$ of R is an HNP ring.*

PROOF. Since $R \subseteq S \subset R_{\mathrm{quo}}$, we know that S is a prime ring with Goldie quotient ring R_{quo}. Suppose a right ideal I of S is given. Then $I \cap R$ is a finitely generated projective right ideal of R and, by 13.4(ii), $I = (I \cap R)S$. Hence I is a finitely generated right ideal of S and it is projective because $(I \cap R)S \cong (I \cap R) \otimes_R S$. Thus S is right Noetherian and right hereditary. Symmetry completes the argument. □

Next we aim to explore the relationship between overrings of R and simple R-modules.

13.6. LEMMA. *All simple right R-modules occur as submodules of $(R_{\mathrm{quo}}/R)_R$.*

PROOF. Let $X = R/M$ be a simple R-module. Since, by convention, $R \neq R_{\mathrm{quo}}$, the socle of R is zero [**McR 01**, 3.3.4]. Therefore M is essential in R. Since M is finitely generated projective, $M \oplus K \cong R^{(n)}$ for some module K and integer n. So $R^{(n)}$ is an essential submodule of the module $N \cong R \oplus K$ and $N/R^{(n)} \cong X$. Thus X embeds in $E(R^{(n)})/R^{(n)} \cong (R_{\mathrm{quo}}/R)^{(n)}$, since $E(R) \cong R_{\mathrm{quo}}$ by 12.1(v); but X is simple, so X embeds in R_{quo}/R. □

This shows that each simple R-module occurs in the socle of R_{quo}/R. Later results [24.4, 24.5] give the multiplicity of each simple module in that socle.

13.7. DEFINITION. Recall, from 12.13, that if $x \in R_{\text{quo}}$ and $\bar{x} = x+R \in R_{\text{quo}}/R$, then $\bar{x}R$ has finite length. Now let $\mathcal{Y}$ be a set of isomorphism classes of simple right R-modules and set

$$R(\mathcal{Y}) = \{x \in R_{\text{quo}} \mid \text{every composition factor of } \bar{x}R \text{ belongs to } \mathcal{Y}\}. \tag{13.7.1}$$

When $\mathcal{Y} = \{Y\}$, a singleton set, we write $R(Y)$ rather than $R(\{Y\})$.

We say that an overring S of R *kills* an R-module X if $X \otimes_R S = 0$.

13.8. THEOREM.

(i) *For each set $\mathcal{Y}$, $R(\mathcal{Y})$ is a ring.*
(ii) *$R(\mathcal{Y})$ kills each simple module in $\mathcal{Y}$ and kills no other simple module.*
(iii) *$R(\mathcal{Y}) \subseteq R(\mathcal{Y}') \Leftrightarrow \mathcal{Y} \subseteq \mathcal{Y}'$.*
(iv) *Every overring S of R equals $R(\mathcal{Y})$ for some unique $\mathcal{Y}$. Indeed, $\mathcal{Y}$ is the set of isomorphism classes of simple composition factors of finitely generated submodules of S/R.*
(v) *$R(\mathcal{Y}) = R_{\text{quo}}$ if and only if $\mathcal{Y}$ is the set of all isomorphism classes of simple right R-modules.*

PROOF. (i) First we prove that $S = R(\mathcal{Y})$ is a ring. Choose $x, y \in R_{\text{quo}}$. Then all composition factors of $\bar{x}R \oplus \bar{y}R$ belong to $\mathcal{Y}$. Hence the same is true of all composition factors of the homomorphic image $\bar{x}R + \bar{y}R \subseteq R_{\text{quo}}/R$, and so of its submodule $(\bar{x} + \bar{y})R$. Thus S is closed under addition.

For closure under multiplication, note that there is an epimorphism $(yR + R)/R \twoheadrightarrow (xyR + xR)/xR$ given by multiplication by x. So the latter module has all its composition factors in $\mathcal{Y}$. The same is then true, in turn, of $(xyR + xR + R)/(xR + R)$, of $xR + R/R$ and hence of $(xyR + xR + R)/R$ and of its submodule $(xyR + R)/R$, as required.

(ii) To see that S kills every $Y \in \mathcal{Y}$, first note that, by 13.6, R_{quo}/R has a submodule $\bar{y}R \cong Y$; and $y \in S$ since $S = R(\mathcal{Y})$. On the other hand, S kills $(S/R)_R$ since S is a localization of R. (Use 13.2 and apply right exactness of $\otimes$ to the short exact sequence $R \hookrightarrow S \twoheadrightarrow S/R$.) Since ${}_RS$ is flat, S also kills the submodule Y of S/R, as desired.

Conversely, suppose that $Y \otimes_R S = 0$. As above, there exists $y \in Q$ such that $(yR + R)/R \cong Y$. Since $Y \otimes_R S = 0$, right exactness of $\otimes$ implies that $(yS + S)/S = 0$. Hence $y \in S$ and therefore $Y \in \mathcal{Y}$. Thus S kills precisely those simple R-modules in $\mathcal{Y}$.

(iii) This is obvious from (ii) and the definition of $R(\mathcal{Y})$.

(iv) Let S be an overring of R, and $\mathcal{Y}$ the set of simple R-modules killed by S. We claim that $S = R(\mathcal{Y})$. To see this, first let $x \in S$. Then $(xS + R)/R$ is killed by S; and so too are all its simple composition factors. Hence they belong to $\mathcal{Y}$ and so $x \in R(\mathcal{Y})$. Thus $S \subseteq R(\mathcal{Y})$. Conversely, let $a \in R(\mathcal{Y})$; so all the simple composition factors of $(aR+R)/R$ belong to $\mathcal{Y}$ and therefore are killed by S. Hence $(aS+S)/S = 0$; i.e. $R(\mathcal{Y}) \subseteq S$. The uniqueness of $\mathcal{Y}$ follows immediately from (ii).

The argument has already shown that $\mathcal{Y}$ contains the set of isomorphism classes of simple composition factors of finitely generated submodules of S/R; and 13.6 shows that each simple isomorphism type in $\mathcal{Y}$ occurs in this set.

(v) This is clear from (iii) and (iv). $\square$

13.9. DEFINITION. We say that an overring S of R is *determined by killing the set* $\mathcal{Y}$ *of simple R-modules* if S is the unique overring of R that kills precisely those simple modules in $\mathcal{Y}$. By 13.8, there always is such a ring, namely $S = R(\mathcal{Y})$.

13.10. EXAMPLE. Recall that each of 11.6 and 11.10 describes an idealizer ring $R = \mathbb{I}(A)$ where both S and R are hereditary Noetherian domains. Note, by 12.1, that $R_{\mathrm{quo}} = S_{\mathrm{quo}}$ and so S is an overring of R. Moreover, by 11.11, $V = (S/R)_R$ is a simple right R-module. Thus we see that $S = R(V)$, the ring determined by killing V.

13.11. COROLLARY. *Every overring $S = R(\mathcal{Y})$ of R is equal to the subring of R_{quo} generated by the rings $R(Y)$ for $Y \in \mathcal{Y}$.*

PROOF. If $Y \in \mathcal{Y}$, then S kills Y and so, by 13.8, $S \supseteq R(Y)$. Therefore $S \supseteq S'$ where S' is the ring generated by all such $R(Y)$. On the other hand, since $R(Y) \subseteq S'$ for each Y, 13.8 shows that S' kills every $Y \in \mathcal{Y}$. Hence $S \subseteq S'$. □

13.12. THEOREM. *Let $S = R(\mathcal{Y})$.*

(a) *If U_R is a simple R-module with $U \notin \mathcal{Y}$, then:*
   (i) *U embeds in $U \otimes S$;*
   (ii) *the image of U in $U \otimes S$ is essential;*
   (iii) *if U' is a simple R-module with $U' \notin \mathcal{Y}$ and $U' \otimes S \cong U \otimes S$ then $U' \cong U$;*
   (iv) *$(U \otimes S)_S$ is simple and $U \otimes S = US$.*

(b) *Every simple S-module has the form $U \otimes S$ for some $U \notin \mathcal{Y}$ which is unique up to isomorphism. The image of U in $U \otimes S$ is its R-module socle.*

PROOF. (a) (i) We know that $U \cong R/M$ for some maximal right ideal M of R. Note that $MS \cap R = R$ or M. Say the former; then $MS/M \supseteq R/M \cong U$. But then $U \otimes S \subset (MS/M) \otimes S = 0$, yet $U \otimes S \neq 0$. Hence $MS \cap R = M$. Thus

$$R/M \cong (R + MS)/MS \subseteq S/MS = (R/M) \otimes S = U \otimes S \cong S/MS \tag{13.12.0}$$

and so $U \subseteq U \otimes S$.

(ii) Now let $a \in S - MS$. Note that $(aR + R/R) \cong R/J$ where $J = \{x \in R \mid ax \in R\}$. Since the composition factors of R/J all belong to $\mathcal{Y}$, then $JS = S$. Suppose that $aJ \subseteq M$; then $aJS \subseteq MS$. Thus $aS \subseteq MS$ and therefore $a \in MS$, contradicting the choice of a. We conclude that $aJ \not\subseteq M$. Thus $aR \cap R \not\subseteq M$ since $aJ \subseteq aR \cap R$. However $M \subseteq (aR + M) \cap R \subseteq R$. Hence $(aR + M) \cap R = R$ and so $aR + MS \supseteq R + MS$. Hence $(R + MS)/MS$ is essential in S/MS.

(iii) This is clear from (ii).

(iv) By (i) we see that $(R + MS)/MS \cong R/M$ and by (ii) that this is contained in every R-submodule of S/MS. Thus $(S + MS)/MS$ is contained in every S-submodule of S/MS; but $S + MS/MS = S/MS$. So S/MS is simple. However $U \otimes S \cong S/MS$ and, by (i), $U \subseteq US \subseteq U \otimes S$. So $U \otimes S = US$.

(b) Let V be a simple S-module, $V \cong S/K$ say for some maximal right ideal K of S. Choose $a \in S - K$. Consider $(aR + K)/K$. This has finite length over R and therefore has a simple R-submodule, U say. Now U cannot belong to $\mathcal{Y}$ because $U \otimes S \twoheadrightarrow US = V \neq 0$. However, $U \otimes S$ is simple, by (a), and so $V \cong U \otimes S$. □

13.13. COROLLARY. *Let $S = R(\mathcal{Y})$. Let W be a simple R-module such that $W \notin \mathcal{Y}$ and $\operatorname{Ext}(Y, W) = 0$ for all $Y \in \mathcal{Y}$. Then $W \cong W \otimes S$ via $w \to w \otimes 1$.*

PROOF. We know that W is an essential submodule of $(W \otimes S)_R$. If possible, choose any $w \otimes x \notin W$ with $x \in S$. Now $(xR + R)/R$ has finite length and its composition factors all come from $\mathcal{Y}$. The same is therefore true of its homomorphic image $((w \otimes x)R + W)/W$. However, this would mean there was a nonsplit extension of W by some $Y \in \mathcal{Y}$. So no such $w \otimes x$ exists. □

14. Right Finite Overrings and Idealizer Rings

The preceding section gives a description of each overring of an HNP ring R in terms of simple modules.[1] We need to be able to determine which of these are right finite overrings; and that is the burden of this section. We give a couple of useful descriptions. The first is a rather straightforward description in terms of idempotent ideals of R. The second concerns the simple modules involved and gives us the connection with idealizer rings.

We start, however, with a result about semiprime Goldie rings.

14.1. DEFINITION. Let T be a semiprime Goldie ring and ${}_TA_T \subseteq T_{\mathrm{quo}}$. We call A *invertible* if there exist ${}_TX_T \subseteq T_{\mathrm{quo}}$ and ${}_TY_T \subseteq T_{\mathrm{quo}}$ such that $XA = T = AY$. When A is invertible, the usual associativity argument shows that $X = Y$ and that X is the unique left inverse and unique right inverse of A. Therefore we write $A^{-1} = X$, and call A^{-1} the *inverse* of A.

14.2. LEMMA. *Let A be a proper invertible ideal of a semiprime Goldie ring T.*

(i) $A^{-1} = A^*_T = ({}_TA)^* \supset T$.
(ii) $O_l(A) = O_r(A) = T$.
(iii) *A_T and ${}_TA$ are progenerators.*
(iv) *Given any right ideal I of T, there is a (1,1)-correspondence between submodules of T/I and those of A/IA given by $J/I \to JA/IA$.*

PROOF. First let I be any right ideal of T such that $I \cap A = 0$. Then $IA = 0$ and so $I = IAA^{-1} = 0$. So A is an essential right ideal of T and thus contains a unit of T_{quo}.

(i) $A^{-1} \subseteq A^*_T$ since $A^{-1}A = T$; and therefore $T = A^{-1}A \subseteq A^*_TA$. However $A^*_TA \subseteq T$ (see 1.5), and hence $A^*_TA = T$. Since A^{-1} is the unique left inverse of A, we get $A^{-1} = A^*_T$. The equality $A^{-1} = ({}_TA)^*$ holds by symmetry. Since A is an ideal of T, then $A^*_T \supseteq T$; and the inclusion is strict because A is invertible.

(ii) Suppose that $xA \subseteq A$ with $x \in T_{\mathrm{quo}}$. Right-multiplying by A^{-1} shows that $x \in T$. Thus $O_l(A) \subseteq T$; and the opposite inclusion is obvious. The corresponding assertion about $O_r(A)$ holds by symmetry.

(iii) It is now a consequence of 12.11 that A_T and ${}_TA$ are finitely generated projective; and the standard identification applied to the equations $AA^* = A^*A = T$ shows that A is a generator on each side.

(iv) The map $J \to JA$ converts any right ideal J between T and I into a right ideal between A and IA; and right multiplication by A^{-1} provides the inverse map. □

14.3. LEMMA. *Let R be an HNP ring.*

(i) *Each nonzero maximal ideal is either idempotent or invertible.*
(ii) *If A is a proper nonzero idempotent ideal, then $(A_R)^* = O_l(A) \supset R$, $({}_RA)^* = O_r(A) \supset R$ and $O_l(A) \neq O_r(A)$.*

[1] In this section R denotes an HNP ring unless the contrary is specified.

(iii) *If A is an invertible ideal then $\lambda(A^n/A^{n+1}) = \lambda(R/A)$ for every n.*
(iv) *If A is a nonzero ideal maximal amongst proper idempotent ideals, then A is a maximal ideal.*

PROOF. (i) Let M be a nonzero maximal ideal of R. Then $M \subseteq (M_R)^*M \subseteq R$ since $1 \in (M_R)^*$. Since M_R is finitely generated projective, its trace ideal $(M_R)^*M$ is idempotent [12.11]. The maximality of M implies that either $(M_R)^*M = M$ or $(M_R)^*M = R$. So if M is not idempotent then $(M_R)^*M = R$ and, by symmetry, $M(_RM)^* = R$; thus M is invertible.

(ii) Clearly $R_{\text{quo}} \supseteq (A_R)^* \supseteq O_l(A)$. Suppose $x \in (A_R)^*$; so $xA \subseteq R$. Hence $xA = xA^2 \subseteq RA = A$, since $A = A^2$. Thus $(A_R)^* \subseteq O_l(A)$. The (1-1)-correspondence provided by 12.12 demonstrates that, since $A \subset R$ then $(A_R)^* \supset R$; so $O_l(A) \supset R$. By symmetry, $(_RA)^* = O_r(A) \supset R$. Finally, $(A_R)^*A \subseteq R$; yet $(_RA)^*A = O_r(A)$ by 12.11. Hence $(_RA)^* \neq (A_R)^*$ and thus $O_r(A) \neq O_l(A)$.

(iii) By 12.14, $\lambda(R/A)_R$ is finite. 14.2(iv), applied to the right ideal A^n, completes the proof.

(iv) Suppose A is not a maximal ideal and let M be a maximal ideal with $A \subset M$. The maximality of A shows that M cannot be idempotent. Hence, by (i), M is invertible and so, by (iii), $\lambda(M^n/M^{n+1}) = \lambda(R/M)$ for every n. Hence $\lambda(R/M^n) = n\lambda(R/M) \geq n$ for every n. Now $A = A^n \subseteq M^n$ for every n, and therefore $\lambda(R/A) \geq \lambda(R/M^n) \geq n$ for every n. This is a contradiction since 12.14 shows that $\lambda(R/A)$ is finite. □

Note that, unlike the invertible case [14.2], for a proper nonzero idempotent ideal A, $O_l(A)$ never equals $O_r(A)$ and each always properly contains R.

14.4. THEOREM. *There is an order-reversing (1,1)-correspondence between right finite overrings S of R (each of which is again an HNP ring, by* 13.5*) and nonzero idempotent ideals I of R given by $S \to (S_R)^*$ and $I \to O_r(I)$.*

PROOF. Let S be a right finite overring. Then $(S_R)^*$ is a nonzero ideal of R and is also a right ideal of S. Hence $(S_R)^* = (S_R)^*S = I$, say. This is the trace ideal which is known to be idempotent since S_R is finitely generated projective [12.11]. Moreover $O_r(I) = I^*$ by 14.3; and $I^* = (S_R)^{**} = S$ by 1.6.

Next let I be a nonzero idempotent ideal of R and let $S = O_r(I)$. There is a regular element, d say, in I and $S \cong dS \subseteq R$. Hence S is a right finite overring. However 14.3 shows that $S = (_RI)^*$; and $(_RI)^{**} = I$, using 1.6 again. □

Next we look at those right finite overrings corresponding to a maximal idempotent ideal in the way described in 14.4. These are, by that theorem, minimal among overrings. An easy illustration of this occurs with $S = A_1(F)$ or $B_1(F)$ and $R = \mathbb{I}(yS)$, as described in 11.10; then yS is a maximal idempotent ideal of R and S is a minimal overring of R.

The next result, whose proof requires some Morita theory, shows how this situation occurs for each basic idealizer. This result and the next require our standing convention that an HNP ring is not equal to its quotient ring.

14.5. THEOREM. *Let S be an HNP ring and $T = \mathbb{I}_S(A)$, a basic idealizer. Then T is an HNP ring, $T_{\text{quo}} = S_{\text{quo}}$, and S is a minimal and right finite overring of T.*

PROOF. We know from 4.19 that T is right Noetherian and right hereditary. By 12.1(i) there is a regular element of S, d say, in A. Then $dS \subseteq T$ and so 12.1(iv) shows that T is a prime Goldie ring and $T_{\text{quo}} = S_{\text{quo}}$; so S is indeed an overring. Then 4.15 shows that S is a minimal overring and 2.3 that S_T is finitely generated.

It remains to show that T is left Noetherian and left hereditary. To achieve this, we will demonstrate a symmetry which is not instantly apparent. Recall from 4.2 that A is an idempotent maximal ideal of T. Consider the ring $S' = O_l(A)$. By 12.9, we know that $S' \cong \text{End}(A_S)$; and A_S is a finitely generated projective generator module (i.e. a progenerator). Thus S and S' are Morita equivalent rings and so, by 11.4, S' is an HNP ring. Moreover

$$T = \mathbb{I}_S(A) = O_l(A) \cap O_r(A) = S \cap S' = \mathbb{I}_{S'}(A).$$

The symmetry we seek, from which the left-handed properties would follow immediately, now requires only that A is an isomaximal generative left ideal of S'.

To see that, first note, from 12.12 above, that there is a lattice anti-isomorphism between the lattice of submodules of $(S/A)_S$ and the lattice of submodules of ${}_S(A^*/S)$. Hence ${}_S(A^*/S)$ is semisimple isotypic; and, since A_S is a progenerator, the same is true of ${}_{S'}(A \otimes_S (A^*/S))$. However,

$$A \otimes_S (A^*/S) \cong (A \otimes_S A^*)/(A \otimes_S S) \cong S'/A.$$

Thus A is an isomaximal left ideal of S'. Also $A^2 = A$ since A_S is generative; and $AA^* = O_l(A) = S'$ since A_S is finitely generated projective. Thus

$$AS' = A(AA^*) = AA^* = S'$$

and the proof is complete. □

14.6. Corollary. *Let S be a ring and $\mathbb{I}(A)$ be a basic idealizer. Then S is an HNP ring if and only if $\mathbb{I}(A)$ is an HNP ring.*

Proof. The theorem deals with the case when S is an HNP ring. Conversely, suppose that $R = \mathbb{I}(A)$ is an HNP ring. Note that A is a nonzero ideal of R and, by 4.14, $S = {}_RA^*$. But ${}_RA^* \cong \{q \in R_{\text{quo}} \mid Aq \subseteq A\} \subseteq R_{\text{quo}}$. So S is an overring of R and so, by 13.5, is an HNP ring. □

An immediate consequence of 14.5 is one direction of the following theorem.

14.7. Theorem. *An overring S of R is minimal and right finite over R if and only if R is a basic idealizer from S.*

Proof. Let S be minimal and right finite over R. We will show that $R = \mathbb{I}_S(A)$ where $A = S_R^*$ and that this is a basic idealizer. We can make our standard identification $A = \{x \in S \mid xS \subseteq R\}$; and this makes clear that A is an ideal of R and a right ideal of S; so $R \subseteq \mathbb{I}_S(A)$. By 12.12, A is an essential left ideal of R (and so contains a regular element, d say), S_R is finitely generated projective and $(S/R)_R$ has finite length. Moreover, 12.14 shows that $(R/A)_R$ has finite length. Hence both $(S/A)_R$ and $(S/A)_S$ have finite length.

Next, since S_R is finitely generated projective, the Internal Dual Basis Lemma 12.11 shows that $S_R S_R^* = O_l(S) = S$; that is, $SA = S$, proving that A is a generative right ideal of S. Now $\mathbb{I}_S(A) \neq S$ since A is a generative right ideal of S and so $R \subseteq \mathbb{I}_S(A) \subset S$. Minimality of S over R implies that $R = \mathbb{I}_S(A)$.

It remains only to show that A is isomaximal in S. Since S is Noetherian, we can choose B maximal among the (necessarily generative) right ideals of S such that $A \subseteq B \subset S$ and $R = \mathbb{I}_S(B)$.

We claim that $(S/B)_S$ has no proper fully invariant submodules. If C/B is any such submodule, then $R = \mathbb{I}_S(B) \subseteq \mathbb{I}_S(C) \subseteq S$ [1.4]. Since B is generative in S, so too is C; and therefore the final inclusion $\mathbb{I}_S(C) \subseteq S$ is proper. Minimality of S over R then shows that $R = \mathbb{I}_S(C)$. The maximality of B shows that $C = B$, a contradiction that proves the claim.

Since $(S/A)_S$ has finite length, so does $(S/B)_S$. Since $(S/B)_S$ has no proper fully invariant submodules, 3.2 asserts that it is isotypic. Thus $R = \mathbb{I}_S(B)$ is a basic idealizer. But then, by 4.2, $B = S_R^*$; that is, $B = A$ as desired. □

14.8. COROLLARY. *An overring S of R is right finite if and only if R is an iterated basic idealizer from S*

PROOF. Let S be a right finite overring of R. Then $(S/R)_R$ has finite length [12.12]. Therefore there is a minimal, necessarily right finite, overring S_1 such that $S \supseteq S_1 \supset R$. By 14.7, R is a basic idealizer from S_1. Now S_1 is again an HNP ring [13.5], and S is a right finite overring of S_1. Moreover

$$\lambda(S/S_1)_{S_1} \leq \lambda(S/S_1)_R < \lambda(S/R)_R.$$

So induction on length completes the proof that R is an iterated basic idealizer from S.

The converse is immediate from 6.2 □

We saw in the preceding section that overrings of an HNP ring R are given by the simple modules they kill, and thus that minimal overrings are determined by a single simple module. The next result describes, in terms of some properties of the simple R-module, when that minimal overring is right finite.

14.9. THEOREM. *(a) Let W be an unfaithful simple right R-module such that $A = \operatorname{ann}(W)$ is idempotent and let $S = O_r(A)$. Then S is a minimal overring of R and is right finite.*

Moreover $R = \mathbb{I}_S(A)$ is a basic idealizer of type $U = [VW]$ for the simple right R-module V defined uniquely, up to isomorphism, by the property that $\operatorname{Ext}_R^1(V, W) \neq 0$. Also $\operatorname{Ext}_R^1(V, W)$ is a vector space of dimension 1 over each of $\operatorname{End}(V)$ and $\operatorname{End}(W)$; and U_S is faithful if and only if V_R is faithful.

(b) Every minimal and right finite overring S of R has this form for some uniquely defined W; and then S is the overring determined by killing V.

PROOF. (a) A is a maximal ideal, by 12.14. Therefore, by 14.4, the overring $S = O_r(A)$ is a minimal right finite overring of R and $A = (S_R)^*$. Hence, as seen in 14.7, $R = \mathbb{I}_S(A)$ is a basic idealizer, say of type $U = [VW']$. We know, from 4.8, that $A = \operatorname{ann} W'$ and so we deduce that $W \cong W'$. Then 5.8 shows that V is the unique simple right R-module with $\operatorname{Ext}_R^1(V, W) \neq 0$; and 5.7 shows that $\operatorname{Ext}_R^1(V, W)$ is a vector space of dimension 1 over each of $\operatorname{End}(V)$ and $\operatorname{End}(W)$. Finally, we consider faithfulness. First suppose that $0 \neq \operatorname{ann}(V_R) = B$. Since R is prime, $0 \neq BA \subseteq \operatorname{ann}(U_S)$. Conversely, suppose that $0 \neq \operatorname{ann}(U_S) = C$. Since $(R_{\text{quo}})_R$ is an essential extension of R, then $0 \neq C \cap R \subseteq \operatorname{ann}(V_R)$.

(b) Let S be some right finite overring of R. Then 14.7 shows that $R = \mathbb{I}_S(A)$ is a basic idealizer, say of type $U = [VW]$; and 4.8 shows that $A = \operatorname{ann} W$. Since A is a right ideal of S we see that $S \subseteq O_r(A)$; but 14.4 shows that $O_r(A)$ is a minimal right finite overring. Hence $S = O_r(A)$; and 4.8 shows that V is the unique simple R-module killed by S. □

15. Successors of Simple Modules; Factor Rings

We introduce a 'successor' relation for certain pairs of simple R-modules[1], a concept that plays a central role in the remainder of our study of HNP rings and their modules. The first application is to show that every proper factor ring of R is an Artinian serial ring, and sometimes a principal ideal ring. (This application uses the results of §§50, 51 on Artinian serial and principal ideal rings.) The section closes by translating this successor relation — when the two simple modules are unfaithful — into a successor relation between their annihilators. This last topic will be continued in §23.

15.1. Definitions. If $\operatorname{Ext}^1_\Lambda(V, W) \neq 0$, for simple modules V, W over some ring Λ, we say that W is a *successor* of V and V is a *predecessor* of W.

Recall from 12.14 that the annihilator of any unfaithful simple right R-module is a maximal ideal and so, by 14.3(i), is either idempotent or invertible. Many properties of unfaithful simple R-modules are proved differently in these two cases.

15.2. Theorem.

(i) *Every unfaithful simple R-module has a unique predecessor.*

(ii) *If a simple module V has an unfaithful successor W, then W is the unique successor to V. Furthermore, in this situation, $\operatorname{Ext}^1_R(V, W)$ is a vector space of dimension 1 over each of $\operatorname{End}(V)$ and $\operatorname{End}(W)$.*

Proof. Fix an unfaithful simple W_R. Thus $WA = 0$ for some maximal ideal $A \neq 0$, necessarily idempotent or invertible. The case when A is idempotent is proved in the more detailed 14.9. The invertible case is a special case of 15.4 below, whose proof we now prepare for. □

We allow R to be a little more general in the next two results.

15.3. Lemma. *Let R be a Noetherian prime ring, M an invertible maximal ideal such that the ring R/M is Artinian and A a right ideal such that R/A has finite length. Then for every positive integer n we have $A/AM^n \cong R/M^n$.*

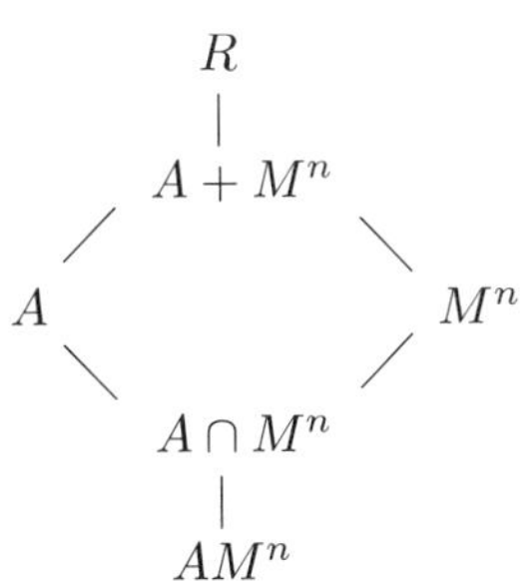

Proof. Since M^n is invertible, it follows that $\lambda(R/A) = \lambda(M^n/AM^n)$. From the submodule diagram we see that $\lambda\big(R/(A + M^n)\big) = \lambda\big((A \cap M^n)/AM^n\big)$. Hence $\lambda(R/M^n) = \lambda(A/AM^n)$ and, in particular, $\lambda(R/M) = \lambda(A/AM)$. Since R/M is a simple Artinian ring, it follows that $A/AM \cong R/M$, a cyclic R-module. Let $J = M/M^n$, the Jacobson radical of R/M^n. Then $(A/AM^n)/(A/AM^n)J \cong A/AM$. Hence A/AM^n is a cyclic module over R/M^n. But any cyclic R/M^n-module whose length equals $\lambda(R/M^n)$ is isomorphic to R/M^n. Thus $A/AM^n \cong R/M^n$ as required. □

15.4. Theorem. *Let R be a Noetherian prime ring, M an invertible maximal ideal such that the ring R/M is Artinian; and let W be the simple R-module such that $WM = 0$.*

(i) *For all n, R/M^n is an Artinian principal ideal ring with radical M/M^n.*

[1]In this section R denotes an HNP ring unless the contrary is specified.

(ii) $\operatorname{Ext}^1_R(W,W)$ *has dimension 1 as a vector space over the endomorphism ring of* W *acting via either variable.*

(iii) W *is its own unique predecessor and successor.*

PROOF. (i) Let A be any right ideal of R containing M^n. Then R/A has finite length and so, by 15.3, A/AM^n is cyclic. The same must be true of its homomorphic image A/M^n. This shows that R/M^n is a right Artinian, principal right ideal ring. Similarly, R/M^n is a left Artinian, principal left ideal ring. Clearly M/M^n is the radical of R/M^n.

(ii) Consider the short exact sequence $M/M^2 \hookrightarrow R/M^2 \twoheadrightarrow R/M$ of right R-modules. Invertibility of M shows that $M/M^2 \neq 0$; and the sequence is nonsplit since $M/M^2 = \operatorname{rad}(R/M^2)$ and the radical never contains nonzero idempotents. So $\operatorname{Ext}^1_R(R/M, M/M^2) \neq 0$. Note that $R/M \cong W^{(t)}$ for some t and $M/M^2 \cong R/M$ by 15.3. Hence $\operatorname{Ext}^1_R(R/M, M/M^2) = \operatorname{Ext}^1_R(W,W)^{(t^2)}$ and so $\operatorname{Ext}^1_R(W,W) \neq 0$.

Let $\Lambda = R/M^2$. Note that if X is any extension of W by W, then $XM^2 = 0$, and hence X is a Λ-module. By viewing Ext^1 as equivalence classes of extensions [53.11], we deduce that $\operatorname{Ext}^1_R(W,W) = \operatorname{Ext}^1_\Lambda(W,W) \neq 0$. Now Λ is an Artinian principal ideal ring, by (i), and hence an Artinian serial ring [51.3]. Therefore the dimension 1 property of $\operatorname{Ext}^1_\Lambda(W,W)$ with respect to its first and second variables, is a special case of 50.16 and 50.12, respectively.

(iii) Since $\operatorname{Ext}^1_R(W,W) \neq 0$, W is a successor and predecessor of itself. Now let X be any simple right R-module such that $X \ncong W$. It remains to show that $\operatorname{Ext}^1_R(W,X) = 0$ and $\operatorname{Ext}^1_R(X,W) = 0$. In each case, we assume there is a nonsplit extension (which, by 5.5, is uniserial and hence cyclic) and obtain a contradiction.

Case 1: $\operatorname{Ext}(W,X)$. Here we have $R \supset B \supset C$ with $R/B \cong W$ and $B/C \cong X$.

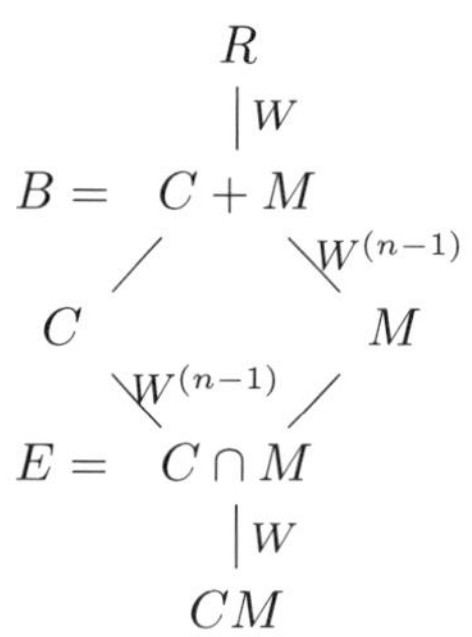

We begin by considering this submodule diagram. Since $R/M \cong W^{(n)}$, and so $R/(C+M) \cong W^{(k)}$ for some k, then $C \subset C+M$. Also $M \subseteq B$ since $WM = 0$. Therefore $C \subset C+M \subseteq C+B = B \subset R$. Thus $C+M = B$ and $R/(C+M) \cong W$ as in the diagram. Since $R/M \cong W^{(n)}$, we can fill in $W^{(n-1)}$ in the two places shown. We have $C/CM \cong R/M \cong W^{(n)}$ by 15.3, so $E/CM \cong W$ as displayed. Since $CM \subset E \subseteq M$, invertibility of M yields $C \subset EM^{-1} \subseteq R$. Thus EM^{-1}/C is a nonzero submodule of R/C annihilated by M; hence R/C has a submodule isomorphic to W. This is the desired contradiction since the unique simple submodule of the uniserial module R/C is $X \ncong W$.

Case 2: $\operatorname{Ext}(X,W)$. Here we have $R \supset B \supset C$ with $R/B \cong X$ and $B/C \cong W$. Since $R/B \cong X$ and $R/M \cong W^{(n)}$, it follows that $B+M = R$. Therefore $B/(B \cap M) \cong R/M$. However, by 15.3, $B/BM \cong R/M$; hence $BM = B \cap M$. Therefore $R/BM \cong R/(B\cap M) \cong R/B \oplus R/M$ which is a semisimple R-module. Moreover, $C \supseteq BM$ since M annihilates $W \cong B/C$. Therefore R/BM maps onto R/C, yielding the contradiction that R/C is semisimple. □

We now return to the situation that R is an HNP ring, noting that the preceding result completes the proof of 15.2. We now provide some examples to demonstrate that the emphasis on unfaithful simple modules is required for the conclusions to be valid.

15.5. EXAMPLE. (i) As described in 11.7, the simple hereditary Noetherian domain $S = B_1(F)$ has infinitely many simple modules, all faithful since S is a simple ring, and $\mathrm{Ext}^1(U, U') \neq 0$ for all simple S-modules U, U'. So this demonstrates that faithful predecessors and successors of faithful simple modules are not unique.

(ii) Now let $R = \mathbb{I}_S(yS)$. We know from 11.10 that this is a basic idealizer and is a hereditary Noetherian domain. Further, if $U = S/yS$ then U_S is simple and, in the notation of 4.5, $U_R = [VW]_R$ for simple R-modules V, W; and, by 11.11, $V = S/R$ is faithful and $W = R/yS$ is unfaithful. We see from 5.7 that $\mathrm{Ext}^1_R(W, W) = 0$. Of course, W is the only unfaithful simple R-module; so W is an unfaithful simple module with no unfaithful successor.

Now let U' be any of the infinitely many simple S-modules. Since $\mathrm{Ext}^1_S(U, U') \neq 0$ then, by 5.3, $\mathrm{Ext}^1_R(W, U') \neq 0$. If $U' \ncong U$ then U'_R is simple [4.4]; and U'_R is faithful since $\mathrm{ann}(U'_R) \subseteq \mathrm{ann}(U'_S)$ which is zero since S is simple. This shows non-uniqueness of faithful successors of the unfaithful simple module W.

15.6. THEOREM. *Each proper factor ring of an HNP ring is an Artinian serial ring.*

PROOF. Let R be the HNP ring and $\Lambda = R/N$ for some ideal $N \neq 0$. The ring Λ is Artinian by 12.14. First note that, for all Λ-modules V, W, the group $\mathrm{Ext}^1_\Lambda(V, W)$ is a subgroup of $\mathrm{Ext}^1_R(V, W)$ — because every Λ-module extension is an R-module extension and the Baer sum of two extensions [see 53.11] is the same in both cases. In fact, $\mathrm{Ext}^1_\Lambda(V, W)$ is an $\mathrm{End}_R(W) = \mathrm{End}_\Lambda(W)$-submodule of $\mathrm{Ext}^1_R(V, W)$, since the action of this endomorphism ring on extensions [see (53.13.1)] is the same in both cases.

Now let V, W be simple Λ-modules, and hence unfaithful simple R-modules. If $\mathrm{Ext}^1_\Lambda(V, W) \neq 0$, then W is uniquely specified by this property, and $\mathrm{Ext}^1_\Lambda(V, W)$ is 1-dimensional over $\mathrm{End}_\Lambda(W)$ [15.2]. Hence, by 50.12, Λ is a right serial ring. Symmetry then shows that Λ is a serial ring. □

This result is improved in 31.1 where it is shown that these serial rings can be constructed from Artinian principal ideal rings.

When simple modules V, W are both unfaithful, the predecessor-successor relation can be translated into a property of their annihilators. The next two results give two ways of doing this.

15.7. PROPOSITION. *Let V, W be unfaithful simple right R-modules, with annihilators M, N respectively. The following are equivalent:*

(i) *W is the successor of V;*
(ii) *$(M \cap N)/MN \neq 0$.*

PROOF. Since the ring R/MN is Artinian, this is a special case of 50.9, together with the fact that (for HNP rings) unfaithful successors are unique [15.2]. □

More detail is available by considering the two types of maximal ideals separately.

15.8. PROPOSITION. *Let V, W be unfaithful simple right R-modules, with annihilators M, N respectively.*

(i) *Suppose that M is idempotent. Then W is the successor of V if and only if $O_r(N) = O_l(M)$; and in that case $W \ncong V$, $M \neq N$, N is idempotent, $O_l(M) \neq R$, and and $O_r(N) \neq R$.*

(ii) *Suppose that N is invertible. Then W is the successor of V if and only if $W \cong V$; and in that case $M = N$ and $O_r(M) = O_l(M) = R$.*

(iii) *Suppose that W is the successor of V. Then either M and N are both idempotent and (i) applies or else $M = N$ is invertible and (ii) applies.*

PROOF. (i) Let $S = O_r(N)$ and $T = O_l(M)$. Note first that $TM \subseteq M$ and so $(T/R)M = 0$. Note also that $T \supset R$ by the left-handed version of 14.4. Therefore T/R is a nonzero right R/M-module and is a direct sum of copies of V. It follows that T is the minimal overring of R determined by killing V [13.8].

Now suppose that W is the successor of V. Then 14.9 shows that $R = \mathbb{I}_S(N)$, a basic idealizer, and that S is the minimal overring determined by killing the simple R-module V. Hence $S = T$; i.e. $O_r(N) = O_l(M)$.

Conversely, suppose that $S = T$. We already have seen that T is the minimal overring of R determined by killing V; so the same is true of $S = O_r(N)$. Therefore, by 14.9, $R = \mathbb{I}_S(N)$ and $\operatorname{Ext}^1_R(V, W) \neq 0$. Thus W is the successor of V.

Finally, the supplementary claims: 4.4 shows that $W \ncong V$; it is then immediate that $M \neq N$; 4.2 shows that N is idempotent; and we have already proved the remaining statements.

(ii) 15.4 shows that W is its own successor and predecessor. So it remains only to show that $O_r(M) = O_l(M) = R$. However, if $xM \subseteq M$ then $xMM^{-1} \subseteq MM^{-1}$; i.e. $xR \subseteq R$ and so $O_l(M) = R$. Symmetry completes the argument.

(iii) We know, by 14.3, that M is either invertible or idempotent. If it is idempotent then (i) applies. Otherwise M is invertible. Then, by 15.4, V is its own unique successor; so $V \cong W$ and $M = N$. Hence (ii) applies. □

16. Uniserial Modules

A predecessor/successor pair of simple modules has a nonsplit extension which is uniserial of length 2. In this section[1] we show how a sequence $W_1, W_2, \ldots, W_n$ such that W_{i+1} is a successor of W_i for each $i \in \{1, \ldots, n-1\}$ leads naturally to uniserial modules of longer length. Our main result states that, when all of these successors are unfaithful, the resulting uniserial module is unique [16.5].

16.1. LEMMA. *Let T be any ring and M_T a module of finite length with a composition series $M = C_0 \supset C_1 \supset \ldots \supset C_n = 0$. The following are equivalent:*

(i) *M is uniserial;*

(ii) *the length 2 subfactor C_i/C_{i+2} is nonsplit for each $i \in \{0, \ldots, n-2\}$.*

PROOF. It is clear that (i) implies (ii). So assume that (ii) holds. We prove (i) by induction on n; and may assume that $n \geq 2$. Since the subfactor C_{n-2}/C_n is nonsplit, 5.5 proves that C_{n-2} is uniserial, proving the case $n = 2$ and allowing us to assume that $n \geq 3$ from now on. By the induction hypothesis, the module M/C_{n-1} is uniserial. We aim to prove that M is uniserial.

First we claim that $\operatorname{soc}(M) = C_{n-1}$. If not, then $X \oplus C_{n-1} \subseteq \operatorname{soc}(M)$ for some simple submodule X of M. But then the simple submodule $(X \oplus C_{n-1})/C_{n-1}$ of the uniserial module M/C_{n-1} must be the socle C_{n-2}/C_{n-1} of M/C_{n-1}. Therefore $X \oplus C_{n-1} = C_{n-2}$, a contradiction since C_{n-2} is uniserial. This proves the claim.

Now let X and Y be nonzero submodules of M. Then $C_{n-1} \cap X \neq 0$ by the previous claim. Since C_{n-1} is simple, this implies that $C_{n-1} \subseteq X$. Similarly

[1] In this section R denotes an HNP ring unless the contrary is specified.

$C_{n-1} \subseteq Y$. Since M/C_{n-1} is uniserial, we have (say) $X/C_{n-1} \subseteq Y/C_{n-1}$ and therefore $X \subseteq Y$, completing the proof. □

16.2. PROPOSITION. *Let R be a right hereditary ring, M_R be a uniserial module of finite length with a composition series $M = C_0 \supset C_1 \supset \ldots \supset C_n = 0$ and X be a simple R-module.*

(i) *If $\operatorname{Ext}^1(C_{n-1}, X) \neq 0$ then there is a uniserial module which is an extension of X by M.*
(ii) *If $\operatorname{Ext}^1(X, M/C_1) \neq 0$ then there is a uniserial module which is an extension of M by X.*

PROOF. (i) The given composition series of M yields a short exact sequence

$$\mathcal{F}: \quad C_{n-1} \overset{\alpha}{\hookrightarrow} M \twoheadrightarrow M/C_{n-1}.$$

By hypothesis, we can choose a nonsplit extension $\mathcal{D}$ of X by C_{n-1}, as shown in the first row of diagram (16.2.1).

$$\begin{array}{lccccc} \mathcal{D}: & X & \hookrightarrow & V & \twoheadrightarrow & C_{n-1} \\ & \downarrow 1 & & \downarrow \gamma & & \downarrow \alpha \\ \mathcal{E}: & X & \hookrightarrow & U & \twoheadrightarrow & M \end{array} \tag{16.2.1}$$

If we apply $\operatorname{Hom}(-, X)$ to the short exact sequence $\mathcal{F}$ we obtain a long exact Ext sequence containing the sequence

$$\ldots \to \operatorname{Ext}^1(M, X) \overset{(\alpha,1)}{\longrightarrow} \operatorname{Ext}^1(C_{n-1}, X) \to 0$$

where the zero occurs because R is hereditary and so $\operatorname{Ext}^2 = 0$. Therefore $\mathcal{D}$ is the image, via $(\alpha, 1)$, of some $\mathcal{E} \in \operatorname{Ext}^1(M, X)$; and $\mathcal{E}$ is nonsplit since it maps onto the nonzero element $\mathcal{D}$ of $\operatorname{Ext}^1(C_{n-1}, X)$.

In the notation of (53.13.1) we have $\mathcal{D} = \alpha\mathcal{E}$; and therefore $\mathcal{D}$ and $\mathcal{E}$ are part of a commutative diagram of the form (16.2.1). Therefore the given composition series of M, together with $\mathcal{E}$, yields a composition series of U. Since $\mathcal{E}$ is nonsplit and M is uniserial, we can apply 16.1 to this composition series of U to conclude that U is uniserial, as desired.

(ii) This is proved similarly. □

16.3. COROLLARY. *Let R be a right hereditary ring and let $W_1, W_2, \ldots, W_n$ be a sequence of simple R-modules such that W_{i+1} is a successor of W_i for each $i \in \{1, \ldots, n-1\}$. Then there is a uniserial right R-module M_n of length n having this sequence of simple modules as its composition factors (from top to bottom).*

PROOF. We assume, by induction on n, that there is such a module M_{n-1} whose composition factors are $W_1, W_2, \ldots, W_{n-1}$. Then 16.2 demonstrates that there is a uniserial module M_n which is an extension of W_n by M_{n-1}. □

We return to the full hypothesis that R is an HNP ring. The case of particular interest here is when each successor in the sequence (i.e. $W_2, \ldots, W_n$) is unfaithful. We show, in 16.5 below, that M_n is then unique up to isomorphism.

16.4. PROPOSITION. *Consider a sequence of simple R-modules $W_1, W_2, \ldots, W_n$ such that W_{i+1} is an unfaithful successor of W_i for each $i \in \{1, \ldots, n-1\}$. Let M_n be a uniserial R-module whose sequence of composition factors (from top to bottom) is $W_1, \ldots, W_n$ (see 16.3). Then for every simple R-module X:*

(i) $\operatorname{Ext}^1(M_n, X) \cong \operatorname{Ext}^1(W_n, X)$ *canonically as left* $\operatorname{End}(X)$*-modules;*
(ii) $\operatorname{Ext}^1(X, M_n) \cong \operatorname{Ext}^1(X, W_1)$ *canonically as right* $\operatorname{End}(X)$*-modules.*

PROOF. Note first that the whole result is trivial when $n = 1$. We establish the result by induction on n. Thus we take $n \geq 2$ and assume the corresponding results for $n - 1$.

(i) Apply $\operatorname{Hom}(-, X)$ to the short exact sequence

$$0 \to W_n \to M_n \to M_n/W_n = M_{n-1} \to 0$$

getting the long exact sequence

$$0 \to \operatorname{Hom}(M_{n-1}, X) \to \operatorname{Hom}(M_n, X) \xrightarrow{\alpha} \operatorname{Hom}(W_n, X)$$
$$\xrightarrow{\delta} \operatorname{Ext}^1(M_{n-1}, X) \xrightarrow{\beta} \operatorname{Ext}^1(M_n, X) \to \operatorname{Ext}^1(W_n, X) \to 0$$

where the final zero comes because R is right hereditary. We need to consider the named maps in the long exact sequence, starting with α. Since X is simple, any map from M_n to X must annihilate the radical of M_n, and hence W_n. Thus α must be the zero map.

Next we turn to the injection δ. Of course, $\operatorname{Hom}(W_n, X)$ is zero unless $X \cong W_n$, in which case it is a 1-dimensional vector space over $\operatorname{End}(X)$. On the other hand, $\operatorname{Ext}^1(M_{n-1}, X) \cong \operatorname{Ext}^1(W_{n-1}, X)$ by the induction hypothesis. However, 15.2 tells us that $\operatorname{Ext}^1(W_{n-1}, X)$ is zero unless $X \cong W_n$, the successor of W_{n-1}, in which case it is a 1-dimensional vector space over $\operatorname{End}(X)$. It follows, in either case, that $\operatorname{Hom}(W_n, X)$ and $\operatorname{Ext}^1(M_{n-1}, X)$ are isomorphic as vector spaces over $\operatorname{End}(X)$ and therefore the injection δ is an isomorphism and $\beta = 0$. The result follows.

(ii) This is proved in an entirely similar fashion to (i), so its proof is omitted. □

The next corollary is the main result of this section.

16.5. COROLLARY. *Let $W_1, W_2, \dots, W_n$ be a sequence of simple R-modules, such that W_{i+1} is the unique unfaithful successor of W_i $i \in \{1, \dots, n-1\}$. Then the uniserial R-module M_n, whose sequence of composition factors, from top to bottom, is $W_1, \dots, W_n$ [given by 16.3], is unique up to isomorphism.*

Indeed, up to R-module isomorphism, M_n is the unique nonsplit extension of W_n by M_{n-1}.

PROOF. We assume the result for $n - 1$; so M_{n-1} is unique. Let M_n' be any other nonsplit extension of W_n by M_{n-1}. We want to show that $M_n \cong M_n'$, and for this it suffices to show that $\operatorname{Ext}^1_R(M_{n-1}, W_n)$ has dimension 1 over the division ring $\operatorname{End}(W_n)$ [53.16]. However, 16.4 shows that $\operatorname{Ext}^1(M_{n-1}, W_n) \cong \operatorname{Ext}^1(W_{n-1}, W_n)$ as $\operatorname{End}(W_n)$-vector spaces; and, since W_n is the unfaithful successor of W_{n-1}, the latter Ext^1 has dimension 1 over $\operatorname{End}(W_n)$ [15.2]. □

17. Right-Left Symmetry: Duality and Transpose

Two types of 'duality' relate the structure of finitely generated right modules over HNP rings to that of finitely generated left modules.[1] The reason we need two types of duality is that every finitely generated right module is the direct sum of a projective module and a module of finite length [12.18]. Concerning the former modules, we have already noted in the Dual Basis Lemma [1.6] that, over any ring, the (ordinary) dual M^* of any finitely generated projective right module M is a

[1]In this section R denotes an HNP ring unless the contrary is specified.

finitely generated projective left module, and that $M \cong M^{**}$, the double dual, via the canonical identification which maps $m \in M$ to the map $M^* \to R$ given by $f \to f(m)$.

In this section we establish a type of duality, called the 'transpose', between right and left modules of finite length. Unlike ordinary duality, this duality depends upon the fact that our ring is an HNP ring.

17.1. DEFINITION. We define the *transpose* of a right R-module X of finite length over an HNP ring to be the left R-module $X^{\mathrm{tr}} = \mathrm{Ext}^1_R(X, R_R)$. Similarly, we define the *transpose* of a left R-module Y of finite length to be the right R-module $Y^{\mathrm{tr}} = \mathrm{Ext}^1_R(Y, {}_RR)$. Finally, for any right or left R-module X, we let $X^{\mathrm{tt}} = (X^{\mathrm{tr}})^{\mathrm{tr}}$.

17.2. LEMMA. *For every finitely generated projective R-module P we have* $\mathrm{udim}(P) = \mathrm{udim}(P^*)$.

PROOF. We know [12.7] that, if $\mathrm{udim}(P) = n$, then $P \cong P_1 \oplus \ldots \oplus P_n$ with each P_i a uniform right ideal of R and so each $P_i^* \neq 0$. Now $P^* \cong P_1^* \oplus \ldots \oplus P_n^*$. Hence $\mathrm{udim}(P^*) \geq \mathrm{udim}(P)$; and so also $\mathrm{udim}(P^{**}) \geq \mathrm{udim}(P^*) \geq \mathrm{udim}(P)$. However, $P^{**} \cong P$. □

The next result shows how we compute X^{tr}. The critical way in which R being an HNP is used is that every presentation $K \hookrightarrow P \twoheadrightarrow X$ of X_R, with P projective, is also a projective resolution of X (because K is necessarily projective).

17.3. THEOREM. *Let X be a right R-module of finite length. Choose any short exact sequence $K_R \hookrightarrow P_R \twoheadrightarrow X$ with P, and hence also K, finitely generated projective. Then $P^* \subseteq K^*$ (by restricting to K each map $f \in P^*$). This yields a short exact sequence $P_R^* \hookrightarrow K_R^* \twoheadrightarrow X^{\mathrm{tr}}$ and X^{tr} is a left R-module of finite length.*

PROOF. Apply $\mathrm{Hom}(-, R)$ to the given short exact sequence. We get the exact sequence:

$$0 \to X^* \to P^* \to K^* \xrightarrow{\partial} \mathrm{Ext}^1_R(X, R) \to \mathrm{Ext}^1_R(P, R) \to \ldots \tag{17.3.1}$$

in which the map $P^* \to K^*$ is by restriction. Since X has finite length and, by our convention [§0.2], R is not simple Artinian, we have $X^* = 0$; and since P is projective, the term on the right is zero. Since X has finite length, $\mathrm{udim}(P) = \mathrm{udim}(K)$ [12.17]. Therefore $\mathrm{udim}(P^*) = \mathrm{udim}(K^*)$ [17.2], and hence X^{tr} has finite length [12.17]. □

The following abbreviated version of 17.3 is the form in which we usually apply it.

17.4. COROLLARY. *Suppose that X_R has finite length and $X \cong P/K$, with P finitely generated projective. Then X^{tr} has finite length and $X^{\mathrm{tr}} \cong K^*/P^*$.* □

17.5. LEMMA. *Let P be a finitely generated projective right R-module and K a submodule having* $\mathrm{udim}(K) = \mathrm{udim}(P)$. *The map $L \to L^*$ gives a lattice anti-isomorphism between the sets of submodules L_R with $K \subseteq L \subseteq P$ and submodules ${}_RM$ with $P^* \subseteq M \subseteq K^*$. The inverse anti-isomorphism is given by $M \to M^*$.*

PROOF. Let $K \subseteq L \subseteq P$. Then the theorem gives embeddings $P^* \subseteq L^* \subseteq K^*$ simply by restriction of maps. As mentioned above, the Dual Basis Lemma [1.6] tells us that we may identify P and its double dual P^{**} via the canonical identification which maps $p \in P$ to the map $P^* \to R$ given by $f \to f(p)$. Of course, whenever $p \in L$, this map is precisely the restriction to P^* of the corresponding map from L^* to R. Thus, under these identifications, $L^{**} = L$ for each L_R. The corresponding facts for ${}_RM$ follow by symmetry and then the result follows immediately. □

17.6. PROPOSITION. *The transpose $X \to X^{\mathrm{tr}}$ is a contravariant additive functor on right modules of finite length.*

PROOF. Let $f\colon X_1 \to X_2$ be a homomorphism of R-modules of finite length and, in the notation of 17.4, write each $X_i \cong P_i/K_i$. The two rows of the first diagram in (17.6.1) display this. Since P_1 is projective, there exists a map g making the second square in this diagram commute, and since the maps $K_i \to P_i$ are inclusions, taking h to be the restriction of g to K_1 makes the first square commute.

(17.6.1)
$$\begin{array}{ccccc} K_1 & \hookrightarrow & P_1 & \twoheadrightarrow & X_1 \\ \downarrow h & & \downarrow g & & \downarrow f \\ K_2 & \hookrightarrow & P_2 & \twoheadrightarrow & X_2 \end{array} \qquad \begin{array}{ccccc} K_1^* & \hookrightarrow & P_1^* & \twoheadrightarrow & X_1^{\mathrm{tr}} \\ \uparrow h^* & & \uparrow g^* & & \uparrow f^{\mathrm{tr}} \\ K_2^* & \hookrightarrow & P_2^* & \twoheadrightarrow & X_2^{\mathrm{tr}} \end{array}$$

Dualizing the first square yields the commutative left-hand square in the second diagram; and this induces a unique map f^{tr} making the second square of this diagram commute.

The theorem on the functorial property of Ext^1 [53.14] states that f^{tr} is independent of the choices of the modules P_i and K_i and maps in the first diagram. Thus, for any right R-homomorphism $f\colon X_1 \to X_2$, f^{tr} is a left R-homomorphism $X_2^{\mathrm{tr}} \to X_1^{\mathrm{tr}}$. Hence the transpose becomes a contravariant functor, as desired. □

We can now state the main result of this section.

17.7. THEOREM. *Let X be a right R-module of finite length.*

(i) *$X^{\mathrm{tt}} \cong X$ for every R-module X of finite length.*
(ii) *The lattice of submodules of X^{tr} is anti-isomorphic to that of X.*
(iii) $\lambda(X^{\mathrm{tr}}) = \lambda(X)$.
(iv) *X^{tr} is semisimple if and only if X is semisimple.*
(v) *X^{tr} is uniserial if and only if X is uniserial.*
(vi) *If $W \hookrightarrow X \twoheadrightarrow V$ is a short exact sequence, then so is the transposed sequence $V^{\mathrm{tr}} \hookrightarrow X^{\mathrm{tr}} \twoheadrightarrow W^{\mathrm{tr}}$. The transposed sequence is split if and only if the original sequence is split.*

PROOF. We use the notation $X = P/K$ and $X^{\mathrm{tr}} = K^*/P^*$ of 17.4.

(i) $X^{\mathrm{tt}} \cong P^{**}/K^{**} = P/K \cong X$, the equality being provided by the identifications $P = P^{**}$ and $K = K^{**}$ in the Dual Basis Lemma [1.6].

(ii) The lattices of submodules of X and X^{tr} are, respectively, isomorphic to the lattices of submodules between P and K and between K^* and P^*; and the latter two lattices are anti-isomorphic by 17.5.

(iii), (iv), (v) Clear from (ii).

(vi) [See 17.6] We may take the given short exact sequence to be:

$$\mathcal{E}\colon L/K \hookrightarrow P/K \twoheadrightarrow P/L\,.$$

Conversely, every configuration of the form $K \subseteq L \subseteq P$ yields a short exact sequence of the form $\mathcal{E}$.

Dualizing the given configuration yields $P^* \subseteq L^* \subseteq K^*$, and hence the short exact sequence $\mathcal{E}^{\mathrm{tr}}$: $L^*/P^* \hookrightarrow K^*/P^* \twoheadrightarrow K^*/L^*$ of transposes, proving the first assertion of (vi). (Among the details to be checked above we note, in particular, that the inclusion map $L \hookrightarrow P$ lifts the inclusion map in $\mathcal{E}$, and so dualizing gives the transpose $K^*/P^* \twoheadrightarrow K^*/L^*$ of the inclusion map in $\mathcal{E}$.)

Finally, $\mathcal{E}$ is split if and only if there exists a submodule $H \subseteq P$ such that $K = H \cap L$ and $P = H + L$. Dualizing these relations gives $P = H^* \cap L^*$ and $K^* = H^* + L^*$; and so $\mathcal{E}$ splits if and only if $\mathcal{E}^{\mathrm{tr}}$ docs. □

17.7(ii) does not give an explicit formula for the anti-isomorphism between the sets of submodules of X and of X^{tr}. Although we do not need it, we give the formula, and a brief version of its proof.

17.8. COROLLARY. *Let X_R have finite length, and let W range over its submodules, yielding pairs of short exact sequences:*

$$W \overset{\alpha}{\hookrightarrow} X \overset{\beta}{\twoheadrightarrow} V \qquad \textit{and} \qquad V^{\mathrm{tr}} \overset{\beta^{\mathrm{tr}}}{\hookrightarrow} X^{\mathrm{tr}} \overset{\alpha^{\mathrm{tr}}}{\twoheadrightarrow} W^{\mathrm{tr}}$$

Then the correspondence $W \longrightarrow \ker(\alpha^{\mathrm{tr}}) = \mathrm{im}(\beta^{\mathrm{tr}})$ is a lattice anti-isomorphism between the sets of submodules of X and of X^{tr}.

PROOF. Using the fact (17.7(vi)) that tr reverses short exact sequences, one can adapt the proof of 17.7(ii). □

17.9. DEFINITION. Let Λ be a ring and V a simple right Λ-module such that $\mathrm{ann}(V) = M$, a maximal ideal such that Λ/M is simple Artinian. The notation V^{op} — *V-opposite* — denotes the unique simple left Λ-module such that $MV^{\mathrm{op}} = 0$. (This defines V^{op} up to isomorphism because V and V^{op} are respectively the unique simple right and left modules over the simple Artinian ring Λ/M.)

Under similar circumstances, we define the opposite, X^{op}, of an appropriate simple left Λ-module, X. Then $(V^{\mathrm{op}})^{\mathrm{op}} \cong V$.

17.10. COROLLARY. *If a simple module W_R is the unfaithful successor of V then V^{tr} is the unfaithful successor of W^{tr}. Moreover $V^{\mathrm{tr}} \cong W^{\mathrm{op}}$.*

PROOF. Since $\mathrm{Ext}^1(V, W) \neq 0$, there is a nonsplit short exact sequence $V \hookrightarrow U \twoheadrightarrow W$ with U of length 2, and hence uniserial [5.5]. Applying tr gives another short exact sequence $W^{\mathrm{tr}} \hookrightarrow U^{\mathrm{tr}} \twoheadrightarrow V^{\mathrm{tr}}$ with U^{tr} again uniserial of length 2 [17.7]. Thus $\mathrm{Ext}^1(W^{\mathrm{tr}}, V^{\mathrm{tr}}) \neq 0$. We will show that if $A = \mathrm{ann}(W)$ then $A = \mathrm{ann}(V^{\mathrm{tr}})$ and hence $V^{\mathrm{tr}} \cong W^{\mathrm{op}}$. Recall that A, being a maximal ideal of R, is either idempotent or invertible [14.3]. We consider each case separately.

First, suppose that A is idempotent. Then $R = \mathbb{I}_S(A)$, a basic idealizer of type $U = [VW]$, for some overring S [14.9]. Therefore $A = S_R^*$ [4.14]; and $(S/R)_R \cong V^{(n)}$ for some $n > 0$ [4.4]. Hence $(V^{(n)})^{\mathrm{tr}} \cong R/S^* \cong R/A$, and so ${}_R(R/A) \cong (V^{\mathrm{tr}})^{(n)}$. Since A is a 2-sided ideal of R, this implies that $AV^{\mathrm{tr}} = 0$, as desired.

Secondly, suppose A is invertible. Then W is its own unique predecessor [15.4], hence $V \cong W$. Since W is the simple right module over the simple Artinian ring R/A we have $R/A \cong W^{(n)}$ for some $n > 0$, and therefore ${}_R(A^*/R) \cong (W^{\mathrm{tr}})^{(n)}$ [17.4]. Since A is invertible, $AA^* = R$. Therefore $AW^{\mathrm{tr}} = AV^{\mathrm{tr}} = 0$, since $W \cong V$. This completes the proof. □

When both simples are unfaithful, 17.10 yields the following additional symmetry result, about opposites rather than transposes, whose proof clarifies the relation between these two notions.

17.11. COROLLARY. *Suppose that V and W are unfaithful simple modules. Then W is the successor of V if and only if V^{op} is the successor of W^{op}.*

PROOF. Suppose first that V, W is a successor pair with W unfaithful. Since V is unfaithful, it has a predecessor (say) X [15.2]. Two applications of 17.10 to the successor sequence X, V, W then yield the successor sequence $W^{\text{tr}}, V^{\text{tr}}, X^{\text{tr}}$. An application of 17.10 to the successor pair V, W yields $V^{\text{tr}} \cong W^{\text{op}}$; and another application to the successor pair X, V yields $X^{\text{tr}} \cong V^{\text{op}}$, as desired. The converse follows by symmetry since $(V^{\text{op}})^{\text{op}} \cong V$. □

17.12. COROLLARY. *Let $f\colon X \to Y$ be a homomorphism of R-modules of finite length.*

(i) $\operatorname{Hom}(X, Y) \cong \operatorname{Hom}(Y^{\text{tr}}, X^{\text{tr}})$ *via* $f \to f^{\text{tr}}$ *(as abelian groups).*
(ii) *If $X = Y$ then the isomorphism in* (i) *is a ring isomorphism (with f acting on the left and f^{tr} acting on the right).*

PROOF. (i) We apply 17.6 to get f^{tr}. Choose a presentation for each of X and Y: say $K_R \hookrightarrow P_R \twoheadrightarrow X$ and $L_R \hookrightarrow Q_R \twoheadrightarrow Y$. The map f can be lifted to a map $g\colon P \to Q$ which maps K to L. This induces a map $Q^* \to P^*$ which maps L^* to K^*. This, in turn, induces the map $f^{\text{tr}}\colon Y^{\text{tr}} \to X^{\text{tr}}$. Repetition of this process returns us to f. Thus we get (i).

(ii) Since the transpose functor is contravariant, it preserves function composition, but reverses the order. This reversal is hidden here because of our convention that f and f^{tr} act on the left and right, respectively, of the right and left modules X and X^{tr}. □

When is the transpose of a simple module faithful? For an example, see immediately below; and for the general answer, see 21.3.

17.13. EXAMPLE. We illustrate the section by once again using the rings $S = A_1(F)$ or $S = B_1(F)$ and $R = \mathbb{I}(yS)$, as in Examples 11.10 and 15.5. We apply the preceding result to the short exact sequence

$$0 \to R/yS \to S/yS \to S/R \to 0$$

which is a nonsplit extension in $\operatorname{Ext}^1_R(S/R, R/yS)$. We note that the R-module duals of S, R, yS are easily checked to be yS, R, ySy^{-1} respectively. So the corresponding transpose sequence of left R-modules is

$$0 \to R/yS \to (ySy^{-1})/yS \to (ySy^{-1})/R \to 0\,.$$

This is, of course, nonsplit from the above result; and $(ySy^{-1})/yS$ is precisely the uniserial left R-module of length 2 obtained when the ring $T = ySy^{-1}$ is idealized with respect to its maximal left ideal yS.

Note that, as predicted by 17.10, the dual of the unfaithful simple successor right R-module is the faithful simple predecessor left R-module and that of the faithful simple predecessor right R-module is the unfaithful simple successor left R-module.

17.14. REMARK (Reason for name 'transpose'). Let R be a principal ideal domain. Then any map $\alpha\colon R_R \to R_R$ equals left multiplication by some unique element a of R. The dual map α^* then equals right multiplication by a.

Now consider a presentation $R_{m\times 1} \overset{\alpha}{\hookrightarrow} R_{n\times 1} \twoheadrightarrow X$ of a right R-module X of finite length, with the elements of the two free modules written as rows. Then α equals left multiplication by some unique $n \times m$ matrix A over R. The dual homomorphism then yields a presentation $R_{1\times n} \overset{\alpha^*}{\hookrightarrow} R_{1\times m} \twoheadrightarrow Y$ where α^* equals right multiplication by the transpose of A and therefore, by 17.4, $Y \cong X^{\mathrm{tr}}$. Moreover, by that same result, the isomorphism class of Y is determined by that of X.

CHAPTER 4

Towers

This chapter introduces towers, which provide a structure on the collection of (isomorphism types of) simple modules over an HNP ring R. A tower is a finite set of nonisomorphic simple modules which is either linearly ordered (and called a faithful tower) or else cyclically ordered (and called a cycle tower). After preliminary work in §18, towers are defined in §19. Then §§20–24 look at examples of towers (via idealizers), at right-left symmetry, at the connection with maximal ideals, and with injective modules. Their crucial role in classifying projective modules will become clearer in later chapters. Because of this, we take time, in §25, to construct examples of HNP rings with preassigned arbitrary numbers and lengths of towers of each type (faithful and cycle).

18. Finiteness Results

In preparation for the definition of towers [19.1] and their associated uniserial modules [19.2] we prove a number of finiteness results.[1] These include a generalization [18.4] of part of the well-known Krull-Akizuki theorem for commutative Noetherian rings.

18.1. Definition. A ring is called *right bounded* if each essential right ideal contains a nonzero two-sided ideal. As usual, omission of the term 'right' or 'left' means that the ring satisfies both the right and left conditions.

In 23.4, we show that an HNP ring is right bounded if and only it is left bounded.

The next few results rely only on the ring being a Noetherian prime ring which satisfies the restricted minimum condition (on both sides) — which is true for any HNP ring [12.16].

18.2. Lemma. *For a Noetherian prime ring T which satisfies the restricted minimum condition, the following are equivalent:*

(i) *T is right bounded;*
(ii) *each simple right T-module is unfaithful;*
(iii) *the annihilator of every right T-module of finite length is nonzero;*
(iv) *T is not right primitive.*

Proof. This is readily checked. □

18.3. Lemma. *Let T be a Noetherian prime ring with restricted minimum condition (on both sides) and suppose that T is right bounded. Let $I_1 \supseteq I_2 \supseteq \dots$ be a descending chain of essential right ideals and let L be an essential left ideal. Then the descending chain $I_1 + L \supseteq I_2 + L \supseteq \dots$ stabilizes.*

[1]In this section R denotes an HNP ring unless the contrary is specified.

PROOF. We can choose nonzero two-sided ideals $B_1 \supseteq B_2 \supseteq \ldots$ such that $B_i \subseteq I_i$ for each i. The descending chain $B_1 + L \supseteq B_2 + L \supseteq \ldots$ of left ideals containing L must stabilize. So there is an integer i such that $B_i + L = B_{i+1} + L = \ldots$. The descending chain $B_i + I_j \supseteq B_i + I_{j+1} \supseteq \ldots$ of essential right ideals containing B_i also must stabilize. So, for some $j \geq i$ we have $B_i + I_k = B_i + I_{k+1}$ for all $k \geq j$. Then, for each $k \geq j$,

$$\begin{aligned} I_k + L &= I_k + B_k + L = I_k + B_i + L \\ &= I_{k+1} + B_i + L = I_{k+1} + B_{k+1} + L = I_{k+1} + L. \end{aligned}$$ □

18.4. THEOREM. *Let T be a Noetherian prime ring with restricted minimum condition on both sides (e.g. any HNP ring) and suppose that T is right bounded. Let S be an overring such that $T \subseteq S \subset T_{\text{quo}}$. Then each simple right S-module has finite length when viewed as a right T-module and is unfaithful over S.*

PROOF. Let A be some maximal right ideal of S. So A is essential in S since $S \neq T_{\text{quo}}$ [12.14]. Hence A contains a regular element of S; and this has the form ac^{-1} for some regular elements $a, c \in T$. Then we see that $a \in A \cap T$. We will show that $(S/aS)_T$ has finite length and hence the same is true of $(S/A)_T$.

For each $i \geq 1$, let $A_i = a^i S \cap T$. We claim that, for some n,

$$aS + a^{-n} A_n = S\ . \tag{18.4.1}$$

The inclusion ($\subseteq$) is obvious. Conversely, choose any element $s \in S$, and write it in the form $s = t_0^{-1} t_1$ with $t_0, t_1 \in T$ and t_0 regular. Then Tt_0 is an essential left ideal of T, and so 18.3 shows that there is an integer n such that $a^n T + Tt_0 = a^{n+1} T + Tt_0$. Hence $a^n = a^{n+1} t_2 + t_3 t_0$ for some $t_2, t_3 \in T$, and therefore $a^n s = a^{n+1} t_2 s + t_3 t_0 s$. Hence $t_3 t_0 s \in a^n S$. However, $t_3 t_0 s = t_3 t_0 t_0^{-1} t_1 \in T$ and so $t_3 t_0 s \in a^n S \cap T = A_n$. Hence $s = a t_2 s + a^{-n} t_3 t_0 s \in aS + a^{-n} A_n$ which completes the proof of (18.4.1).

From (18.4.1), we deduce the following T-module isomorphisms:

$$\frac{S}{aS} = \frac{aS + a^{-n}A_n}{aS} \cong \frac{a^{-n}A_n}{a^{-n}A_n \cap aS} \cong \frac{A_n}{A_n \cap a^{n+1}S}\ .$$

Note that $A_n \cap a^{n+1} S$ contains the regular element a^{n+1} of T. However, by the restricted minimum condition, the T-module $T/a^{n+1}T$ has finite length, and hence so does its submodule $A_n/a^{n+1}T$. Therefore $A_n/(A_n \cap a^{n+1} S)$ has finite length over T, and hence so does S/aS. Thus each simple right S-module has finite length as a right T-module. By 18.2, it is therefore unfaithful over T and hence over S. □

We now return to the case of an HNP ring R.

18.5. PROPOSITION. *Let S be an overring such that $R \subseteq S \subset R_{\text{quo}}$. Then any unfaithful simple right S-module is uniserial of finite length over R.*

PROOF. Let U be an unfaithful simple right S-module. By 13.12, we know that $U \cong W \otimes_R S$ for some simple right R-module W, and that W is essential in U_R. Note that since $\text{ann}_S(U)$ is a nonzero ideal, then $B = \text{ann}_S(U) \cap R \neq 0$; so W_R and U_R are both R/B-modules. However, R/B is an Artinian serial ring since it is a proper factor ring [15.6]. Therefore U must be a direct sum of uniserial R/B-modules [50.19]; but the simple module W is essential in U. Hence U is uniserial and so its composition length is at most the length of R/B. □

18.6. COROLLARY. *Let R be right bounded and S be an overring such that $R \subseteq S \subset R_{\text{quo}}$. Then S is right bounded and each simple right S-module is uniserial of finite length over R.*

PROOF. By 18.4, each simple right S-module is unfaithful. Then 18.5 applies. □

The next two lemmas deal with sequences of simple modules. Note, however, that the terms 'successor' and 'predecessor' are reserved for the notion defined in 15.1 — when the pair of simple modules have a nonsplit extension.

18.7. LEMMA. *There is no infinite sequence $W_1, W_2, \ldots$ of distinct unfaithful simple right R-modules such that each W_{i+1} is the predecessor of W_i $(i \geq 1)$.*

PROOF. Suppose such a sequence exists. First we reduce the problem to the case when R is right bounded. Recall that we say that an overring S of R 'kills' a simple right module W if $W \otimes_R S = 0$ [13.7]. Let $\mathcal{Y}$ be the set of all simple R-modules not in the given infinite sequence and let $S = R(\mathcal{Y})$, the localization which kills all members of $\mathcal{Y}$ and kills no other simple right R-modules [13.8].

Then 13.12 tells us that the simple right S-modules are precisely the modules $W_i \otimes_R S$. It also shows that $S \neq R_{\text{quo}}$, since $\mathcal{Y}$ does not include all (isomorphism classes of) simple right R-modules. We know that W_{i+1} is the predecessor of W_i. Thus $\text{Ext}_R^1(Y, W_i) = 0$ for all $Y \in \mathcal{Y}$. Hence 13.13 asserts that $W_i \cong W_i \otimes_R S$ via the map $w \to w \otimes 1$. Thus W_i has a natural S-module structure and, clearly, is unfaithful over S. Thus every simple right S-module is unfaithful. Combined with 13.5, this shows that S is a right bounded HNP ring.

We know from 13.2 that ${}_RS$ is flat and $S \otimes_R S \cong S$ via multiplication. Then 2.5(i) says that $\text{Hom}_R = \text{Hom}_S$ and 4.13(iii) that $\text{Ext}_R^1 = \text{Ext}_S^1$ on S-modules. So we deduce that the sequence of W_i, when viewed as simple right S-modules, retains the property that each W_{i+1} is the predecessor of W_i. This completes our reduction to the right bounded case.

Thus we assume now that R is right bounded, and its simple right modules consist of the W_i. Note that W_1 does not have a successor because the only possibility for a successor would be some W_j, and this would contradict the uniqueness of predecessors and successors stated in 15.2.

Let $\mathcal{Y} = \{W_2, W_3, \ldots\}$ and let $S = R(\mathcal{Y})$, the overring of R that kills precisely the simple right modules in $\mathcal{Y}$, that is, every W_i except W_1. Then 13.12 asserts that $S \subset R_{\text{quo}}$ and $X = W_1 \otimes_R S$ is a simple right S-module. Hence, by 18.6, X_R has finite length, say m. We show that this leads to a contradiction.

Choose any integer $n > m$. There is a uniserial right R-module U whose composition factors from top to bottom are $W_n, \ldots, W_1$ [16.3]. The composition factors of U/W_1 are elements of $\mathcal{Y}$, and hence are killed by S. Therefore $(U/W_1) \otimes_R S = 0$. Hence tensoring the short exact sequence $W_1 \hookrightarrow U \twoheadrightarrow U/W_1$ with the flat module ${}_RS$ yields a short exact sequence showing that $X = W_1 \otimes_R S \cong U \otimes_R S$ as S-modules, and hence as R-modules. This shows that $U \otimes_R S$ has length m as an R-module. On the other hand, the natural map $U \to U \otimes_R S$ is an embedding of R-modules since S does not kill the socle W_1 of the uniserial R-module U. Thus $U \otimes_R S$ has length at least n over R; and this is the desired contradiction. □

We continue to reserve the terms successor and predecessor for the notion defined in 15.1.

18.8. LEMMA. *There is no infinite sequence $W_1, W_2, \dots$ of distinct unfaithful simple right R-modules such that each W_{i+1} is the successor of W_i $(i \geq 1)$.*

PROOF. If there were such a sequence, then the infinite sequence $W_1^{\mathrm{op}}, W_2^{\mathrm{op}}, \dots$ would have the property that each W_i^{op} is unfaithful and each W_{i+1}^{op} is the predecessor of W_i^{op} [17.11], contradicting the left-hand version of 18.7. □

19. Towers of Simple Modules

In this section[1] we introduce the notion of a tower, providing a structure to the collection of (isomorphism classes of) simple modules. Since towers occur in almost every section that follows, we define this concept and its related concepts in considerable detail. This section makes frequent use, without explicit reference, of the facts [15.2] that every unfaithful simple module has a unique predecessor; and if a simple module has an unfaithful successor, then that is the unique successor.

19.1. DEFINITIONS. We call a finite sequence $W_1, W_2, \dots, W_n$ of (isomorphism types of) simple right R-modules a *tower of length n* (or an *R-tower*, if more precision is needed) if the following conditions hold.

(i) For all $i > 1$, W_i is the unique unfaithful successor of W_{i-1}.
(ii) $W_i \not\cong W_j$ when $i \neq j$.
(iii) $W_1, W_2, \dots, W_n$ is maximal with respect to conditions (i) and (ii); that is, it is not a portion of a longer such sequence.

Thus, in every tower only one member, namely W_1, can possibly be faithful.

We call the tower $W_1, W_2, \dots, W_n$ a *faithful tower* if W_1 is faithful. We call it a *cycle tower* if W_1 is the unfaithful successor of W_n. In 19.3 we show that every simple module belongs to a unique tower of one of these two types — unique except, in the case of cycle towers, for cyclic permutations.

Trivial cases. We call a tower 'trivial' if it consists of a single simple module. Thus, a faithful tower is trivial if it consists of a single faithful simple module that has no unfaithful successor. It is possible for a faithful simple module with no unfaithful successor to have a faithful successor [see 15.5]. However if a faithful simple module has an unfaithful successor, then that is its unique successor [15.2].

A trivial cycle tower consists of a single unfaithful simple module that is its own unique predecessor (otherwise we could extend the tower, by 15.2), and hence its own unique successor.

Caution: Let $W_1, W_2, \dots, W_n$ be a cycle tower. Then so is the sequence $W_2, W_3, \dots, W_n, W_1$. By abuse of terminology, *we regard these as the same tower.* Thus we usually regard cycle towers as cyclicly ordered sets.

We call W_1 and W_n respectively the *top* and *base* of the tower $W_1, W_2, \dots, W_n$. Thus, for cycle towers, the meanings of 'top' and 'base' depend on the arbitrarily chosen starting point of the (cyclicly ordered) sequence that enumerates it.

The reason for the terms 'top', 'base', and 'tower' is the following result.

19.2. LEMMA. *Given any tower $W_1, \dots, W_n$ there is a uniserial R-module U, unique up to isomorphism, whose composition factors, from top to bottom are $W_1, W_2, \dots, W_n$. In particular, every n-term cycle tower has n non-isomorphic uniserial modules of length n associated with it.*

---

[1]In this section R denotes an HNP ring unless the contrary is specified.

PROOF. This follows immediately from 16.5. □

19.3. THEOREM. *Every simple right R-module belongs to a unique tower.*

PROOF. First we concentrate attention on the set of (isomorphism types of) unfaithful simple R-modules. Given any unfaithful simple right R-module W, we can form a chain of unfaithful simple modules as follows: first pick an unfaithful successor or unfaithful predecessor of W provided there is one; and then repeatedly pick unfaithful successors or unfaithful predecessors of simples already chosen provided these exist. We recall [15.2] that each member of a consecutive pair of these unfaithful simple modules uniquely defines the other. So the (possibly doubly infinite) chain obtained this way is uniquely specified by W.

Suppose this chain involves the repetition of some isomorphism class. So the chain contains a sequence $W_1, W_2, \ldots, W_{n+1}$ of unfaithful simple right R-modules, with $W_1, \ldots, W_n$ distinct, consecutive pairs being predecessor and successor of each other, and $W_{n+1} \cong W_1$. The uniqueness of predecessor and successor given by 15.2 tells us that W_2 is the successor of W_{n+1}, etc. We deduce that $W_1, \ldots, W_n$ is a cycle tower and that the chain consists of this cycle tower repeated indefinitely in each direction. Thus W belongs to a uniquely specified cycle tower.

Next suppose there is no repetition in the chain. We know from 18.7 that any sequence $W_1, W_2, \ldots$ of distinct unfaithful simple right R-modules such that each W_{i+1} is the predecessor of W_i must be finite; and we know from 18.8 the same is true for sequences $W_1, W_2, \ldots$ of distinct unfaithful simple right R-modules such that each W_{i+1} is the successor of W_i. So the chain constructed from W must be finite; say it is $W_1, W_2, \ldots, W_n$. Then W belongs to this sequence, each W_{i+1} is the unfaithful successor of W_i and the sequence cannot be extended by an unfaithful simple module in either direction without repetition. 15.2 tells us that W_1, like every other unfaithful simple R-module, has a predecessor, W_0 say. Since the sequence cannot be extended by unfaithful simple modules, W_0 must be faithful. Then the sequence $W_0, \ldots, W_n$ forms a faithful tower.

Finally, consider any faithful simple module, V say. There are two possibilities: either V is the predecessor of some unfaithful simple module, W say; or it is not. In the latter case, by definition, V itself forms a trivial faithful tower. Suppose the former is the case. Then W cannot belong to a cycle tower because it that case its (unique) predecessor is unfaithful. Therefore $W = W_1$ is the first unfaithful term in a faithful tower $V, W_1, \ldots, W_n$. □

This shows that the set of towers of an HNP ring R forms a well-defined structure on the set of (isomorphism classes of) simple R-modules. This structure is one of the main building blocks of the module theory presented here. We now show it is preserved under Morita equivalence.

19.4. THEOREM. *Let R, S be Morita equivalent HNP rings. Then the correspondence between the categories of right R-modules and right S-modules preserves the tower structure; i.e. it preserves the towers, their lengths and their (cycle or faithful) types.*

PROOF. Let V_R, W_R be simple. Since the given Morita equivalence, say μ, is an equivalence and preserves the lattice of submodules of a module, μ provides a bijection between (isomorphism classes of) simple R-modules and simple S-modules. Further, this bijection preserves faithfulness since that is a Morita invariant property [**Anderson and Fuller 92**, 21.6(4)]. Finally, we note that $\mathrm{Ext}^1_R(V,W) \neq 0 \Leftrightarrow \mathrm{Ext}^1_S(\mu(V),\mu(W)) \neq 0$ because $\mathrm{Ext}^1_R(V,W) \neq 0$ is equivalent to the existence of a short exact sequence $W \hookrightarrow U \twoheadrightarrow V$ in which U is uniserial of length 2. □

20. Idealizers and Tower Examples

Let $\mathcal{A}$ be a multichain in an HNP ring S. This section[1] begins by describing precisely the relationship between the tower structure of S and that of the multichain idealizer $R = \mathbb{I}_S(\mathcal{A})$ [§8]. This is used to provide some straightforward examples of faithful towers and cycle towers of arbitrary lengths in HNP rings. These examples have only finitely many nontrivial towers. (§25 is devoted to the construction of an HNP ring with an arbitrary, possibly infinite, collection of nontrivial faithful and cycle towers.)

We repeatedly use the facts that every multichain idealizer R from a ring S can be obtained as an iterated basic idealizer from S and vice versa [8.6]. Consequently (by repeated use of 14.6), R is an HNP ring if and only if S is; and $R_{\mathrm{quo}} = S_{\mathrm{quo}}$ [14.5].

Our first lemma slightly sharpens 6.4, for HNP rings.

20.1. LEMMA. *Let S be an HNP ring, R an iterated basic idealizer from S, and U_S a simple module. Thus, by 6.4, U_R is uniserial of finite length over R, say with R-composition series $U = X_0 \supset X_1 \supset \ldots \supset X_t \supset X_{t+1} = 0$. Then:*

(i) *X_i/X_{i+1} is unfaithful for all $i \geq 1$; and*
(ii) *X_0/X_1 is faithful if and only if U_S is faithful.*

PROOF. (i) This is is part of 6.4.

(ii) By 12.1(vii), U_S is unfaithful if and only if the same is true of U_R. Now if U_R is unfaithful then so, of course, is its factor X_0/X_1. Conversely, if X_0/X_1 is unfaithful then each composition factor of U_R is unfaithful and the product of their annihilators, in appropriate order, annihilates U and is nonzero since R is prime. Hence U_S is unfaithful. □

The next result makes explicit the relation between a multichain $\mathcal{A}$ of S and and the towers in the resulting idealizer $R = \mathbb{I}_S(\mathcal{A})$.

20.2. THEOREM. *Let S be an HNP ring with a multichain $\mathcal{A}$ of type $\{U_1,\ldots,U_t\}$ and $R = \mathbb{I}(\mathcal{A})$.*

(i) *The towers of R are precisely those of S but with each $(U_i)_S$ replaced by the composition factors, from top to bottom, of the uniserial module $(U_i)_R$.*
(ii) *Each R-tower is a cycle tower or, respectively, a faithful tower if and only if the same is true of the S-tower from which it is obtained.*

PROOF. By 8.9, each $(U_i)_R$ is uniserial of length $|\mathcal{A}_i| + 1$ where $\mathcal{A}_i$ is the component basic chain of type U_i in $\mathcal{A}$. We know from 6.4 that the simple R-modules are precisely those of S but with the U_i replaced by their R-composition factors. We note, from 20.1, that the new simple modules are all unfaithful except for the top R-composition factor of each U_i which is faithful over S.

Moreover, by referring to 6.10, the reader will see that the extension group of each pair of simple R-modules is determined by that of a prescribed pair of simple

[1] In this section R denotes an HNP ring unless the contrary is specified.

S-modules. In particular, each pair of simple S-modules which remain simple over R retains the same extension group; each consecutive pair of R-composition factors of one of the U_i is a predecessor and unfaithful successor; the top composition factor of U_i is an unfaithful successor if and only U_i was over S; and likewise each simple S-module with no unfaithful successor (or, if it is one of the U_i, its bottom R-composition factor) retains that property over R. Finally, 20.1 implies that the new sequence of simple modules is a cycle tower, respectively a faithful tower, if and only if the original tower was so. □

Next, some elementary facts which will be of frequent use in providing examples of basic chains and multichains.

20.3. LEMMA. *Let T be a ring which is not right Artinian and X be a faithful simple right T-module. Then $X^{(n)}$ is cyclic for all $n \geq 0$.*

PROOF. We proceed by induction on n. So suppose $T/K \cong X^{(n-1)}$ for some K. Since T is not right Artinian, $K \neq 0$; and since X is faithful, $\cap\{M \mid T/M \cong X\} = 0$. Hence there exists M with $T/M \cong X$ and $M \not\supseteq K$. Then $M + K = T$ and so $T/(M \cap K) \cong T/M \oplus T/K \cong X^{(n)}$, as required. □

It is an easy exercise to extend this to show that if P is a non-Artinian projective T-module then $X^{(n)}$ is a homomorphic image of P for any n. (This is also a consequence of 49.12.)

20.4. LEMMA. *Let T be a ring, U a simple T-module and A a right ideal with $T/A \cong U^{(n)}$ for some $n > 0$. Then there is a chain of right ideals $T \supset A_1 \supset A_2 \supset \ldots \supset A_n = A$ such that $T/A_i \cong U^{(i)}$ for each $i \leq n$.*

PROOF. Lift to T a composition series for T/A. □

We now give the promised examples.

20.5. EXAMPLE. There is an HNP ring R with any prescribed finite number t of nontrivial faithful towers with prescribed lengths, say $n_1, \ldots, n_t$.

PROOF. Let S be a simple HNP ring having at least t isomorphism types of simple modules; for example either of the Weyl Algebras $A_1(F)$ or $B_1(F)$ with F a field of characteristic 0 [see 11.8, 11.7]. The simplicity of S ensures that all towers in S are trivial and faithful.

Choose any t non-isomorphic simple S-modules $U_1, \ldots, U_t$. It is enough to find, for each i, a basic chain $\mathcal{A}_i$ of type U_i and length n_i. For then, letting $\mathcal{A}$ be the union of the component basic chains $\mathcal{A}_i$ and letting $R = \mathbb{I}(\mathcal{A})$ yields the desired HNP ring [20.2].

To obtain $\mathcal{A}_i$, note, by 20.3, that $U_i^{(n_i)}$ is cyclic, say $U_i^{(n_i)} \cong S/A$. Then 20.4 provides a chain of length n_i which is a basic chain since each nonzero right ideal of S is generative. □

20.6. EXAMPLE. There is an HNP ring R with any prescribed finite number t of nontrivial cycle towers with prescribed lengths, say $n_1, \ldots, n_t$.

PROOF. Let $S = M_m(D)$ where $m > n_i$ for each i and where D is a commutative Dedekind domain having at least t distinct maximal ideals [e.g. $D = \mathbb{Z}$]. By 19.4, Morita equivalent rings have the same tower structures. Thus S has at least t towers, each of which is a trivial cycle tower. Fix t distinct such towers, letting

U_i be the simple module in each. As in the previous example, it is enough to find, for each i, a basic chain $\mathcal{A}_i$ of type U_i and length n_i. Therefore we may fix our attention on some $U = U_i$, $n = n_i$.

Let $H = \mathrm{ann}_S(U)$; so H is an invertible maximal ideal. We have $S/H \cong U^{(m)}$ and $n < m$. Hence 20.4 provides a chain $S \supset A_1 \supset \ldots \supset A_m = H$ in which each $S/A_i \cong U^{(i)}$. Note that each A_i with $i < m$ is generative, since H is a maximal ideal. So the basic chain $\mathcal{A}_i$ made up of the right ideals $A_1, \ldots, A_n$ is as required. □

So far we have not given an example of an HNP ring with both cycle and faithful towers.

20.7. EXAMPLE. There is an HNP ring R with a cycle tower of prescribed length (say) $n > 1$ and a faithful tower of prescribed length $s > 1$.

PROOF. Let $D = \Phi[x, \sigma]$, the skew polynomial ring over a field Φ, with σ an automorphism of Φ of infinite order and multiplication given by $\alpha x = x\alpha^\sigma$ ($\alpha \in \Phi$). We know, from [**Jacobson 43**, p. 29], that D is a principal left and principal right ideal domain (basically, the division algorithm works). Evidently $x^n D = Dx^n$ for all n although, of course, x is not central in D. Indeed [**Jacobson 43**, p. 38], these are the only nonzero ideals of D; and clearly they are invertible.

Since all right and all left ideals of D are free modules, D is an HNP ring. Next, we show that its towers are all trivial; and that it has exactly one cycle tower and at least one faithful tower. The lone maximal ideal $xD = Dx$ is invertible and is a maximal right ideal; hence D/xD is the unique trivial cycle tower. No nontrivial tower can exist, because any such tower would contain an unfaithful simple module whose annihilator was a maximal ideal distinct from xD. In particular, $D/(x-1)D$ is a trivial faithful tower.

The rest of this proof is mostly a repetition of parts of the previous two examples, so we give less detail than previously.

Let $S = M_m(D)$ where $m > n$. By Morita equivalence [19.4], S is an HNP ring with the same tower structure as D; so its towers are all trivial, there is exactly one cycle tower and at least one faithful tower. The lone member of the cycle tower is U where $U^{(m)} \cong M_m(D/xD)$. The argument of the final paragraph of the proof of 20.6 shows that S has a basic chain $\mathcal{A}$ of length n and type U. Consider one of the faithful towers of S; say of type U'. As in the final paragraph of the proof of 20.5, we see that S has an s-term basic chain $\mathcal{A}'$ of type U'. Now apply 20.2 to the multichain idealizer ring $R = \mathbb{I}(\mathcal{A} \cup \mathcal{A}')$. □

21. Right Versus Left Towers

In this section[1] we compare the right and left tower structures of an HNP ring. Some consequences fill gaps in the previous work on right-left symmetry. Recall that X^{tr} denotes the transpose of an R-module X of finite length [17.1].

21.1. DEFINITION. For a tower $\mathcal{T} = W_1, \ldots, W_n$, let $\mathcal{T}^{\mathrm{tr}} = W_n^{\mathrm{tr}}, \ldots, W_1^{\mathrm{tr}}$.

21.2. PROPOSITION. *The function $\mathcal{T} \to \mathcal{T}^{\mathrm{tr}}$ is a (1,1)-correspondence between the set of towers of simple right R-modules and the set of towers of simple left R-modules. Moreover, it preserves the type (faithful or cycle) of every tower.*

[1]In this section R denotes an HNP ring unless the contrary is specified.

PROOF. Recall that 17.10 shows that if W is the unfaithful successor of V then V^{tr} is the unfaithful successor of W^{tr}. Repeated use of this shows that the sequence of transposes is part of a tower. Since, by 17.7, $V^{\mathrm{tt}} \cong V$, the sequence must be an entire tower. Suppose the original tower is a cycle tower; so W_1 is the unfaithful successor of W_n. Then W_n^{tr} is the unfaithful successor of W_1^{tr}. Hence the sequence of transposes of a cycle tower is a cycle tower; and symmetry gives the converse. □

An immediate consequence of 21.2 answers the question, left open in the section on transposes, of when the transpose of a simple module is faithful.

21.3. COROLLARY. *Let W_R be a simple module. Then W^{tr} is faithful if and only if W occurs as the bottom element of a faithful tower. (In particular, both W and W^{tr} are faithful if and only the tower containing W is trivial and faithful.)* □

There is a simpler formula for the unfaithful members of $\mathcal{T}^{\mathrm{tr}}$. Recall that W^{op} denotes the opposite of an unfaithful simple R-module W [17.9].

21.4. COROLLARY. *Let $\mathcal{T} = W_1, \ldots, W_n$ be a tower.*
(i) *If $\mathcal{T}$ is a faithful tower, then $\mathcal{T}^{\mathrm{tr}} = W_n^{\mathrm{tr}}, W_n^{\mathrm{op}}, W_{n-1}^{\mathrm{op}}, \ldots, W_2^{\mathrm{op}}$.*
(ii) *If $\mathcal{T}$ is a cycle tower, then $\mathcal{T}^{\mathrm{tr}} = W_1^{\mathrm{op}}, W_n^{\mathrm{op}}, W_{n-1}^{\mathrm{op}}, \ldots, W_2^{\mathrm{op}}$.*

Remark. Note that (ii) states that the transpose of a cycle tower $\mathcal{T}$, viewed as a cycle, is simply the reversed cycle of opposites of the members of $\mathcal{T}$.

PROOF. By 17.10, $W_i^{\mathrm{tr}} = W_{i+1}^{\mathrm{op}}$ whenever W_i is unfaithful and $i \leq n-1$; and for cycle towers, it is also true that $W_n^{\mathrm{tr}} = W_1^{\mathrm{op}}$. □

21.5. DEFINITION. For each unfaithful simple R-module W, the ring $R/\operatorname{ann}(W)$ is simple Artinian; so by definition [9.2] R has finite rank $\rho(R, W)$ at W. By the *rank sequence* of a tower $\mathcal{T}$, we mean the sequence of ranks $\rho(R, W_i)$ of its unfaithful simple modules, from top to bottom.

21.6. COROLLARY. *Let $\mathcal{T} = W_1, \ldots, W_n$ be a tower. Then, for each $i \leq n-1$, $\rho(R, W_i^{\mathrm{tr}}) = \rho(R, W_{i+1})$.*
If $\mathcal{T}$ is a cycle tower, we also have $\rho(R, W_n^{\mathrm{tr}}) = \rho(R, W_1)$.

PROOF. Evidently, $\rho(R, W^{\mathrm{op}}) = \rho(R, W)$ for every unfaithful simple W. The stated result follows from this and 21.4. □

21.7. COROLLARY. *Let $\mathcal{T}$ be a tower. The rank sequence of $\mathcal{T}^{\mathrm{tr}}$ is the reverse of that of $\mathcal{T}$.*

PROOF. This is an immediate consequence of 21.6. □

22. Cycles of Maximal Ideals; Multiplicative Ideal Theory

Before the introduction of towers in [HNP1], many properties of cycle towers were phrased in terms of maximal ideals. In this section[1] we show the precise connection between these two aspects; and we do likewise for the unfaithful members of faithful towers.

We also describe the analogues for noncommutative HNP rings of one of the best-known properties of commutative Dedekind domains, the unique factorization of ideals as products of maximal ideals.

[1]In this section R denotes an HNP ring unless the contrary is specified.

22.1. PROPOSITION.

(i) *Let* $W_1, \ldots, W_n$ *be a sequence of unfaithful simple right* R*-modules and let* $A_i = \operatorname{ann}(W_i)$ *for each* i*. Then* $O_l(A_i) = O_r(A_{i+1})$ *for* $i = 1, \ldots, n-1$ *if and only if the* W_i *all belong to the same tower and each* W_i *is the predecessor of* W_{i+1}.

(ii) *If* $\mathcal{C}$ *is a nontrivial tower then the annihilators of its unfaithful simple modules are idempotent maximal ideals.*

(iii) *Each idempotent maximal ideal is the annihilator of some unfaithful simple module in a nontrivial tower.*

PROOF. (i),(ii) Both parts are interpretations for towers of 15.8 which showed in particular that the hypothesis that $O_l(A_i) = O_r(A_{i+1})$ is equivalent to the assertion that the unfaithful simple module W_i is the predecessor of W_{i+1}.

(iii) 14.9 states that if M is an idempotent maximal ideal and W is the simple module it annihilates then W is an unfaithful successor; thus W belongs to a nontrivial tower. □

Combining (ii) and (iii) yields:

22.2. COROLLARY. *There is a (1,1)-correspondence between idempotent maximal ideals and (isomorphism classes of) simple unfaithful modules belonging to nontrivial towers.* □

22.3. COROLLARY. *Every (possibly doubly infinite) sequence* $\ldots, A_i, A_{i+1}, \ldots$ *of distinct nonzero maximal ideals* A_i *of* R *such that* $O_l(A_i) = O_r(A_{i+1})$ *for each* i *is necessarily finite.*

PROOF. Let W_i be the simple right R-module annihilated by A_i. By 22.1, each W_i is the predecessor of W_{i+1} and hence all of the W_i are contained in a single tower. However, all towers are finite. □

22.4. DEFINITION. A *cycle of maximal ideals* is a sequence $A_1, \ldots, A_n$ of distinct nonzero maximal ideals such that $O_l(A_i) = O_r(A_{i+1})$ for each i and $O_l(A_n) = O_r(A_1)$. Evidently, $A_2, \ldots, A_n, A_1$ is also a cycle; and we regard these as being the same cycle of maximal ideals. Thus we regard a cycle as a cyclically ordered set.

If W_i denotes the simple R-module such that $W_iA_i = 0$, then the cyclicly ordered sequence $W_1, \ldots, W_n$ is called the cycle tower *associated with* the cycle of maximal ideals $A_1, \ldots, A_n$, because of 22.6(i) below.

22.5. REMARK. The indexing of the maximal ideals in a cycle, as given above, is the reverse of what has been conventional in the literature. This makes it mesh smoothly with the indexing of the members of the tower; see 22.6(i). Note, however, that if we were dealing with towers of simple left modules, such a reversal would not be needed.

22.6. COROLLARY. *Let* $W_1, \ldots, W_n$ *be a sequence of unfaithful simple right* R*-modules and let* $A_i = \operatorname{ann}(W_i)$ *for each* i*. Then:*

(i) $A_1, \ldots, A_n$ *is a cycle of maximal ideals if and only if* $W_1, \ldots, W_n$ *is a cycle tower of simple modules.*

(ii) *Two cycles of maximal ideals coincide if they overlap.*

PROOF. (i) This is immediate from 22.1.

(ii) This is clear from (i) since 19.3 shows that a cycle tower is uniquely specified by any member. □

There is a similar correspondence for faithful towers.

22.7. COROLLARY. *Let $W_1, \dots, W_n$ be a sequence of unfaithful simple right R-modules and let $A_i = \operatorname{ann}(W_i)$ for each i. Then $W_1, \dots, W_n$ is the sequence of unfaithful simple modules in some faithful tower if and only if (i) $O_l(A_i) = O_r(A_{i+1})$ for each i; and (ii) the sequence $A_1, \dots, A_n$ cannot be extended at either end while preserving condition (i).*

PROOF. This is immediate from 22.1. □

22.8. LEMMA. *Let $A_1, A_2, \dots, A_n$ be nonzero maximal ideals of R, let $X = \cap A_i$ and let V be a simple right R-module.*

(i) *The only maximal ideals containing X are the A_i.*
(ii) *$VX = 0$ if and only if $VA_i = 0$ for some i.*
(iii) *If $A_1, A_2, \dots, A_n$ is a cycle of maximal ideals then V belongs to the associated cycle tower if and only if $VX = 0$.*

PROOF. (i) Let P be a maximal ideal containing X. Then $P \supseteq A_1A_2 \dots A_n$ and so, since maximal ideals are prime, $P \supseteq A_i$ for some i.

(ii) Suppose $VX = 0$. Note that $X \supseteq A_1A_2 \dots A_n \neq 0$. Thus $\operatorname{ann}_R(V)$ is a nonzero, and hence maximal, ideal containing X; and so, by (i), $\operatorname{ann}(V) = A_i$ for some i. The converse is trivial.

(iii) Clear from (ii). □

We now apply what is proved above, to study multiplicative ideal theory.

22.9. PROPOSITION. *Let $A_1, A_2, \dots, A_n$ be a cycle of maximal ideals. Then $X = \cap A_i$ is an invertible ideal.*

PROOF. 15.8(ii) gives this when $n = 1$; so we may assume that $n \geq 2$. Evidently, $X^*X \supseteq X$; we want to show that $X^*X = R$.

Let $W_1, \dots, W_n$ be the associated cycle tower of simple modules. Thus every $W_iX = 0$. Each W_i is the unfaithful successor of W_{i-1} (letting $W_0 = W_n$). Therefore there is a uniserial module of length 2, $U_i = [W_{i-1}W_i]$ with top and bottom composition factors W_{i-1}, W_i respectively. Since $W_{i-1}X = 0$ we have $U_iX = W_i$ or 0. But the latter possibility implies that the uniserial module U_i of length 2 is a module over the semisimple ring R/X, a contradiction. Thus $U_iX = W_i$. The uniserial module U_i is cyclic, and so there is a surjection $R \twoheadrightarrow U_i$. Right multiplying by X yields a surjection $X \twoheadrightarrow W_i$. Since R/A_i is a direct sum of copies of W_i, this shows that there is a nonzero map $X \to R/A_i$. Projectivity of X implies that this lifts to a map $X \to R$ whose image is not contained in A_i. We deduce that $X^*X \not\subseteq A_i$ for each i. But the A_i are the only maximal ideals containing X [22.8]. Therefore $X^*X = R$.

The left-handed version of what has just been proved is that $XX' = R$, where X' is the dual of the left R-module X. Associativity of the product X^*XX' then shows that $X' = X^*$. □

22.10. DEFINITION. An ideal C of which some power is idempotent is called an *eventually idempotent* ideal.

22.11. PROPOSITION. *Let C be a proper ideal of R such that the set of maximal ideals containing C does not include a complete cycle. Then C is eventually idempotent. In more detail: $C^n = C^{2n}$ where n is the number of maximal ideals containing C.*

PROOF. By 15.6, the ring R/C^{2n}, like every proper factor ring of R, is an Artinian serial ring. Hence, by Nakayama's Theorem (50.19), R/C^{2n} is the direct sum of some finite set $\mathcal{U}$ of uniserial modules.

Consider any $U \in \mathcal{U}$. Each composition factor of U is annihilated by C^{2n}, and hence by C. Thus U has at most n distinct composition factors. Moreover, consecutive composition factors of the uniserial R-module U need to be predecessor and successor. Therefore U can have no repeated composition factor unless it has (at least) a whole cycle tower of composition factors, which our hypothesis on C rules out.

Thus U has length $\leq n$, and hence $UC^n = 0$. Since $R/C^{2n} = \oplus\{U \in \mathcal{U}\}$ we deduce that $(R/C^{2n})C^n = 0$, and hence $C^n \subseteq C^{2n}$. The opposite inclusion is obvious. □

22.12. COROLLARY. *An ideal is a maximal invertible ideal (i.e. maximal in the set of invertible ideals) if and only if it is the intersection of a cycle.*

PROOF. 22.9 shows that the intersection of a cycle is invertible and 22.11 shows it is maximal amongst invertible ideals, since larger ideals are all eventually idempotent. Conversely, suppose that I is a maximal invertible ideal. Then 22.11 shows that the set of maximal ideals containing I includes a complete cycle; and the intersection X of the maximal ideals in this cycle is an invertible ideal, by 22.9. The maximality of I tells us that $I = X$. □

22.13. LEMMA. *Let M, N be nonzero ideals of R, and suppose that no tower contains a composition factor of both R/M and R/N. Then $MN = M \cap N = NM$.*

PROOF. By the definition of a tower, $\operatorname{Ext}^1(X, Y) = 0 = \operatorname{Ext}^1(Y, X)$ for every composition factor X of R/M and Y of R/N. Therefore the lemma is a special case of 50.8. □

22.14. LEMMA. *Let C be a nonzero right ideal of R, and $I \neq R$ an invertible ideal. Then $C \supset CI$.*

PROOF. Since I is invertible, one sees that $\lambda(R/I) = \lambda(I^n/I^{n+1})$ for each n. Hence $R/\cap I^n$ cannot have finite length and so $\cap I^n = 0$, by the restricted minimum condition for R [12.14]. Now suppose that the lemma is false. Then $C = CI$ and hence $C = CI^n \subseteq I^n$ for every n. It follows that $C \subseteq \cap I^n = 0$, a contradiction. □

22.15. THEOREM.

(i) *Each nonzero ideal of R is the product EI of an eventually idempotent ideal E and an invertible ideal I.*
(ii) *Each invertible ideal is a (possibly empty) product of maximal invertible ideals.*
(iii) *The set of invertible ideals generates a free abelian group under multiplication, and the maximal invertible ideals form a basis of this group.*
(iv) *The intersection of a finite set of maximal ideals is invertible if and only if it is the union of a finite set of cycles.*

PROOF. (i) Let A be any nonzero ideal. If the set $\mathcal{X}$ of maximal ideals containing A does not contain a cycle, then A is eventually idempotent [22.11], and hence the desired factorization is $A = AR$. Suppose, therefore, that $\mathcal{X} \supseteq \mathcal{C}$ for some cycle $\mathcal{C}$. Then the intersection C of the ideals in $\mathcal{C}$ is invertible [22.9] and $A \subseteq C$.

Then $R \supseteq AC^{-1} \supset A$, where the final inclusion is strict by 22.14. By Noetherian induction, we may suppose that $AC^{-1} = EY$ for some eventually idempotent ideal E and some invertible ideal Y. Then $A = E(YC)$ as required.

(ii) A similar Noetherian induction establishes this.

(iii) Let X_1, X_2 be two distinct maximal invertible ideals. Then each X_i is the intersection of all of the maximal ideals in the cycle of maximal ideals that contains X_i [22.12]. It follows that no tower of simple modules contains a composition factor of both R/X_1 and R/X_2. Therefore $X_1X_2 = X_1 \cap X_2 = X_2X_1$ [22.13].

The crux of the rest of the proof is to prove the following claim. If

$$I_1I_2 \dots I_m = J_1J_2 \dots J_n$$

for maximal invertible ideals I_μ and J_ν, then I_1 equals J_ν for some ν, and hence we can cancel it from both sides and repeat the process, concluding that the two products are the same except for the order of occurrence of the factors.

As mentioned above, each of these maximal invertible ideals is the intersection of the maximal ideals which contain it; and they form a cycle. Choose a maximal ideal $M_1 \supseteq I_1$. Then $J_1J_2 \dots J_n = I_1I_2 \dots I_m \subseteq M_1$ and so, since M_1 is prime, some $J_\nu \subseteq M_1$. Hence, by 22.8, M_1 is one of the maximal ideals in the cycle whose intersection is J_ν. Thus the cycles whose intersections are I_1 and J_ν have a maximal ideal in common, and hence are the same cycle, completing the proof of the claim. We omit the remaining simple details in the proof of (iii).

(iv) Let the set of maximal ideals be $\mathcal{Y}$ and the intersection be Y. If $\mathcal{Y}$ does not include a complete cycle, then 22.11 tells us that Y is eventually idempotent, and so is not invertible. Suppose, then, that $\mathcal{Y}$ does include a complete cycle $\mathcal{C}$ say, and let X be the intersection of the members of $\mathcal{C}$. Then X is invertible and $X \supseteq Y$. Then $R \supseteq YX^{-1} \supset Y$, where the strict inclusion holds because X is invertible [22.14]. Now YX^{-1} is the intersection of the maximal ideals in $\mathcal{Y} - \mathcal{C}$. By induction, we may assume that YX^{-1} is invertible if and only if $\mathcal{Y} - \mathcal{C}$ is a union of complete cycles; so the result follows. □

23. Types of HNP Rings

In this section[1] we review various types of HNP rings that have been studied in the literature, and show how they can be defined in terms of their towers of simple modules.

23.1. Definition. An HNP ring R is said to have *enough invertible ideals* if each nonzero ideal contains an invertible ideal. Note that this is a left-right symmetric definition.

23.2. Proposition. *The following are equivalent.*

(i) *R has enough invertible ideals.*
(ii) *Every faithful tower of simple right R-modules is trivial.*
(iii) *Every unfaithful simple right R-module is annihilated by a maximal invertible ideal.*

Proof. (ii)⇔(iii) Statement (ii) is equivalent to the assertion that every unfaithful simple module belongs to a cycle tower. The equivalence of this with (iii) is given by 22.8 together with 22.12.

[1]In this section R denotes an HNP ring unless the contrary is specified.

(i)⇒(iii) Choose an unfaithful simple module W. By (i), the annihilator of W contains an invertible ideal J. We have $J = I_1 I_2 \dots I_n$, a product of maximal invertible ideals [22.15]. Since W is simple, some I_ν annihilates W, and therefore W belongs to a cycle tower [22.8].

(iii)⇒(i) Let A be a nonzero ideal; so $(R/A)_R$ has finite length. Therefore, by (iii), each composition factor of R/A is annihilated by an invertible ideal. Hence there is an invertible ideal I such that $(R/A)I = 0$; i.e. $A \supseteq I$, as desired. □

Recall that R is *right bounded* if every essential right ideal contains a nonzero ideal or equivalently, since R is an HNP ring, if every right R-module of finite length has a nonzero annihilator [18.1, 18.2].

23.3. COROLLARY. *Every right bounded HNP ring has enough invertible ideals.*

PROOF. Since R has no faithful simple right modules, it has no faithful towers. Therefore the desired conclusion is part of 23.2. □

23.4. PROPOSITION.

(i) *R is right bounded if and only if every tower of simple right modules is a cycle tower.*
(ii) *If R is right bounded then R is also left bounded.*

PROOF. (i) If R is right bounded, then it has no faithful simple right modules. Thus every tower of simple right modules can only be a cycle tower. Conversely, suppose that all towers are cycle towers. Then R has has no faithful towers and so no faithful simple modules. Hence it is right bounded.

(ii) By part (i), R is right bounded if and only if every tower of simple right modules is a cycle tower. However, by 21.2, the transpose functor gives a bijection between the set of towers of simple right modules and the set of towers of simple left modules, preserving their type (faithful or cycle). □

23.5. DEFINITION. An HNP ring in which every nonzero ideal is invertible is called a *Dedekind prime ring*. This clearly is a left-right symmetric condition. A Dedekind prime ring which is an integral domain is called a *noncommutative Dedekind domain.*

23.6. PROPOSITION. *The following are equivalent:*

(i) *R is a Dedekind prime ring.*
(ii) *Every tower of simple right modules is trivial.*
(iii) *Every nonzero ideal is a progenerator right module.*
(iv) *No nonzero maximal ideal is idempotent. (Equivalently, every nonzero maximal ideal is invertible.)*

PROOF. We already know, by 14.3, that the two conditions in (iv) are equivalent.

(i)⇒(iii) This was proved in 14.2.

(iii)⇒(iv) Let M be a nonzero maximal ideal. Since M_R is a progenerator, the direct sum of some number of copies of M maps onto R, and therefore $M^*M = R$ (using the standard identifications in 12.10). If $M^2 = M$, then left-multiplying by M^* would imply the contradiction $M = R$. Thus (iv) holds.

(iv)⇒(ii) Let V be an unfaithful simple R-module and M the maximal ideal such that $VM = 0$. By hypothesis, M is invertible and maximal. Hence, by

22.15(iv), M itself forms a cycle of maximal ideals. Hence, using the correspondence in 22.6 between cycles of simple modules and cycles of maximal ideals, V is a trivial cycle tower. Thus (ii) holds.

(ii)$\Rightarrow$(i) If M is an idempotent maximal ideal then, by 22.1(iii), the unfaithful simple module it annihilates belongs to a nontrivial tower, contradicting (ii). So we see that every nonzero maximal ideal is invertible.

Now let $A \neq 0$ be a proper ideal. Then $A \subseteq M$ for some maximal ideal M, and then $A \subset AM^{-1} \subseteq R$, where the first inclusion is strict by invertibility of M [22.14]. By Noetherian induction, we may assume that AM^{-1} is invertible. Then so is $A = (AM^{-1})M$ as desired. □

23.7. REMARK. One immediate consequence is that a maximal D-order R over a commutative Dedekind domain D, as described in 11.5, is a Dedekind prime ring. For if R had an idempotent maximal ideal A say, then, by 14.4, $O_r(A) = S$ would strictly contain R and be finitely generated over R and hence over D.

We will see later, in 28.17, that an HNP ring R has only finitely many nontrivial towers if and only if R is a multichain idealizer from a Dedekind prime ring. With that in mind, we note the following related result which is an immediate consequence of the (1,1)-correspondence between idempotent maximal ideals and unfaithful members of nontrivial towers given by 22.2.

23.8. COROLLARY. *R has only finitely many nontrivial towers if and only if R has only finitely many idempotent maximal ideals.* □

24. Injective Modules

This section[1] applies what we know about towers to study the structure of injective modules over HNP rings.

We begin by recalling some results from the literature which, combined, make clear that to understand injective modules over a Noetherian ring, T say, it is enough to understand the indecomposable injective modules or, equivalently, the uniform injective modules.

24.1. THEOREM. *Let T be a right Noetherian ring.*

(i) *Every injective T-module is a direct sum of indecomposable injective modules.*
(ii) *Every direct sum of injective T-modules is injective.*
(iii) *An injective T-module is indecomposable if and only if it is uniform.*
(iv) *Two direct sums of indecomposable injective T-modules are isomorphic if and only if the same isomorphism classes occur in both direct sums, with the same multiplicities.*

PROOF. (i) [**Goodearl-Warfield 89**, 4.20].
(ii) [**Goodearl-Warfield 89**, 4.19].
(iii) [**Goodearl-Warfield 89**, 4.20].
(iv) Combine [**Facchini 75**, 2.12, 2.25, 2.26]. □

In the case of a Noetherian prime ring, there are some immediate examples of indecomposable injective modules.

[1]In this section R denotes an HNP ring unless the contrary is specified.

24.2. Lemma. *Let T be a Noetherian prime ring. Then each T_{quo}-module is an injective T-module and is indecomposable precisely when it is a simple T_{quo}-module.*

Proof. 12.1 tells us that T_{quo} is injective over T. The rest is clear. $\square$

The remaining indecomposable injective R-modules are described next.

24.3. Corollary.

(i) *The indecomposable injective R-modules that are not R_{quo}-modules coincide with the injective envelopes $E(W)$, W_R simple.*

(ii) *Every $E(W)$ with W_R simple is isomorphic to a direct summand of the injective R-module R_{quo}/R.*

Proof. (i) Let X be an indecomposable injective R-module. Choose any nonzero finitely generated submodule $S \subseteq X$. Then S, like every finitely generated R-module, has a decomposition $S = P \oplus T$ where P is projective and T has finite length [12.18]. Uniformity of X [24.1] then implies that $S = P$ or $S = T$.

Suppose that $S = T$. Then S, and hence X, has a simple submodule W, and hence X contains an injective envelope $E(W)$ of W as a direct summand. Indecomposability of X then shows that $X = E(W)$, as desired.

The remaining possibility is that S is projective. Then P is isomorphic to a uniform right ideal of R since, by 12.7, each finitely generated projective R-module is a direct sum of uniform right ideals. So we may suppose that $P \subseteq R$. Since P_R is a uniform right ideal of R, 12.4(iii) asserts that PR_{quo} is a simple right R_{quo}-module. Note that PR_{quo} is an essential extension of P_R: for each nonzero element $p \in PR_{\text{quo}}$ has the form $p = ab^{-1}$ with $a \in U$ and b a regular element of R; and then $0 \neq a \in pR \cap U$. Thus $P \subseteq PR_{\text{quo}} \subseteq X$. The fact [24.2] that PR_{quo} is injective over R and X_R is indecomposable shows that $PR_{\text{quo}} = X$, and so X is a simple right R_{quo}-module.

(ii) Consider any $E(W)$ with W_R simple. We already know that R_{quo}/R is injective [12.2] and contains a copy of every simple W [13.6]. Hence the injective module $E(W)$ is isomorphic to a direct summand of R_{quo}/R. $\square$

Henceforth, therefore, we concentrate on the injective envelopes of simple R-modules, beginning with the multiplicity of each $E(W)$ as a direct summand of R_{quo}/R — equivalently the multiplicity of W itself as a direct summand of $\operatorname{soc}(R_{\text{quo}}/R)$.

24.4. Theorem. *Let V be a simple right R-module.*

(i) *If V^{tr} is faithful, then $(R_{\text{quo}}/R)_R$ contains the direct sum of infinitely many copies of V.*

(ii) *Let V^{tr} be unfaithful and let $n = \rho(R, V^{\text{tr}})$. Then $(R_{\text{quo}}/R)_R$ contains the direct sum of n, and no more than n, copies of V.*

Proof. (i) Choose a positive integer n. Then $(V^{\text{tr}})^{(n)}$ is cyclic, by 20.3; say $(V^{\text{tr}})^{(n)} \cong R/L$ for some left ideal L. Taking transposes and using 17.7 yields $V^{(n)} \cong L^*/R \subseteq (R_{\text{quo}}/R)_R$.

(ii) Let L range over all left ideals of R such that ${}_R(R/L)$ is a direct sum of copies of V^{tr}, and let $n(L)$ be the number of terms in this direct sum. Now $R/\operatorname{ann}(V^{\text{tr}}) \cong M_n(D)$ for some division ring D where $n = \rho(R, V^{\text{tr}})$. So the maximum possible value of $n(L)$ is n. Taking transposes, as in the proof of part (i), completes the proof. $\square$

24.5. COROLLARY. *The following are equivalent for a simple R-module V.*

(i) $(R_{\text{quo}}/R)_R$ *contains the direct sum of infinitely many copies of* V.
(ii) V *is the bottom element of a faithful tower.*

PROOF. Combine 24.4 with 21.3. □

We now turn to the structure of $E(W)$ itself. Note that the notation soc^n and soc_n for the nth socle and n^{th} socle layer is defined in 50.1.

24.6. PROPOSITION. *For every simple R-module W, E(W) satisfies the DCC on finitely generated submodules. Hence* $E(W) = \cup_{i=1}^{\infty} \operatorname{soc}^i\big(E(W)\big)$.

PROOF. Since R_{quo}/R maps onto $E(W)$ [24.3] it suffices to prove the proposition for R_{quo}/R in place of $E(W)$. This holds since the lattice of such submodules is anti-isomorphic to the lattice of finitely generated essential left ideals of the left Noetherian ring R [12.12]. □

Note that $E(W)$ does not always satisfy the DCC on all submodules; see Example 24.12.

24.7. LEMMA. *Let S be an overring of R and M an S-module.*

(i) M_S *is injective if and only if* M_R *is injective.*
(ii) *The injective envelope* $E(M_S)$ *is also the injective envelope* $E(M_R)$.

PROOF. (i) $S \otimes_R S \cong S$ via multiplication and ${}_RS$ is flat [13.2]. Therefore this result is a special case of 4.13 (ii).

(ii) Since $E(M_S)$ is an essential extension of M_S and $R_{\text{quo}} = S_{\text{quo}}$, we see that $E(M_S)$ is an essential extension of M_R, and therefore $E(M_S) \subseteq E(M_R)$. On the other hand, $E(M_S)$ is an injective R-module, by part (i), and is therefore a direct summand of $E(M_R)$. Since $E(M_R)$ is an essential extension of M_R, this implies that $E(M_S) = E(M_R)$. □

24.8. LEMMA. *Let V, W be simple R-modules. Then:*

(i) $\operatorname{soc}\big(E(W)/W\big) \cong V^{(k)} \oplus X$ *where k is the dimension of* $\operatorname{Ext}^1(V, W)$ *over* $\operatorname{End}(V)$ *and no direct summand of X is isomorphic to V.*
(ii) $E(W)/W$ *is the injective hull of its socle.*

PROOF. (i) This is a special case of 50.3(ii).

(ii) $E(W)/W$ is injective because R is right hereditary [12.2], and is the injective envelope of its socle because $E(W)$ satisfies the DCC for finitely generated submodules [24.6]. □

24.9. THEOREM. *Let* $W_1, W_2, \ldots, W_n$ *be a sequence of simple R-modules such that* W_{i+1} *is the (necessarily unique) unfaithful successor of* W_i *for* $1 \le i \le n-1$. *Let* M_n *be the uniserial R-module of length n whose sequence of composition factors, from top to bottom, is* $W_1, \ldots, W_n$ *[16.5]. Then:*

(i) $E(M_n) = E(W_n)$*; and*
(ii) M_n *is the unique submodule of* $E(W_n)$ *of length n.*

PROOF. (i) Let E be the injective envelope of W_n. Since M_n is an essential extension of W_n, we have $E(M_n) = E$.

(ii) This is obvious if $n = 1$. Suppose $n \ge 2$ and that the result is true for $n-1$. Let Y be a submodule of E having length n. Then $Y \supset W_n$ and Y/W_n is a submodule of E/W_n having length $n-1$. We know from 24.8(ii) that E/W_n is the injective envelope of $\operatorname{soc}(E/W_n)$. If we can show that $\operatorname{soc}(E/W_n) \cong W_{n-1}$ then it will follow from our induction hypothesis that M_n/W_n is the unique submodule of E/W_{n-1} of length $n-1$ and so $Y = M_n$.

So, consider $\operatorname{soc}(E/W_n)$. We know from 15.2 that $\operatorname{Ext}^1(W_{n-1}, W_n)$ is 1-dimensional over $\operatorname{End}(W_n)$ and hence, by 24.8(i), $\operatorname{soc}(E/W_n) \cong W_{n-1} \oplus X$ where X is semisimple and has no direct summands isomorphic to W_{n-1}. Suppose that $X \neq 0$. Choose any simple submodule, V say, of X and let U be its inverse image in E. Then U is an essential extension of W_n and so $\operatorname{Ext}^1(V, W_n) \neq 0$. Since W_{n-1} is the predecessor of W_n, we deduce that $V \cong W_{n-1}$, a contradiction to the conditions on X. Hence $X = 0$ and $\operatorname{soc}(E/W_n) \cong W_{n-1}$ as required. □

Our most complete result about $E(W)$ occurs in the case that W belongs to a cycle tower.

24.10. COROLLARY. *Let W be a simple R-module that is a member of a cycle tower, which we can take to be $W_1, \ldots, W_n = W$. Then $E(W)$ is an Artinian uniserial module of infinite length. Its sequence of composition factors, listed from the bottom up, consists of infinitely many repetitions of the sequence $W_n, W_{n-1}, \ldots, W_1$.*

PROOF. For each positive integer k, let M_{kn} be the uniserial module whose composition factors, listed from top to bottom, consist of k repetitions of the sequence $W_1, \ldots, W_n$ [16.5]. Each M_{kn} is an essential extension of W and therefore can be considered to be a submodule of $E(W)$. In fact, each M_{kn} is the unique submodule of $E(W)$ of length kn [24.9] and therefore equals $\operatorname{soc}^{kn}(E(W))$, the $(kn)^{\text{th}}$ iterated socle. Since $E(W) = \cup_{i=1}^{\infty} \operatorname{soc}^i(E(W))$ [24.6], we see that $E(W) = \cup_{k=1}^{\infty} M_{kn}$, completing the proof. □

If W belongs to a faithful tower, there is a uniserial submodule of $E(W)$ which has an easily described structure, given in the next corollary. Several examples that follow show that, above that submodule, the structure of $E(W)$ can get quite complicated.

24.11. COROLLARY. *Let W be a simple R-module that is a member of a faithful tower, and let the beginning of that tower be $W_1, W_2, \ldots, W_n = W$; so W_1 is the faithful top. Let M_n be the uniserial R-module whose sequence of composition factors, listed from top to bottom, is $W_1, W_2, \ldots, W_n = W$ [16.5]. Then M_n is the unique submodule of $E(W)$ of length n.*

PROOF. This is a special case of 24.9. □

We now give some examples of what happens above the uniserial bottom part of $E(W)$, when W belongs to a faithful tower. We begin with rings whose towers are trivial. One extreme situation involves the ring $B_1(F)$ described in 11.7.

24.12. EXAMPLE. Let $R = B_1(F)$ where F has characteristic zero. Then for every simple R-module W and every integer $n > 1$, the socle layer $\operatorname{soc}_n(E(W))$ [50.1] is the direct sum of infinitely many copies of each of the infinitely many simple R-modules.

PROOF. We begin with the case $n = 2$. As mentioned in 11.7, R has infinitely many simple modules and the dimension of $\operatorname{Ext}^1(V, W)$ over $\operatorname{End}(V)$ is infinite for all simple R-modules V, W. It follows from 24.8 that $\operatorname{soc}_2(E(W)) = \operatorname{soc}(E(W)/W)$

is the direct sum of infinitely many copies of each of the infinitely many simple R-modules.

Now suppose that $n > 2$ and that we know the stated result for $n-1$. Let X be any simple R-module. By our induction hypothesis, X is a direct summand of $\operatorname{soc}_{n-1}(E(W)) = \operatorname{soc}(E(W)/\operatorname{soc}^{n-2}(E(W))$. Since R is hereditary, homomorphic images of injectives are injective [12.2]; so $E(W)/\operatorname{soc}^{n-2}(E(W))$ is injective. Thus $E(X)$ is a direct summand of $E(W)/\operatorname{soc}^{n-2}(E(W))$. Therefore $\operatorname{soc}(E(X)/X)$ is a direct summand of $\operatorname{soc}(E(W)/\operatorname{soc}^{n-1}(E(W))) = \operatorname{soc}_n(E(W)$. However, we know from the preceding paragraph that $\operatorname{soc}\big(E(X)/X\big)$ is the direct sum of infinitely many copies of each of the infinitely many simple R-modules. □

A less extreme example comes from the ring $A_1(F)$ described in 11.8.

24.13. Example. Let $R = A_1$. Then R has a simple module W such that:

(i) $\operatorname{soc}\big(E(W)/W\big)$ is the direct sum of finitely many copies of infinitely many, but not all, simple R-modules; and
(ii) for every nonnegative integer n there is a simple R-module V such that $\operatorname{soc}\big(E(W)/W\big)$ contains a copy of $V^{(n)}$ but not of $V^{(n+1)}$.

Proof. We apply 24.8 several times. Recall from 11.8 that, for all simple R-modules V, W, the F-dimension of $\operatorname{Ext}^1(V,W)$ over $\operatorname{End}(V)$ is finite. Thus no simple module occurs infinitely often in the direct sum in part (i).

We also know [11.8] that the elements of $A_1 = A_1(F)$ consist of polynomials in two noncommuting variables x, y and that, whenever $p, q \in F[y]$, the R-modules $R/(x+p)R$ and $R/(x+q)R$ have the following properties: (a) they are simple modules; (b) they are isomorphic if and only if $p = q$; (c) the F-dimension of $\operatorname{Ext}^1\big(R/(x+p)R, R/(x+q)R\big)$ equals the degree of $p - q$; and (d) F is the endomorphism ring of each of these modules. Fixing q and letting p vary gives (ii). □

The preceding examples, when combined with results about idealizers, provide further examples of interest which show that similar behaviour can occur in nontrivial faithful towers, above the necessarily uniserial bottom part.

24.14. Example. Let n be a positive integer. There exists an HNP ring R, a module U_R, and a simple module W_R such that:

(i) U is the unique submodule of $E(W)$ of length n, U is uniserial and its sequence of composition factors, from top to base, is a faithful tower;
(ii) $\operatorname{soc}_R\big(E(W)/U\big)$ is the direct sum of infinitely many simple R-modules.

Proof. Let $S = B_1(F)$, where F has characteristic 0, and let U_S be a (necessarily faithful) simple module. Then by 24.12, $\operatorname{soc}_S\big(E(U_S)/U\big)$ is the direct sum of infinitely many copies of every simple S-module.

By 20.5, there is a HNP ring $R \subseteq S$ such $R_{\text{quo}} = S_{\text{quo}}$ such that U_R is uniserial of length n and its sequence of composition factors is a faithful R-tower. Let $E = E(U_S)$. Then we also have $E = E(U_R)$ [24.7]. Since U_R is uniserial of finite length, with socle W we also have $E = E(W_R)$. As observed above, $\operatorname{soc}_S(E/U)$ is a direct sum of infinitely many simple S-modules. Since every simple S-module has finite length as an R-module [6.4] we see that $\operatorname{soc}_R(E/U) = \operatorname{soc}_R\big(\operatorname{soc}_S(E/U)\big)$ which is an infinite direct sum of simple R-modules, as desired. □

Another example of interest comes by applying a similar argument to the simple HNP ring C described in 11.9. We recall from there that its simple modules W are all injective: so $E(W) = \mathrm{soc}(E(W)) = W$.

24.15. EXAMPLE. Let n be a positive integer. There exists an HNP ring R and a module U_R such that:

(i) U is uniserial of length n, its sequence of composition factors, from top to base, is a faithful tower; and
(ii) U is injective.

PROOF. Let S be a simple HNP (such as C) which has an injective simple module U. If $n = 1$ we are done. Suppose therefore that $n > 1$. By 20.5 there is an HNP ring $R \subseteq S$ such that $R_{\text{quo}} = S_{\text{quo}}$ with a module U as described in (i). Moreover, U_R is injective because U_S is injective [24.7]. □

25. Examples of Towers: Arbitrary Case

The purpose of this section is to construct an HNP domain having an arbitrarily complicated example of tower structure in the sense that one can specify the number (finite or infinite) of nontrivial faithful towers and cycle towers desired, together with the length and rank sequence of each.

As with the examples in §20, the construction involves the use of idealizer subrings. The detailed construction involves several different aspects which are dealt with in subsections. After an introductory example, there follow some general results about certain directed limits of HNP domains; and then two basic examples which form the building blocks of the ultimate example. That example is then constructed and its properties established. All that is required in the aftermath is that such an example exists; so the reader might skip all but the statement of the main result, 25.27, on a first reading.

Introductory example

This short subsection shows one superficially possible, but failing, route for obtaining an example with infinitely many nontrivial towers. The ring $S = M_2(\mathbb{Z})$ has infinitely many invertible maximal ideals p_iS where p_i is the ith prime. One can idealize S at the maximal right ideal $e_{11}S + e_{22}p_iS$ getting the subring

$$R_i = \begin{bmatrix} \mathbb{Z} & \mathbb{Z} \\ p_i\mathbb{Z} & \mathbb{Z} \end{bmatrix}$$

which has a cycle tower of length 2. An example with an arbitrary finite number of nontrivial towers is obtainable by intersecting that number of R_i for distinct values of i. So one might hope to get infinitely many such cycle towers by taking the intersection of the rings R_i as i runs through the natural numbers. However, that gives the ring

$$T = \begin{bmatrix} \mathbb{Z} & \mathbb{Z} \\ 0 & \mathbb{Z} \end{bmatrix}$$

which is neither prime nor hereditary!

Directed unions

Because of the difficulties this example exhibits, the construction we are going to describe follows a different, longer route involving direct limits of HNP domains.

25.1. NOTATION. We start with a directed, partially ordered set Π and a collection of subrings $\{R_\pi \mid \pi \in \Pi\}$ of an integral domain S such that $R_\pi \subseteq R_\mu$ and R_μ is flat as a left R_π-module whenever $\pi \leq \mu$. Let $H = \varinjlim\{R_\pi \mid \pi \in \Pi\}$.

We fix this notation for this subsection in which we will show that, under certain circumstances, properties shared by the R_π are inherited by H.

25.2. LEMMA. *Suppose that:*

(i) *for each $\pi \in \Pi$ and each simple R_π-module X, there is an integer $n(X)$ and some $\rho \geq \pi$ such that $\lambda(X \otimes_{R_\pi} R_\mu) \leq n$ for all $\mu \geq \rho$;*
(ii) *each R_π is right Noetherian with restricted minimum condition.*

Then H is right Noetherian with restricted minimum condition.

PROOF. Suppose that $f \in H$ and $f \neq 0$. Then $f \in R_\pi$ for some π. Therefore, $\lambda(R_\pi/fR_\pi)$ is finite; say it is t. By (i), for each simple composition factor X_i of R_π/fR_π, there is a $\rho \geq \pi$ such that there is a fixed bound on the length of $X_i \otimes R_\mu$ for $\mu \geq \rho$. An induction on t, using the left flatness of R_μ over R_π, shows that there is a $\rho \geq \pi$ and an integer $n(\pi, f)$ such that $\lambda(R_\mu/fR_\mu) \leq n$ for all $\mu \geq \rho$.

Suppose that $\lambda(H/fH) > n$. Then we can choose $a_1, a_2, \ldots, a_{n+1} \in H$ such that $a_j \notin a_1H + \ldots a_{j-1}H + fH$ for each $j \in \{1, 2, \ldots, n+1\}$. Choose $\mu \geq \rho$ so that $a_1, a_2, \ldots, a_{n+1} \in R_\mu$. Then $a_j \notin a_1R_\mu + \ldots a_{j-1}R_\mu + fR_\mu$ for each j; and so $\lambda(R_\mu/fR_\mu) \geq n+1$ which contradicts the choice of n. We deduce that $\lambda(H/fH) \leq n$. Hence H is right Noetherian with restricted minimum condition. $\square$

25.3. LEMMA. *Suppose that, in addition to the conditions (i) and (ii) of* 25.2*:*

(iii) *each R_π is right hereditary.*

Then H is right hereditary.

PROOF. Let B be any right ideal of H. By 25.2 above, we know that B is finitely generated. Choose a finite generating set for B and then $\pi \in \Pi$ so that the generating set belongs to R_π. Evidently $B = (B \cap R_\pi)H$. Now H is left flat over R_π since it is the direct limit of left flat R_π-modules. Hence $(B \cap R_\pi) \otimes_{R_\pi} H \cong (B \cap R_\pi)H$ via multiplication. However, since $B \cap R_\pi$ is projective over R_π, so too is $(B \cap R_\pi) \otimes_{R_\pi} H$ over H. $\square$

Recall from 9.6 that if there are rings $R \subseteq S$ and finite length modules $U_R \subseteq V_S$ then we say a composition series $0 \subset U_1 \subset \ldots \subset U_n = U$ fully embeds in a composition series $0 \subset V_1 \subset \ldots \subset V_n = V$ if, for each i, $U_i = V_i \cap U$.

25.4. LEMMA. *Suppose that, in addition to the conditions (i), (ii) and (iii) above:*

(iv) *for each π, there is a uniserial R_π-module U_π of fixed finite length, n say;*
(v) *for each $\mu \geq \pi$, $U_\mu \supseteq U_\pi$ and the composition series of U_π fully embeds in that of U_μ.*

Let $U = \cup(U_\pi)$. Then U is a uniserial H-module of length n, the ith member of its composition series is the union of the ith members of the composition series of the U_π and the composition series of each U_π fully embeds in that of U.

If, further, for every μ, no two composition factors of U_μ are isomorphic, then the same is true of U.

PROOF. Let the composition series of U_π be written

$$0 = U_\pi[0] \subset U_\pi[1] \subset U_\pi[2] \subset \ldots \subset U_\pi[n] = U_\pi$$

and, for each i, let $U[i] = \cup U_\pi[i]$. First, we show that $U[i] \cap U_\pi = U_\pi[i]$ for each π and each i. For let $x \in U[i]$. Then $x \in U_\mu[i]$ for some μ which we can choose so that $\pi \leq \mu$. Therefore $x \in U_\mu[i] \cap U_\pi = U_\pi[i]$ by the fully embedded property. We note the obvious consequence that the submodules $U[i]$ are distinct.

We next check that each factor $U[i+1]/U[i]$ is a simple H-module. To see this, choose any two elements in $U[i+1] - U[i]$. They both belong to some U_μ and so belong to $U_\mu[i+1] - U_\mu[i]$. Hence each is a multiple of the other by an element of R_μ and so by an element of H. Hence the factor is simple over H, as desired. This shows that the $U[i]$ form a composition series in which the composition series of each U_π is fully embedded.

To see that U is uniserial, choose any two elements $v_1, v_2 \in U$. Then both belong to some U_μ. That module is uniserial; so one of the v_i is a multiple of the other, by some element of R_μ and so by an element of H. We deduce that U is uniserial.

Finally, suppose there is an isomorphism

$$\alpha : U[i+1]/U[i] \to U[j+1]/U[j]$$

for some $i \neq j$. This restricts to a nonzero map

$$\alpha_\mu : U_\mu[i+1]/U_\mu[i] \to U_\mu[j+1]/U_\mu[j]$$

which must, necessarily, be an isomorphism over R_μ. □

Basic examples

The ultimate example at which we aim is constructed in a fashion which combines two distinct methods of obtaining towers, one dealing with cycle towers and the other with faithful towers. In this subsection we describe separately two constructions of HNP domains and establish some of their properties. Each construction involves a polynomial ring over a division ring. In each of the two cases, we describe a basic chain idealizer and a tower whose rank sequence is specified.

25.5. DEFINITION. The *tower rank* of a tower $\mathcal{T}$ in an HNP ring R is defined to be the sum of the members of the rank sequence of $\mathcal{T}$ [see 21.5]; that is, $\Sigma\rho(R, W_i)$ as W_i ranges through the unfaithful members of $\mathcal{T}$.

25.6. LEMMA. *Suppose that a cycle tower of specified rank sequence (and hence specified length, $n+1$ say, and tower rank, m say) is desired. Choose any division ring E and form the division ring of fractions $F = E(y_1, \ldots, y_m)$ where the y_i are indeterminates commuting with elements of E and with each other. Let σ be the automorphism of $F = E(y_1, \ldots, y_m)$ which leaves E unaffected but maps y_i to y_{i+1} for each i (taking $y_{m+1} = y_1$). Next form the skew polynomial ring $E(y_1, \ldots, y_m)[x; \sigma]$ with $ax = x\sigma(a)$ for each $a \in E(y_1, \ldots, y_m)$ and let D be its quotient division ring. Finally, consider the ring $D[z]$ with z a commuting indeterminate. Then:*

(i) *$U = D[z]/(z-x)D[z]$ is a simple $D[z]$-module,* $\operatorname{ann}(U) = (z^m - x^m)D[z]$ *and the factor ring $D[z]/(z^m - x^m)D[z]$ is a full $m \times m$-matrix ring over the fixed division ring $F^\sigma(x)$;*

(ii) *we can choose a basic chain $\mathcal{C}$ of length n of isomaximal right ideals of type U such that the basic chain idealizer $R = \mathbb{I}(\mathcal{C})$ is a hereditary Noetherian* *domain having one nontrivial tower, which is a cycle tower with the desired rank sequence;*

(iii) *R can also be constructed as the idealizer of a basic chain of isomaximal left ideals of type $U' = D[z]/D[z](z-x)$.*

PROOF. (i) Since U is 1-dimensional over D, it is simple. Since $z^m - x^m$ has $z - x$ as a factor, $(z^m - x^m)D[z] \subseteq (z-x)D[z]$; and since $z^m - x^m$ is central in $D[z]$, then $(z^m - x^m)D[z]$ is a two-sided ideal and the factor by it is m-dimensional over D. It is easy to check, using 1.3, that $\operatorname{End}(U_{D[z]}) \cong D^\sigma = F^\sigma(x)$.

One can check (see [**Cohn 77**, p. 47]) that F is m-dimensional over the fixed sub-division ring F^σ and so $F(x)$ is m-dimensional over $F^\sigma(x)$.

(Alternatively, we can form the skew group ring $D[\sigma]$ which, by [**McR 01**, 7.8.12], is simple. It follows that the standard Morita context given by [**McR 01**, 7.8.3, 7.8.5] is a Morita equivalence. Hence, by [**McR 01**, 7.8.13], $D[\sigma] \cong M_m(D^\sigma)$ and D is m-dimensional over D^σ.)

The density theorem then tells us that $D[z]/\operatorname{ann}(U) \cong M_m(D^\sigma)$ and that, as a right $D[z]$-module, $D[z]/\operatorname{ann}(U) \cong U^{(m)}$. It follows that $\operatorname{ann}(U) = (z^m - x^m)D[z]$.

(ii) Since $D[z]/ann(U)$ is an $m \times m$ matrix ring, 9.4 provides the basic chain and the desired rank sequence; 20.2 shows that R has a single nontrivial tower, which is a cycle tower; and the comments preceding 20.1 make clear that R is an HNP domain.

(iii) We can view $R/(z^m - x^m)D[z]$ as a basic chain idealizer from the simple Artinian ring $D[z]/(z^m - x^m)D[z]$; and 6.2 (viii) tells us that this can be obtained similarly from a basic chain of left ideals. □

25.7. LEMMA. *Suppose that a faithful tower of specified rank sequence (and hence specified length, $n+1$ say) is desired. Choose any division ring E and form the division ring of fractions $E(t,y)$ where t and y are indeterminates commuting with elements of E and with each other. Let σ be the automorphism of $E(t,y)$ which leaves $E(t)$ unaffected but maps y to yt. Next form the skew polynomial ring $E(t,y)[x;\sigma]$ with $ax = x\sigma(a)$ for each $a \in E(t,y]$ and let D be its quotient division ring. Finally, consider the ring $D[z]$ with z a commuting indeterminate. Then:*

(i) *$U = D[z]/(z-x)D[z]$ is a faithful simple $D[z]$-module;*

(ii) *we can choose a basic chain $\mathcal{C}$ of length n of isomaximal right ideals of type U such that the idealizer $R = \mathbb{I}(\mathcal{C})$ is a hereditary Noetherian domain having one nontrivial tower, which is a faithful tower of length $n+1$ with the desired rank sequence;*

(iii) *if $aD[z]$ is the minimal member of $\mathcal{C}$, then R is also the idealizer of a basic chain of left ideals in $aD[z]a^{-1}$ of type $aD[z]a^{-1}/aD[z]a^{-1}a(z-x)a^{-1}$.*

PROOF. (i) Let $D[z]/(z-x)D[z] = M$. Now $(z-x)D[z]$ is a maximal right ideal since $z - x$ has degree 1; and so M is simple. Suppose M is unfaithful, with annihilator A say. Then $A = aD[z]$ for some central element

$$a = z^s + z^{s-1}a_1 + \cdots + a_s$$

with each a_i in the centre of D, which is $k(t)$ where k is the centre of E. Now $A \subseteq (z-x)D[z]$, so

$$a = (z-x)(z^{s-1} + z^{s-2}b_1 \cdots + b_{s-1}).$$

Comparing coefficients and then multiplying each resulting equation by a power of x gives the equations:

$$x^{s-1}(b_1 - x) = x^{s-1}a_1$$
$$x^{s-2}(b_2 - xb_1) = x^{s-2}a_2$$
$$\dots$$
$$x(-xb_{s-1}) = xa_s.$$

Their sum is

$$-x^s = x^{s-1}a_1 + x^{s-2}a_2 + \cdots + xa_s$$

thus making x algebraic over $k(t)$. This contradiction shows that M is faithful.

(ii) 20.3 and 20.4 provide us with a basic chain for which 9.5 gives the desired rank sequence. 20.2 shows that R has a single nontrivial tower, which is a faithful tower; and the comments preceding 20.1 make clear that R is an HNP domain.

(iii) Let $T = \mathbb{I}_{D[z]}(aD[z])$. We know, e.g. from the proof of 14.5, that T is also the idealizer of the isomaximal left ideal $aD[z] = aD[z]a^{-1}a$ of the ring $aD[z]a^{-1}$. Then, as in the proof of 25.6(iii), we deduce that R is the idealizer of a basic chain of isomaximal left ideals of $aD[z]a^{-1}$. This chain has $aD[z] = aD[z]a^{-1}a$ as its minimal member. However, $aD[z]a^{-1}a \subseteq aD[z]a^{-1}a(z-x)a^{-1}$ which, by symmetry, is a maximal left ideal of $aD[z]a^{-1}$. So we deduce that the chain has the type claimed. □

It may well help to point out that (iii) tells us that R can be obtained from an isomorphic copy of $D[z]$ in precisely the same way, but using left ideals. Thus each of the two basic chain idealizer rings described above has a symmetric construction.

Construction — Sketch

Before the construction of the example having an arbitrary collection of faithful and cycle towers, it may help the reader to have a sketch first of how the example is built. It starts with a field, K say, and an index set Δ which indexes the towers wanted.

Next we construct some division rings. First, for each $\delta \in \Delta$, we choose (in the manner of 25.6 or 25.7) finitely many indeterminates (with some form of twist) which, when adjoined to K, generate an Ore domain whose division ring of fractions we call $D(\delta)$. Next, for each subset Δ' of Δ, we adjoin to K all the indeterminates chosen for each $\delta \in \Delta'$, making the indeterminates for δ commute with those for any $\delta' \neq \delta$. The resulting Ore domain again has a quotient division ring, denoted by $D(\Delta')$. If $\delta \in \Delta' \subseteq \Delta$ then $D(\delta) \subseteq D(\Delta') \subseteq D(\Delta)$.

Finally we introduce one further commuting indeterminate z. The ring H we seek will be a subring of the principal ideal domain $D(\Delta)[z]$. To describe H we first choose, as in 25.6 and 25.7, a special basic chain $\mathcal{C}_\delta$ in $D(\delta)[z]$ in such a way that $R(\delta) = \mathbb{I}(\mathcal{C}_\delta)$ has a tower of the type and rank sequence desired. Moreover, the basic chain is chosen in such a way that it remains a basic chain when extended to $D(\Delta')[z]$ for any subset Δ' such that $\delta \in \Delta' \subseteq \Delta$. Moreover, distinct δ in Δ' produce basic chains of different simple type. Therefore if Δ' is finite, then in $D(\Delta')[z]$ there is a multichain comprising one basic chain for each $\delta \in \Delta'$. We define $R(\Delta')$ to be the idealizer of this multichain; and the ring H we seek is the union of all $R(\Delta')$ as Δ' ranges over the finite subsets of Δ. In fact, H may also be obtained by intersecting the idealizers of each chain $\mathcal{C}_\delta$ extended to $D(\Delta)[z]$, thus echoing the 'failed' example which began this section.

Detailed construction and notation: division rings and polynomials

We now seek to combine the two basic examples to produce arbitrary numbers of nontrivial towers of each type in a single example. In order to do so we will require some *fixed notation* which we build up in this and the succeeding subsection. This subsection deals with a specific division ring D containing a rich collection of sub-division rings D' and with the relationship between the simple modules in the polynomial rings $D[z]$ and $D'[z]$.

25.8. NOTATION. Suppose we are given two arbitrary sets Λ and Γ and we wish to have a nontrivial faithful tower of length $n(\lambda)$ for each $\lambda \in \Lambda$, and one nontrivial cycle tower of length $n(\gamma)$ for each $\gamma \in \Gamma$ — and in each case we predetermine the rank sequence desired. Thus, if we let Δ be the disjoint union of Λ and Γ, then we want a tower of length $n(\delta)$ and specified rank sequence for each $\delta \in \Delta$. We will eventually be using the results above about direct limits; and the partially ordered set Π introduced in 25.1 will be the collection of all finite subsets of Δ.

We start with a division ring K of the form $K = k(t)$, rational functions in one commuting indeterminate t over a division ring k. (Ultimately, K and k will be fields. But it will help to allow the more general case for now.) For each $\lambda \in \Lambda$, we introduce an indeterminate y_λ and for each $\gamma \in \Gamma$ we introduce $n(\gamma)$ indeterminates $y_{(\gamma,1)}, \dots, y_{(\gamma,n(\gamma))}$ where $n(\gamma)$ is the tower rank required; and these indeterminates commute with each other and with the elements of K. Form the polynomial ring in these indeterminates. Any two elements in this polynomial ring belong to a polynomial ring in finitely many indeterminates which is an Ore domain. They thus have a common right multiple; so the whole polynomial ring too is an Ore domain which has a division ring of rational functions $F = K(y_\lambda, y_{(\gamma,i)} \mid \lambda \in \Lambda, \gamma \in \Gamma, i \leq n(\gamma))$.

Next, for each λ, define an automorphism $\sigma(\lambda)$ of F mapping $y_\lambda \to t y_\lambda$ and leaving K and all the other indeterminates invariant. Similarly, for each γ, define an automorphism $\sigma(\gamma)$ of F mapping $y_{(\gamma,i)} \to y_{(\gamma,i+1)}$ (setting $y_{(\gamma,n(\gamma)+1)} = y_{(\gamma,1)}$) and fixing K and all the other indeterminates. Now form a skew polynomial ring in new indeterminates $\{x_\lambda, x_\gamma\}$ for all $\lambda \in \Lambda$ and $\gamma \in \Gamma$ making these indeterminates commute with each other and with K and with all other indeterminates except that, for each $i \in \{1, \dots, n(\gamma)\}$,

$$y_\lambda x_\lambda = x_\lambda y_\lambda t \quad \text{and} \quad y_{(\gamma,i)} x_\gamma = x_\gamma y_{(\gamma,i+1)}.$$

Thus x_λ induces $\sigma(\lambda)$ on F and x_γ induces $\sigma(\gamma)$. As in the preceding paragraph, we see that this skew polynomial ring is an Ore domain. Let $D(\Lambda, \Gamma) = D(\Delta)$ (or $D_K(\Delta)$ if the extra precision is required) be its quotient division ring.

Note that, given any $\Delta' \subseteq \Delta$, there is a sub-division ring $D(\Delta') \subseteq D(\Delta)$. We will, in what follows, be making use of these division rings and their inter-relationships. We note first the following fact.

25.9. LEMMA. *If $\Delta' \subseteq \Delta$ and $K' = D(\Delta - \Delta')$ then $D_K(\Delta) = D_{K'}(\Delta')$.* □

We now adopt the convention that although Δ itself is allowed to be finite or infinite — and in the ultimate example at which we aim, it will be infinite — *any subset Δ_i with a suffix is automatically finite.* However, most of the work involved concerns the finite case because we aim to use 25.2 – 25.4 with Π being the collection of finite subsets of Δ. This work involves the inter-relationship between $D(\Delta_1)$ and $D(\Delta)$ especially in the case when $\Delta_1 = \{\delta\}$, a singleton; we write $D(\{\delta\}) = D(\delta)$.

Next we fix the notation $S(\Delta) = D(\Delta)[z]$ where z is a commuting indeterminate. (Our eventual example will be a subring of the principal ideal domain $S(\Delta)$.)

We need some results about the behaviour of some simple modules and chains under extension. The first result is quite general and easy.

25.10. LEMMA. *Let $D' \subseteq D$ be division rings and consider the principal ideal domains $D'[z] \subseteq D[z]$ with z a commuting indeterminate.*

(a) (i) *Let M be a nonzero right ideal of $D'[z]$. Then M has a monic polynomial generator, say of degree n; and then $D'[z]/M$ is n-dimensional over D' and $D[z]/MD[z]$ is n-dimensional over D.*

(ii) *If $n = 1$ then $M = (z - x)D'[z]$ for some $x \in D'$, M is maximal in $D'[z]$ and $MD[z]$ is maximal in $D[z]$.*

(b) *Let $x \in D'$ and $U = D'[z]/(z - x)D'[z]$.*

(i) *Let A be a right ideal of $D'[z]$. Then $D'[z]/A$ has a composition series of length r with all composition factors isomorphic to U if and only if $A = f(z)D'[z]$ with $f(z)$ a monic polynomial in z of degree r which factorizes as the product of r terms of the form $z - b$ with $D'[z]/(z - b)D'[z] \cong U$.*

(ii) *Suppose that $D'[z]/A \cong U^{(r)}$ and so $A = f(z)D'[z]$ as in (i). Then $V = D[z]/(z - x)D[z]$ is simple and $D[z]/f(z)D[z] \cong V^{(r)}$.*

PROOF. (a)(i), (ii) Straightforward.

(b)(i) We proceed by induction on r. First suppose that A is a maximal right ideal with $D'[z]/A \cong U$. Since U is 1-dimensional over D', then A has a monic linear generator, by (a). Thus $A = (z - b)D'[z]$ for some $b \in D'$. This establishes the result in the case $r = 1$

Say $D'[z]/A \cong U^{(r)}$. Then there is a maximal right ideal M such that $A \subseteq M$ and $D'[z]/M \cong U$. Hence $M = (z - b)D'[z]$ and $A = (z - b)A'$ where A' is a right ideal such that $M/A \cong D'[z]/A'$. By induction, $A' = g(z)D'[z]$ with $g(z)$ a product of monic polynomials of the required type; so the same is true of $f(z)$. The converse is clear.

(ii) We see from (i) that $f(z)$ has the type of factorization described and so $D[z]/f(z)D[z]$ has a composition series whose factors are all isomorphic to V. Since $D[z]$ is free over $D'[z]$, tensoring preserves the submodules and their intersections. $\square$

For each $\delta \in \Delta$, we now *choose and fix* the simple right $S(\delta)$-module $U(\delta) = S(\delta)/(z - x_\delta)S(\delta)$ and a basic chain $\mathcal{C}(\delta) \in S(\delta)$ of type $U(\delta)$ and length $n(\delta) - 1$, as described in the basic examples; see 25.6 or 25.7 depending upon whether $\delta \in \Gamma$ or $\delta \in \Lambda$.

25.11. LEMMA.

(i) *If $\delta \in \Delta$ then $\mathcal{C}(\delta)S(\Delta)$ is a basic chain in $S(\Delta)$ having the same length and rank sequence as $\mathcal{C}(\delta)$ and the type (faithful or unfaithful) of the simple module remains unaltered.*

(ii) *If, further, $\delta' \in \Delta$ with $\delta' \neq \delta$ then $\mathcal{C}(\delta')$ and $\mathcal{C}(\delta)$ are basic chains of different simple type.*

PROOF. (i) Note, by 25.9, that we can construct $S(\Delta)$ from $S(\Delta - \{\delta\})$ by using a different base division ring in the basic example involving δ. Since the simple type is fixed by that construction of that basic example, the type remains unaltered. The rest is clear from 25.10.

(ii) We need to prove that $S(\Delta)/(z-x_\delta)S(\Delta) \not\cong S(\Delta)/(z-x_{\delta'})S(\Delta)$. This is clear if one of δ, δ' belongs to Λ and the other to Γ since then one is unfaithful and the other faithful. If both belong to Γ, the simple modules have different annihilator ideals and so are nonisomorphic. So it only remains to deal with the case when both belong to Λ and the two simple modules are faithful.

Suppose there is an isomorphism

$$\theta : S(\Delta)/(z-x_\delta)S(\Delta) \to S(\Delta)/(z-x_{\delta'})S(\Delta)$$

with $\delta \neq \delta'$. Then $\theta([1+(z-x_\delta)S(\Delta)]) = [t+(z-x_{\delta'})S(\Delta)]$ where $t \in D(\Delta)$. It follows that $t(z-x_\delta) = (z-x_{\delta'})r$ with $r \in S(\Delta_1)$. Comparing coefficients gives $tx_\delta = x_{\delta'}t$ and hence $tx_\delta^m = x_{\delta'}^m t$ for all m. We can write $t = cd^{-1} = e^{-1}f$ where each of c, d, e, f is a polynomial over k in $\{u, x_\delta, w_\delta \mid \delta \in \Delta\}$. Then $fx_\delta^m d = ex_{\delta'}^m c$. However, in this equation, the degree in x_δ of the right-hand side is independent of m unlike that of the left-hand side. This contradiction completes the proof. □

Detailed construction and notation: idealizer rings.

We continue with the notations and conventions established in the preceding subsection; *but henceforth we restrict to the case that K is a field.* We now begin to examine basic chain and multichain idealizer subrings of the various rings $S(\Delta_i)$ as Δ_i ranges through the finite subsets of Δ.

Let $\Delta_1 \subseteq \Delta_2 \subseteq \Delta' \subseteq \Delta$ (so, by our convention, Δ_1 and Δ_2 are finite and Δ' may be finite or infinite). Define $\mathcal{C}(\Delta_1 : \Delta')$ to be the union of the basic chains $\mathcal{C}(\delta)S(\Delta')$ as δ varies through Δ_1. (So $\mathcal{C}(\Delta_1 : \Delta')$ is a multichain since Δ_1 is finite.) Note, in particular, that $\mathcal{C}(\Delta_1 : \Delta') \subseteq \mathcal{C}(\Delta_2 : \Delta')$. Define $R(\Delta_1 : \Delta') = \mathbb{I}_{S(\Delta')}(\mathcal{C}(\Delta_1 : \Delta'))$. We write $R(\Delta_1) = R(\Delta_1 : \Delta_1)$.

We note some straightforward inter-relationships between these idealizer rings.

25.12. LEMMA. *If $\Delta_1 \subseteq \Delta_2 \subseteq \Delta' \subseteq \Delta''$ then $R(\Delta_2 : \Delta') \subseteq R(\Delta_1 : \Delta'')$.*

PROOF. If $x \in R(\Delta_2 : \Delta')$ then $xAS(\Delta') \subseteq AS(\Delta')$ for each $A \in \mathcal{C}(\Delta_2)$ and hence, a fortiori, for each $A \in \mathcal{C}(\Delta_1)$. Therefore $xAS(\Delta')S(\Delta'') \subseteq AS(\Delta')S(\Delta'')$; i.e. $xAS(\Delta'') \subseteq AS(\Delta'')$. Hence $x \in R(\Delta_1 : \Delta'')$, as required. □

A less obvious, but crucial, relationship comes as (i) of the next result. It tells us that the set of subrings $R(\Delta_i)$, as Δ_i ranges through the finite subsets of Δ, is a directed partially ordered set under containment. (We note that it is in the proof of this result that K being a field is used.)

25.13. LEMMA.

(i) *If $\Delta_1 \subseteq \Delta_2$ then $R(\Delta_1) \subseteq R(\Delta_2)$.*
(ii) *If $\Delta_1 \cup \{\delta\} \subseteq \Delta_2$ then $R(\Delta_1) \subseteq R(\delta : \Delta_2) \subseteq R(\delta : \Delta)$.*

PROOF. (i) Let $T = R(\Delta_1 : \Delta_2)$. One sees, from 25.12, that $R(\Delta_2) \subseteq T$ and $R(\Delta_1) \subseteq T$.

Write Δ_2 as a disjoint union $\Delta_2 = \Delta_1 \cup \Delta_3$, say. For each $\delta \in \Delta_3$ there is a basic chain $\mathcal{C}(\delta)$ of $D(\delta)[z]$; and $R(\Delta_2) = [\cap_{\delta\in\Delta_3}\mathbb{I}_{S(\Delta_2)}(\mathcal{C}(\delta)S(\Delta_2))] \cap T$. Note that the elements in $D(\delta)[z]$ commute with those in $D(\Delta_1)$ and so with those in $R(\Delta_1)$. Therefore the elements of $R(\Delta_1)$ commute with the monic generators, given by 25.9, of the right ideals of $\mathcal{C}(\delta)$; and so $R(\Delta_1) \subseteq \mathbb{I}_{S(\Delta_2)}(\mathcal{C}(\delta)S(\Delta_2))$ for all $\delta \in \Delta_3$. Hence $R(\Delta_1) \subseteq R(\Delta_2)$.

(ii) We see from (i) that $R(\Delta_1) \subseteq R(\Delta_2)$ and, from 25.12, that $R(\Delta_2) \subseteq R(\delta : \Delta_2) \subseteq R(\delta : \Delta)$. □

We now note the following facts about direct limits involving the finite subsets of Δ.

25.14. LEMMA. *Let Δ_i range over the finite subsets of Δ. Then:*

(i) $S(\Delta) = \varinjlim(S(\Delta_i))$;
(ii) $R(\delta : \Delta_i) = R(\delta : \Delta) \cap S(\Delta_i)$;
(iii) $R(\delta : \Delta) = \varinjlim(R(\delta : \Delta_i))$;
(iv) $\varinjlim(R(\Delta_i)) = \cap_{\delta \in \Delta} R(\delta : \Delta)$.

PROOF. (i) This is immediate from the definitions.

(ii) Evidently $R(\delta : \Delta_i) \subseteq R(\delta : \Delta) \cap S(\Delta_i)$. On the other hand, suppose that $r \in R(\delta : \Delta) \cap S(\Delta_i)$. That tells us that $r \in S(\Delta_i)$ and $rB_j S(\Delta_i) \subseteq B_j S(\Delta_i)$ for each $B_j \in \mathcal{C}(\delta)$. Let b_j be a monic generator, of degree h say, of the principal right ideal B_j. Then we see that $rb_j = b_j s_j$ for some $s_j \in \Delta$; and so $b_j s_j \in S(\Delta_i)$. We need only show that $s_j \in \Delta_i$. This is a straightforward induction on the degree of s_j as a polynomial in z. (For if the leading term of s_j is $z^m d_m$ then the leading term of $b_j s_j$ is $z^{h+m} d_m$. Hence $d_m \in \Delta_i$ and so $b_j(s_j - z^m d_m) \in S(\Delta_i)$ and $s_j - z^m d_m$ has smaller degree.)

(iii) This follows easily from (i) and (ii).

(iv) Since $R(\Delta_i) \subseteq R(\delta : \Delta_i)$ we see from (iii) that $\varinjlim(R(\Delta_i)) \subseteq \cap_{\delta \in \Delta} R(\delta : \Delta)$. On the other hand, let $r \in \cap_{\delta \in \Delta} R(\delta : \Delta)$. Then $r \in S(\Delta_i)$ for some i. Choose any $\delta \in \Delta_i$. Then, by (iii), $r \in R(\delta : \Delta_i)$; and this holds for each $\delta \in \Delta_i$. Hence $r \in \cap_{\delta \in \Delta_i} R(\delta : \Delta_i) = R(\Delta_i)$. □

25.15. DEFINITION. In the light of (iv) above, we define

$$H = \varinjlim(R(\Delta_i)) = \cap_{\delta \in \Delta} R(\delta : \Delta).$$

This ring H is the example we have been aiming at. It is noteworthy that (iv) tells us that, like the earlier 'failed' example, H is the intersection of a collection of basic chain idealizers from the PID $S(\Delta)$ and so, in an obvious sense, is the idealizer of an infinite multichain. Of course, we have yet to show it has any of the desired properties. In order to do so, we will be using the earlier results about direct limits. Therefore, we will need to check the conditions involved in 25.2, 25.3 and 25.4. We will be using flatness in the following type of case.

25.16. LEMMA. *Let B be an integral domain and A be a hereditary Noetherian domain with $A \subseteq B$. Then every A-submodule M of B is flat.*

PROOF. Each finitely generated A-submodule of M is torsion-free and hence projective. Now M is the direct limit of these projective submodules and so is flat. □

25.17. PROPOSITION. *Let $\Delta_1 \subseteq \Delta' \subseteq \Delta$ with, as usual, Δ_1 finite.*

(i) *$R(\Delta_1 : \Delta')$ is a hereditary Noetherian integral domain and $U(\delta) \otimes S(\Delta')$ has length $n(\delta)$ over $R(\Delta_1 : \Delta')$.*
(ii) *$R(\Delta_1 : \Delta')$ is a left flat $R(\Delta_1)$-module.*

(iii) *Let X be a simple $R(\Delta_1)$-module. Then there is a $\Delta_2 \supseteq \Delta_1$ and an integer $n(X)$ dependent only on X such that the length of $X \otimes_{R(\Delta_1)} R(\Delta_3)$ is bounded above by $n(X)$ for all $\Delta_3 \supseteq \Delta_2$.*

PROOF. (i) Note that $\mathcal{C}(\Delta_1 : \Delta')$ is a multichain in $S(\Delta')$, one component of which is $\mathcal{C}(\delta)S(\Delta')$. The result follows from 14.5 combined with 6.8.

(ii) By (i), $R(\Delta_1)$ is an HNP domain. Now apply 25.16.

(iii) Given a simple right $R(\Delta_1)$-module, X, choose a maximal right ideal M with $R(\Delta_1)/M \cong X$. Pick a nonzero element $f \in M$. It is enough to show that there is a fixed bound on the length of

$$R(\Delta_1)/fR(\Delta_1) \otimes_{R(\Delta_1)} R(\Delta_3) \cong R(\Delta_3)/fR(\Delta_3)$$

for $\Delta_3 \supseteq \Delta_2$; for this will also be a bound on the length of $X \otimes_{R(\Delta_1)} R(\Delta_3)$.

Note next that, as modules over $R(\Delta_1)$, since $S(\Delta_1)/fR(\Delta_1)$ has finite length and $fS(\Delta_1)/fR(\Delta_1) \cong S(\Delta_1)/R(\Delta_1)$, then

$$\lambda(R(\Delta_1)/fR(\Delta_1)) = \lambda(S(\Delta_1)/fS(\Delta_1)).$$

A similar statement holds with Δ_3 replacing Δ_1 for any $\Delta_3 \supseteq \Delta_1$. So it will be enough to find $\Delta_2 \supseteq \Delta_1$ such that the length of $S(\Delta_3)/fS(\Delta_3)$ over $R(\Delta_3)$ is bounded for all $\Delta_3 \supseteq \Delta_2$.

Consider the element $f \in S(\Delta_1)$. It is a polynomial in z; and the length of $S(\Delta_1)/fS(\Delta_1)$ over $S(\Delta_1)$ is determined by the extent to which f factorizes; and this, in turn, is limited by the degree of f as a polynomial. So we may choose some $\Delta_3 \supseteq \Delta_1$ to maximize the length of $S(\Delta_3)/fS(\Delta_3)$ over $S(\Delta_3)$. Without loss of generality, we assume that $\Delta_3 = \Delta_1$.

We argue differently now for two types of simple composition factor U of $S(\Delta_1)/fS(\Delta_1)$. If $U \otimes_{S(\Delta_1)} S(\Delta_2) \cong S(\Delta_2)/(z - x_\delta)S(\Delta_2)$ for some $\Delta_2 \supseteq \Delta_1$ then $U \otimes_{S(\Delta_1)} S(\Delta_3)$ has length $n(\delta)$ over $R(\Delta_3)$ for all $\Delta_3 \supseteq \Delta_2$ by (i). We will assume Δ_2 is picked large enough to handle all such U. The other possibility is that $U \otimes_{S(\Delta_1)} S(\Delta_3) \ncong S(\Delta_3)/(z - x_\delta)S(\Delta_3)$ for all $\Delta_3 \supseteq \Delta_1$; and in that case $U \otimes_{S(\Delta_1)} S(\Delta_3)$ is a simple $R(\Delta_3)$-module.

Thus we see that the length of $S(\Delta_3)/fS(\Delta_3)$ over $R(\Delta_3)$ is bounded by the sum of all the $n(\delta)$ for all the simple composition factors of $S(\Delta_1)/fS(\Delta_1)$ of the former type, and 1 for the latter type. This provides the bound required. □

We can now obtain the first main result about H.

25.18. THEOREM. *H is a hereditary Noetherian domain.*

PROOF. First, note that, because of the symmetry described in 25.6 and 25.7, we need only prove that H is right Noetherian and right hereditary. However, that is immediate since 25.17 provides all the ingredients required to apply 25.2 and 25.3. □

The towers

We next aim towards ensuring that H contains a collection of towers as desired.

We require some results concerning the behaviour of the simple $S(\delta)$-module $U(\delta) = S(\delta)/(z - x_\delta)S(\delta)$ when extended to $S(\Delta_1)$ (where $\delta \in \Delta_1$) and viewed as an $R(\Delta_1)$-module or as an $R(\delta : \Delta_1)$-module. Note first that $U(\delta) \otimes_{R(\delta)} S(\Delta_1) \cong U(\delta) \otimes_{S(\delta)} S(\Delta_1)$ since $U(\delta)$ is an $S(\delta)$-module and $S(\delta) \otimes_{R(\delta)} S(\delta) \cong S(\delta)$. So we may write $U(\delta) \otimes S(\Delta_1)$ without ambiguity.

25.19. LEMMA.

(i) *$U(\delta) \otimes S(\Delta_1)$ is simple as an $S(\Delta_1)$-module and uniserial of length $n(\delta)$ over $R(\delta : \Delta_1)$ and, with the same composition series, over $R(\Delta_1)$. Over each of these idealizer rings, no two composition factors are isomorphic.*

(ii) *If $\Delta_1 \subseteq \Delta_2$ then $U(\delta)\otimes S(\Delta_1) \subseteq U(\delta)\otimes S(\Delta_2)$; and the composition series of $U(\delta) \otimes S(\Delta_1)$ over $R(\delta : \Delta_1)$ embeds fully in that of $U(\delta) \otimes S(\Delta_2)$ over $R(\delta : \Delta_2)$.*

PROOF. (i) We have already seen, in 25.11, that $U(\delta) \otimes S(\Delta_1)$ is simple as an $S(\Delta_1)$-module. Since the basic chain involved in forming the idealizer subring $R(\delta : \Delta_1)$ is of type $U(\delta) \otimes S(\Delta_1)$ and length $n(\delta) - 1$ we see that $U(\delta) \otimes S(\Delta_1)$ is uniserial of length $n(\delta)$ over $R(\delta : \Delta_1)$. Since $R(\Delta_1)$ is a multichain idealizer, one component of which was used in forming $R(\delta : \Delta_1)$, we know that the submodule structure of $U(\delta) \otimes S(\Delta_1)$ over $R(\Delta_1)$ is the same as over $R(\delta : \Delta_1)$.

(ii) Note that the rings $S(\Delta_1) \subseteq S(\Delta_2)$ and the simple module $U(\delta) \otimes S(\Delta_1)$ satisfy the conditions laid down in 9.8. The result follows immediately. □

This extends to $S(\Delta)$.

25.20. COROLLARY. *$U(\delta) \otimes S(\Delta)$ is simple over $S(\Delta)$. It is uniserial of length $n(\delta)$ over $R(\delta : \Delta)$ and, with the same composition series, over H. No two composition factors are isomorphic (over $R(\delta : \Delta)$ or over H).*

PROOF. View $U(\delta) \otimes S(\Delta)$ as the union of the $U(\delta) \otimes S(\Delta_i)$. The preceding result establishes the conditions to apply 25.4 in turn to the collections of rings $S(\Delta_i)$, $R(\delta : \Delta_i)$ and $R(\Delta_i)$ which have, as their respective unions, $S(\Delta)$, $R(\delta : \Delta)$ and H. □

It will help to have a result about faithful flatness. We make use of the fact, see e.g. [**McR 01**, 7.2.3], that if a left R-module L is flat then it is faithfully flat provided that $IL \neq L$ for all proper right ideals I of R. Of course, it is enough to check this for each maximal right ideal. Equivalently, since ${}_RL$ is flat, one need only check that $X \otimes L \neq 0$ for each simple right R-module X.

25.21. PROPOSITION. *For all finite subsets Δ_i of Δ:*

(i) *$S(\Delta)$ is left faithfully flat over $S(\Delta_i)$;*

(ii) *H is left faithfully flat over $R(\Delta_i)$;*

(iii) *$R(\Delta_i : \Delta)$ is left faithfully flat over $R(\Delta_i)$.*

PROOF. (i) Since $S(\Delta_i)$ is an HNP domain, 25.16 shows that $S(\Delta)$ is flat as a left $S(\Delta_i)$-module. To see it is faithfully flat, it is enough to show that if M is a maximal right ideal of $S(\Delta_i)$ then $MS(\Delta) \neq S(\Delta)$. This is clear since the monic polynomial in $S(\Delta_i)$ which generates M is not a unit in $S(\Delta)$.

(ii) Let X be a simple $R(\Delta_i)$-module. It may be a composition factor of $U(\delta) \otimes S(\Delta_i)$ for some $\delta \in \Delta_i$. In that case, we see from above that X embeds in the corresponding H-composition factor of $U(\delta) \otimes S(\Delta)$; and so $X \otimes_{R(\Delta_i)} H \neq 0$. Otherwise, X must be a simple $S(\Delta_i)$-module in which case $X \otimes_{S(\Delta_i)} S(\Delta) \neq 0$. But

$$\begin{aligned} X \otimes_{S(\Delta_i)} S(\Delta) &\cong X \otimes_{R(\Delta_i)} S(\Delta_i) \otimes_{S(\Delta_i)} S(\Delta) \\ &\cong X \otimes_{R(\Delta_i)} S(\Delta) \cong X \otimes_{R(\Delta_i)} H \otimes_H S(\Delta). \end{aligned}$$

Hence $X \otimes_{R(\Delta_i)} H \neq 0$ in each case.

(iii) Similar. □

Recall that Δ is the disjoint union of Λ and Γ.

25.22. THEOREM. *H contains a nontrivial faithful tower of length $n(\lambda)$ for each $\lambda \in \Lambda$ and a cycle tower of length $n(\gamma)$ for each $\gamma \in \Gamma$; and there are no further nontrivial towers in H.*

PROOF. Let $U = U(\delta)$ and $U' = U'(\delta) = U \otimes_{S(\delta)} S(\Delta)$; then 25.20 describes the structure of U' over each of the rings $S(\Delta)$, $R(\delta : \Delta)$ and $R(\Delta)$. In particular, over $S(\Delta)$ it is simple; and over H it is uniserial of length $n(\delta)$ and no two of its simple composition factors are isomorphic.

Now U' is unfaithful over $S(\Delta)$ if and only if it is unfaithful over H since the two rings share the same quotient ring [12.1(vii)]. Hence the composition factors of U' are, at least, a part of a nontrivial tower. We now consider separately two cases.

(a) Suppose that U is an unfaithful simple $S(\delta)$-module and so U is also a complete cycle tower over $R(\delta)$. There is a nonsplit extension Q of $U_{S(\delta)}$ by itself which is uniserial of length $2n(\delta)$ over $R(\delta)$; and the same is true over $R(\Delta_i)$ for each Δ_i containing δ. Hence, by 25.4, $Q \otimes S(\Delta)$ is uniserial of length $2n(\delta)$ over H. However, its composition factors are those of U' repeated twice. It follows that the composition factors of U' form a complete cycle tower for H.

(b) Suppose that U is a faithful simple $S(\delta)$-module and so U is also a complete faithful tower over $R(\delta)$. The same is true for $U \otimes S(\Delta_i)$ for each Δ_i containing δ. We know, from 25.11, that U' is a faithful simple $S(\Delta)$-module and so, from above, U' is a faithful H-module. However, its composition factors over H coincide with those over $R(\delta : \Delta)$, all of which are unfaithful except for the top one. Using 12.1(vii) again, we deduce that all but the top factor are unfaithful over H. We have thus shown that the composition factors of U' over H form a top section of a faithful tower.

It remains to show that this is actually a complete tower and that there are no further towers. Both will be established if we can show that if there are two simple H-modules P, Q such that $\operatorname{Ext}(P, Q) \neq 0$ and Q is unfaithful then they come from such an extension over some $R(\Delta_i)$. Given such a P, Q, there exist right ideals A, B of H such that $H \supset A \supset B$ and H/B is a nonsplit extension of Q by P. Choose $\Delta_i \subset \Delta$ so that $R(\Delta_i)$ contains generating sets, as right ideals of H, of A and B and $\operatorname{ann}(Q)$. Then $R(\Delta_i) \supset (A \cap R(\Delta_i)) \supset (B \cap R(\Delta_i))$. Faithful flatness of H over $R(\Delta_i)$ implies that each factor is simple and the bottom one is unfaithful since $\operatorname{ann}(Q) \cap R(\Delta_i) \neq 0$. The fact that $H/B \cong (R(\Delta_i)/(B \cap R(\Delta_i))) \otimes H$ implies that $R(\Delta_i)/(B \cap R(\Delta_i))$ must be nonsplit; and this completes the argument. □

The rank sequences

Finally, we aim towards demonstrating that each nontrivial tower constructed has the rank sequence desired.

25.23. NOTATION. Suppose we go through the same construction which gave us H except that one specified basic chain, say $\mathcal{C}(\delta_0)$, is replaced by an empty chain. This modified construction gives a ring, H' say, having all the same features as H except those specifically concerned with δ_0.

25.24. LEMMA.

(i) *H' is an HNP ring which lies between H and S.*

(ii) *$H' = \cap\{R(\delta : \Delta) \mid \delta \neq \delta_0\}$ and $H = R(\delta_0 : \Delta) \cap H'$.*

(iii) $S(\delta_0) \subseteq H'$.

PROOF. (i) (ii) We see from 25.18 that H' is an HNP domain and 25.14(iv) shows that H' is the intersection of all $R(\delta : \Delta)$ other than $R(\delta_0 : \Delta)$. The rest is then clear.

(iii) This follows from the observation that the elements of $S(\delta_0)$ commute with each generator of each member of the chains used in the construction.

□

We need to consider the simple module $U(\delta_0)$.

25.25. LEMMA.

(i) *If M is a maximal right ideal of $S(\delta_0)$ such that $S(\delta_0)/M \cong U(\delta_0)$ then MH' is a maximal right ideal of H'.*

(ii) *The original basic chain $\mathcal{C}(\delta_0)$ blows up to a basic chain $\mathcal{C}(\delta_0) \otimes H'$ which has the same length and the same rank sequence as had $\mathcal{C}(\delta_0)$.*

PROOF. (i) M is cyclic, say $M = mS(\delta_0)$. Consider $R(\Delta_1 : \Delta_1)/mR(\Delta_1 : \Delta_1)$. Now $S(\Delta_1)/mS(\Delta_1)$ is simple over $S(\Delta_1)$ by 25.11. It therefore remains simple over $R(\Delta_1)$. Consider the two series of $R(\Delta_1)$-modules

$$S(\Delta_1) \supseteq R(\Delta_1) \supseteq mR(\Delta_1) \quad \text{and} \quad S(\Delta_1) \supseteq mS(\Delta_1) \supseteq mR(\Delta_1).$$

Now $mS(\Delta_1)/mR(\Delta_1) \cong S(\Delta_1) \supseteq R(\Delta_1)$. So the length of $R(\Delta_1)/mR(\Delta_1)$ equals that of $S(\Delta_1)/mS(\Delta_1)$, which is 1. Since H'/mH' is the limit of the $R(\Delta_1)/mR(\Delta_1)$, we deduce from 25.4 that it is, as claimed, simple.

(ii) This follows immediately. □

Let R' be the idealizer, in H', of this basic chain.

25.26. PROPOSITION. *$R' = H$.*

PROOF. Note that each member of $\mathcal{C}(\delta_0)$ is cyclic. Pick one, generated by p, say. Then p also generates the corresponding chain member in $\mathcal{C}(\delta_0) \otimes H'$. Let $r \in R'$. Then $rp = ps$ for some $s \in H'$ and hence $r \in \mathbb{I}(pS)$. This applies to each chain member; so $r \in R(\delta_0 : \Delta)$. However, we saw in 25.24 that $H = R(\delta_0 : \Delta) \cap H'$ so $R' \subseteq H \subseteq H'$.

Now R' is the idealizer in H' of the basic chain $\mathcal{C}(\delta_0) \otimes H'$; and hence, by 8.14, H is the idealizer in H' of a subchain of that chain. Thus the towers of H are the same as those of H' except that the simple H'-module $U(\delta_0) \otimes H'$ is uniserial over H and its simple composition factors form the only additional non-trivial tower. Note that

$$U(\delta_0) \otimes_{S(\delta_0)} H' \otimes_{H'} S(\Delta) \cong U(\delta_0) \otimes_{S(\delta_0)} S(\Delta).$$

Therefore this tower is the one of length $n(\delta_0)$ given by 25.22. Hence the subchain of the basic chain $\mathcal{C}(\delta_0) \otimes H'$ whose idealizer is H is, in fact, the whole chain [8.19]. Therefore $H = R'$. □

25.27. THEOREM. *The hereditary Noetherian domain H defined in* 25.15 *has precisely the nontrivial towers described in* 25.8, *with each having the length and rank sequence specified there.*

PROOF. We continue with the notation of the preceding result. The rank sequence of the nontrivial H-tower whose associated uniserial module is the simple H'-module $U(\delta_0) \otimes H'$ is, by 9.5, given precisely by the rank sequences in the basic chain concerned. And that, by 25.25, is the same as the rank sequence of the basic chain in $S(\delta_0)$ which was organized to be as required.

Since δ_0 was an arbitrary member of Δ this applies to all nontrivial towers of H. □

26. Notes on Chapters 3 and 4

The story of HNP rings[1] starts with the description in [**Steinitz 1911**] of finitely generated modules over full rings of algebraic integers, a description that remains intact over commutative Dedekind domains. The connection with projective modules was made in [**Kaplansky 52**, Lemma 3], by showing that invertible ideals are projective modules. He did so without mentioning the term 'projective', which was not then in use.

The noncommutative version is derived in part from the study of maximal orders over Dedekind domains (in central simple algebras over the field of fractions of that Dedekind domain) [**Reiner 75**]. These maximal orders were shown to be HNP rings; and the more general class of hereditary orders, which are also HNP rings, was investigated by [**Harada 64**], [**Jacobinski 71**] and others.

At the same time, Goldie's Theorem characterized more general classes of orders in semisimple Artinian rings, no longer requiring them to be finitely generated over their centre. Dedekind prime rings [23.5, 23.6] play a role similar to that of maximal orders and once again they are HNP rings. Much earlier than Goldie's Theorem, in [**Asano 39**] and [**Asano 50**] (the earlier results being reported, to some extent, in [**Jacobson 43**]), Asano worked along a related direction. Despite not having Goldie's characterization of orders in simple Artinian rings, Asano was studying such orders whose ideals, or right ideals, had properties similar to those of Dedekind domains. In retrospect, it was shown (see [**Michler 69**], [**Griffith-Robson 70**]) that the rings being studied were precisely bounded Dedekind prime rings.

[**Robson-Small 74**] demonstrated the precise relationship with the 'classical' case: maximal orders and hereditary orders are precisely those Dedekind prime rings, respectively HNP rings, which are finitely generated over their centres (equivalently, their simple Artinian quotient rings are finite dimensional over their centres; equivalently, they satisfy a polynomial identity).

The investigation of Dedekind prime rings and HNP rings was, of course, guided by the properties of Dedekind domains. Thus their ideal structure was investigated in [**Robson 68**] and [**Michler 69**] and then in [**Eisenbud-Robson 70a**] and [**Eisenbud-Robson 70b**]. The latter papers also proved several module-theoretic results showing some similarities to Dedekind domains and the classical theory; but they also began to show some distinctive features in this more general setting, with the major distinctions arising because there can be faithful simple modules over these more general rings.

At the same time and thereafter, many papers were published advancing the understanding of an HNP ring R and its ideal and module structure. We will mention only some of these; namely those containing results which appear pivotal to the overall study and hence have been covered in these two chapters; that is, those which deal with overrings and factor rings of R, and their connections with the successor relations between simple modules involving towers.

---

[1] In this section R denotes an HNP ring unless the contrary is specified.

§11. [**Chatters 72**] showed [11.1] that a hereditary Noetherian ring splits as a direct sum of Artinian and prime rings, thus splitting the study of hereditary Noetherian rings and their modules into the prime and Artinian cases. Both the ring and module structure, in the Artinian case, turn out to be entirely different from that in the prime case, and essentially unrelated to that of HNP rings. So they play no role in this book. For the structure of an Artinian hereditary ring, see [**Harada 66**] or [**McR 01**, 5.4.7].

§12. [**Webber 70**] showed that R satisfies the restricted minimum condition, [12.16]. In effect, this reduced the study of finitely generated R-modules to projective (equivalently, torsion-free) modules and indecomposable modules of finite length, with the latter all being cyclic (but not necessarily uniserial) when R is a Dedekind prime ring.

§13. [**Kuzmanovitch 72**] showed that each overring of an HNP ring R is again an HNP ring [13.5]. [**Goodearl 74**] then demonstrated the (1,1)-correspondence between overrings of R and collections of simple right R-modules [13.8]. We note, in relation to 13.7, that Goodearl's terminology focuses on those simple modules *not* killed by S. The same paper also showed [13.6] that each simple module belongs to the socle of R_{quo}/R.

The proofs in this section are very different to the originals, being developed directly rather than depending on general torsion theory.

§14. [**Robson 72**] showed that a basic idealizer is an HNP ring if and only if the original ring is one [14.6] and, using that, demonstrated that idealizers could be applied as a major tool in the study of HNP rings and in the construction of counterexamples. This paper also proved that each basic idealizer slices one simple into two parts and leaves all other simple modules intact. This tool now appears indispensable and is used extensively here.

§15. [**Eisenbud-Griffith 71**] proved that factor rings are serial [15.6].

§17. The transpose duality was introduced by [**Zaks 74**]. A similar, but not exactly the same, notion of 'transpose' plays a prominent role in the theory of modules over finite dimensional algebras A: Given any short projective resolution $Q \overset{\alpha}{\to} P \twoheadrightarrow X$ of a right A-module X, with Q and P finitely generated, one calls the cokernel of α^*: $P^* \to Q^*$ a 'transpose' of X. (Compare this with our 17.4.) However, this transpose is a functor defined on mod-A modulo projective direct summands, rather than on mod-A itself, as in our situation.

The formula $V^{\text{tr}} \cong W^{\text{op}}$ [17.10] comes from [**Hodges 87**, p. 755(4)].

§18. [**Lenagan 73**, 2.2] proved the unusual chain condition 18.3 involving both left and right ideals. He used it to prove his extension [**Lenagan 73**, 2.3], of the well-known commutative Krull-Akizuki theorem, to bounded Noetherian prime rings with the restricted minimum condition [18.4].

18.7 was proved, for bounded HNP rings, by Lenagan in the course of his proof of [**Lenagan 73**, 3.3]. Then [**Goodearl-Warfield 79**, Cor 21] removed the boundedness hypothesis, essentially proving the finiteness of towers. (See the discussion of §22.)

§19. [**Goodearl-Warfield 79**, §1] proved, using some idealizer results from [**Robson 72**], that every unfaithful simple R-module has a unique predecessor; and whenever a simple module V has an unfaithful successor W, then W is the unique successor (faithful or not). Using the aforementioned finiteness results of

Lenagan, they effectively partitioned all simple R-modules into towers, although they had no need to formalize this, as was required in [**Levy-Robson 99a**].

§22. Recall that every maximal ideal of R is either idempotent or invertible [14.3]. The desire to distinguish between these roles led to the study of certain pairs of nonzero idempotent maximal ideals which was the first step leading to the eventual definition of towers. To show the formal connection, via hindsight, recall [15.8] that if V, W are unfaithful simple modules whose annihilators are the idempotent maximal ideals M, N respectively then $O_r(N) = O_l(M)$ if and only if $\text{Ext}^1(V, W) \neq 0$. [**Eisenbud-Robson 70b**] defined a 'cycle of maximal ideals' to be a sequence $A_1, \ldots, A_n$ of distinct nonzero idempotent maximal ideals such that $O_r(A_i) = O_l(A_{i+1})$ for each i and $O_r(A_n) = O_l(A_1)$. (This numbering is the reverse of the numbering we have adopted for a cycle.) They then proved [their 2.5 and 2.6] that the intersection of a cycle is a maximal invertible ideal (maximal amongst the invertible ideals), and every maximal invertible ideal arises in this way. They considered a single invertible maximal ideal to be the trivial case of this. (See our 22.1 and 22.12.) This improved the earlier result [**Harada 64**, Proposition 7]. They also noted that R has a right finite overring S that is a Dedekind prime ring if and only if R has only a finite number of idempotent maximal ideals. (S would then be a Dedekind closure of R in the terms used in our 30.2.) They asked whether infinitely many idempotent maximal ideals can actually occur. See the discussion of §25 below.

§23. Lenagan's purpose in [**Lenagan 73**] was to prove that every bounded HNP ring has enough invertible ideals [his 3.3; our 23.2].

§24. The results 24.4 and 24.5 on the multiplicity of a simple W_R as a direct summand of $\text{soc}(R_{\text{quo}}/R)$ — equivalently, the multiplicity of $E(W)$ as a direct summand of R_{quo}/R — seem to be new. We were surprised that finiteness of this multiplicity is not determined by whether or not W is faithful.

The structure of $E(W)$ is due to [**Goodearl-Warfield 79**].

§25. Finiteness of towers converts the question of whether R can have infinitely many idempotent maximal ideals, to whether R can have infinitely many non-trivial towers. The answer to this had to await [**Stafford-Warfield 84**] and [**Stafford-Warfield 85**] which proved that R can have infinitely many non-trivial cycle towers or infinitely many non-trivial faithful towers. That result is improved here by 25.27 which shows that suitable HNP rings R can have arbitrarily many towers of prescribed lengths and types (cycle or faithful).

CHAPTER 5

Integral Overrings

The term *integral* has a well-established meaning in commutative ring theory. In particular, for a commutative integral domain D with a field of fractions D_{quo}, an overring A (so with $D \subseteq A \subseteq D_{\text{quo}}$) is an *integral extension* if and only if, for each $a \in A$, the subring $D[a]$ generated by D and a is a finitely generated D-module.

This chapter extends this idea to a right integral overring S of an HNP ring R. The first three sections deal with simple modules. The first determines which overrings S are right integral over R in terms of the set of simple right R-modules killed by S. The next shows that all simple R-modules, including those killed by S, occur as R-composition factors of simple S-modules. Finally we show that, amongst right integral overrings S of R, each is determined by the simple R-modules onto which it maps.

Then §30 concerns maximal right integral overrings of R: these are Dedekind prime rings called 'Dedekind right closures' of R. The chapter ends with an application giving a more precise description of the factor rings of R which helps show that not all Artinian serial rings arise in this way.

27. Simple Modules Killed

27.1. DEFINITIONS. [1] An overring S of R is called a *right integral overring of* R if, for each finite subset F of S, the overring $R\langle F\rangle$ generated by R together with F is a finitely generated right R-module (i.e. is a right finite overring of R). (27.11 shows that we could restrict F to be a one-element set in this definition without changing the notion.)

Recall, from 13.8, that every overring of R has the form $S = R(\mathcal{Y})$ where $\mathcal{Y}$ is the set of simple R-modules Y killed by S (i.e. $Y \otimes_R S = 0$), and the correspondence $\mathcal{Y} \to R(\mathcal{Y})$ is an inclusion-preserving bijection between sets of (isomorphism classes of) simple R-modules and the set of overrings of R. Anticipating the main result of this section [27.9], we call a set $\mathcal{Y}$ of simple R-modules *integral* if it satisfies the following two conditions:

(27.1.1) (i) $\mathcal{Y}$ does not contain any entire cycle tower; and
(ii) $\mathcal{Y}$ does not contain the base of any faithful tower.

One immediate consequence of $\mathcal{Y}$ being integral is that none of its members belong to trivial towers.

Our first lemma isolates the reason for the two exclusions mentioned in (27.1.1).

27.2. LEMMA. *Let $S = R(\mathcal{Y})$ where either:*

(i) *$\mathcal{Y}$ is a cycle R-tower; or*

[1] In this section R denotes an HNP ring unless the contrary is specified.

(ii) $\mathcal{Y} = \{V\}$, *the one-element set consisting of the base element* V *of some faithful* R*-tower.*

Then S *is not right integral over* R*. Indeed* S *is finitely generated over* R *as a ring, but is not right finite over* R*.*

PROOF. First we show that S is finitely generated over R as a ring. Recall that 13.11 asserts that S is generated as a ring by the finite set of overrings $R(W)$ corresponding to each simple isomorphism type $W \in \mathcal{Y}$. Hence, to show that S is finitely generated over R as a ring, it is enough to demonstrate this for each $R(W)$. However, the inclusion-preserving correspondence between overrings of R and sets of simple R-modules given by 13.8 shows that each $R(W)$ is a minimal overring of R and so is generated over R as a ring by any element in $R(W) - R$.

To show that S_R is not finitely generated different arguments are used for the two cases.

(i) Fix some $W \in \mathcal{Y}$. We know [24.3] that R_{quo}/R contains a copy of $E(W)$. We identify $E(W)$ with that copy. By 24.11, $E(W)$ is an Artinian uniserial module of infinite length and each cyclic submodule C of $E(W)$ has finite length with composition factors all belonging to $\mathcal{Y}$. By the definition of $R(\mathcal{Y})$ in 13.7, this tells us that $C \subseteq R(\mathcal{Y})/R$. Hence $E(W) \subseteq R(\mathcal{Y})/R$. However any finitely generated submodule of R_{quo}/R has finite length [12.12]. So $E(W_R)$ is not finitely generated; and hence the same is true of $R(\mathcal{Y})$.

(ii) The first paragraph of this proof demonstrates that $S = R(V)$ is a minimal overring. By 14.9(b), if $R(V)$ were right finite over R, then V would have an unfaithful successor module W. But since V is the base element of a faithful tower, it has no such successor. □

27.3. DEFINITION. Let U be an R-module and S an overring of R. We will say that U *is naturally an* S*-module* when the natural map $U \to U \otimes_R S$ is an isomorphism. This gives U an S-module structure, and is natural in the sense that, for any S-module U, the natural map $U \to U \otimes_R S$ is indeed an isomorphism [13.4].

27.4. LEMMA. *Let* $S = R(\mathcal{Y})$ *where* $\mathcal{Y}$ *is an integral set of simple* R*-modules. Suppose* W_R *is simple and either:*

(i) W *is not in the same* R*-tower as any element of* $\mathcal{Y}$*; or more generally,*
(ii) $W \notin \mathcal{Y}$*, and* W *is not the unfaithful successor of any element of* $\mathcal{Y}$*.*

Then W *is naturally a simple* S*-module.*

PROOF. We wish to show that $W \cong W \otimes_R S$ via $w \to w \otimes 1$; and 13.13 asserts that this is so provided that $\text{Ext}^1_R(Y, W) = 0$ for every $Y \in \mathcal{Y}$.

Suppose, to the contrary, that $\text{Ext}^1_R(Y, W) \neq 0$ for some $Y \in \mathcal{Y}$. Recall that, since $\mathcal{Y}$ is integral, Y is not the base of a faithful tower. Therefore Y has an unfaithful successor, X say, in its tower. However, 15.2 asserts that X is the unique successor (faithful or unfaithful); so we deduce that $W \cong X$, which contradicts the hypothesis. □

27.5. LEMMA. *Let* $\mathcal{Y} = \mathcal{Y}_1 \cup \mathcal{Y}_2$*, a disjoint union of integral sets of simple* R*-modules. Suppose that no element of* $\mathcal{Y}_2$ *is the unfaithful successor of any element of* $\mathcal{Y}_1$*. Then:*

(i) $\mathcal{Y}$ *is an integral set of simple* R*-modules;*
(ii) *each simple* R*-module in* $\mathcal{Y}_2$ *is naturally a simple* $R(\mathcal{Y}_1)$*-module; and*

(iii) $R(\mathcal{Y}) = R(\mathcal{Y}_1)(\mathcal{Y}_2)$.

PROOF. (i) This is clear from the definition of an integral set of simple modules.

(ii) Choose an arbitrary $W \in \mathcal{Y}_2$ and apply 27.4(i).

(iii) Let $T = R(\mathcal{Y}_1)(\mathcal{Y}_2)$. We need to show that $R(\mathcal{Y}) = T$; and to do this, it suffices to show that both rings kill the same simple R-modules [13.8]. Let W_R be simple and $W' = W \otimes_R R(\mathcal{Y}_1)$.

First suppose that $W \otimes_R T = 0$. If $W' = 0$ then W is one of the simple R-modules killed by $R(\mathcal{Y}_1)$; that is, $W \in \mathcal{Y}_1 \subseteq \mathcal{Y}$, and so $W \otimes_R R(\mathcal{Y}) = 0$. On the other hand, if $W' \neq 0$, then W' is a simple $R(\mathcal{Y}_1)$-module; and in this case

$$W' \otimes_{R(\mathcal{Y}_1)} T \cong W \otimes_R R(\mathcal{Y}_1) \otimes_{R(\mathcal{Y}_1)} T \cong W \otimes_R T = 0.$$

Therefore W' is one of the simple $R(\mathcal{Y}_1)$-modules killed by T; that is, $W' \in \mathcal{Y}_2$. By 13.4, $\otimes_R = \otimes_{R(\mathcal{Y}_1)}$ when restricted to $R(\mathcal{Y}_1)$-modules; and so

$$\begin{aligned} 0 = W' \otimes_R R(\mathcal{Y}) &= W \otimes_R R(\mathcal{Y}_1) \otimes_R R(\mathcal{Y}) \\ &= W \otimes_R R(\mathcal{Y}_1) \otimes_{R(\mathcal{Y}_1)} R(\mathcal{Y}) = W \otimes_R R(\mathcal{Y}). \end{aligned}$$

Thus, in either case, $W \otimes_R R(\mathcal{Y}) = 0$.

Conversely, suppose that $W \otimes_R R(\mathcal{Y}) = 0$. Then $W \in \mathcal{Y} = \mathcal{Y}_1 \cup \mathcal{Y}_2$. If $W \in \mathcal{Y}_1$ then $W \otimes_R R(\mathcal{Y}_1) = 0$, and so, since $R(\mathcal{Y}_1) \subseteq T$, then $W \otimes_R T = 0$. If on the other hand, $W \in \mathcal{Y}_2$, then W is a simple $R(\mathcal{Y}_1)$-module and so we have $W \otimes_R T = W \otimes_R R(\mathcal{Y}_1)(\mathcal{Y}_2) = W \otimes_{R(\mathcal{Y}_1)} R(\mathcal{Y}_1)(\mathcal{Y}_2) = 0$, as required. □

27.6. COROLLARY. *Let $\mathcal{A}$ be the collection of all R-towers. Write $\mathcal{A}$ as the disjoint union of two subsets $\mathcal{A}_1 \cup \mathcal{A}_2$. Let $\mathcal{Y}$ be an integral set of simple R-modules and let $\mathcal{Y}_i$ be those simple modules in $\mathcal{Y}$ which belong to towers in $\mathcal{A}_i$. Then each $\mathcal{Y}_i$ is an integral set and $R(\mathcal{Y}) = R(\mathcal{Y}_1)(\mathcal{Y}_2) = R(\mathcal{Y}_2)(\mathcal{Y}_1)$.*

PROOF. It is immediate that the hypotheses of 27.5 are satisfied with the $\mathcal{Y}_i$ in either order. □

The next two results provide a critical link between idealizers and the topic of the present chapter.

27.7. LEMMA. *Let S be an HNP ring and R an iterated basic idealizer from S. Let Y be a simple R-module which is naturally a simple S-module. Suppose that Y has an unfaithful successor Z over R. Then Y has an unfaithful successor X over S and Z is the top composition factor of X when viewed as a uniserial finite length R-module.*

PROOF. Recall [12.1] that an S-module M is faithful if and only if M_R is faithful.

We proceed via induction on the number of basic idealizing steps. So let $R \subseteq T \subset S$ with T being the first basic idealizer on the path from S down to R. Of course, Y is naturally a simple T-module too. So the induction hypothesis ensures that Y has an unfaithful successor Z' over T and Z is the top composition factor of Z'_R. Thus we may assume that $T = R$ and $Z' = Z$; i.e. R is a basic idealizer from S. Then the simple S-modules coincide with the simple R-modules except that one simple S-module, U say, is cut into a uniserial R-module $[VW]$ of length 2. If Z is a simple S-module then we see that Z is the successor of Y_S and the result is proven. Otherwise $Z \cong V$ and U is the successor of Y (because, by 5.3,

$\operatorname{Ext}(Y, W) = 0$ and $\operatorname{Ext}^1_R(Y, V) \cong \operatorname{Ext}^1_S(Y, U)$. Since $V \cong Z$ is unfaithful so too is U (because W always is unfaithful), and the result is proven. □

27.8. PROPOSITION. *Let $S = R(\mathcal{Z})$ where $\mathcal{Z}$ is a finite integral set of simple R-modules. Then R is an iterated basic idealizer from S and S_R is finitely generated.*

PROOF. We prove this by induction on the number of modules in $\mathcal{Z}$.

(i) Suppose $\mathcal{Z} = \{V\}$ and $S = R(V)$. Since $\mathcal{Z}$ is integral, V has an unfaithful successor W in the same tower as V (but not in $\mathcal{Z}$). This tower is thus nontrivial; so $A = \operatorname{ann}_R(W)$ is idempotent [22.1]. Hence, 14.9 shows that $O_r(A) = S$, that R is a basic idealizer from S and that S_R is finitely generated.

(ii) Suppose that $\mathcal{Z}$ contains $n > 1$ elements. Since $\mathcal{Z}$ is integral, we can choose $Y \in \mathcal{Z}$ so that Y has no predecessor in $\mathcal{Z}$. Let $\mathcal{Z}'$ denote $\mathcal{Z} - \{Y\}$ and let $T = R(\mathcal{Z}')$. Since $\mathcal{Z}'$ has fewer members that $\mathcal{Z}$, we may assume that R is an iterated basic idealizer from T. By 18.5(ii), Y is naturally a simple S-module and so a simple T-module and, by 18.5(iii), $S = T(Y)$. So the result follows from (i) provided that $\{Y\}$ is an integral set for T.

We know that Y_R must have an unfaithful successor $Z \ncong Y$, since it belongs to some nontrivial R-tower and cannot be the base. It then follows from 27.7 that Y_T has an unfaithful successor X which has Z as its top R-composition factor; so $Y \ncong X$. Thus $\{Y\}$ is indeed an integral set for T. □

27.9. THEOREM. *An overring $S = R(\mathcal{Y})$ of R is right integral over R if and only if $\mathcal{Y}$ is an integral set of simple R-modules.*

PROOF. Suppose first that the set $\mathcal{Y}$ of simple R-modules is not integral. Thus $\mathcal{Y}$ contains either some entire cycle tower $\mathcal{T}$ or else the base element V of some faithful tower. Let $\mathcal{Z} = \mathcal{T}$ or $\mathcal{Z} = \{V\}$ respectively. Since $\mathcal{Y} \supseteq \mathcal{Z}$ then $S = R(\mathcal{Y}) \supseteq R(\mathcal{Z})$; but 27.2 shows that $R(\mathcal{Z})$, and hence also S, is not integral over R.

Conversely, suppose that $\mathcal{Y}$ is an integral set of simple R-modules, and choose a finite subset $F \subseteq R(\mathcal{Y})$. We want to show that $R\langle F\rangle$ is right finite over R. Since F is a finite set, the R-submodule $(FR + R)/R \subseteq R_{\text{quo}}/R$ has finite length [12.12]. Therefore its set $\mathcal{Z}$ of composition factors is finite and, by the definition of $R(\mathcal{Y})$ [13.7], $\mathcal{Z} \subseteq \mathcal{Y}$. The same definition also ensures that $R(\mathcal{Z}) \supseteq R\langle F\rangle$. By 27.8, $R(\mathcal{Z})$ is right finite over R and the result follows. □

27.10. COROLLARY. *Let S be an overring of R; so $S = R(\mathcal{Y})$ for some set $\mathcal{Y}$. Then the following conditions are equivalent:*

(i) *S is right finite over R;*
(ii) *$\mathcal{Y}$ is a finite integral set of simple R-modules;*
(iii) *R is an iterated basic idealizer from S.*

PROOF. (i)⇒(ii) Since S is right finite over R, then $(S/R)_R$ has finite length [12.12], and hence its set $\mathcal{Z}$ of composition factors is finite. Thus 13.8(iv) shows that $\mathcal{Z} = \mathcal{Y}$. Therefore $\mathcal{Y}$ is finite; and, by 27.9, $\mathcal{Y}$ is an integral set.

(ii)⇒(iii) This is part of 27.8.

(iii)⇒(i) This is part of 6.2(i). □

The next result shows, as mentioned in 27.1, that one could define integrality of overrings in terms of the behaviour of single elements rather than finite sets.

27.11. PROPOSITION. *Let S be an overring of R with the property that $R\langle x\rangle$ is right finite for each $x \in S$. Then S is a right integral overring.*

PROOF. Suppose, to the contrary, that S is not right integral. We then know, from 27.9, that $S = R(\mathcal{Y})$ for some $\mathcal{Y}$ which is not an integral set. We will show that there is an element $x \in S$ such that $R\langle x\rangle$ is not right finite. We will consider two separate cases.

First suppose that $\mathcal{Y}$ includes the base V of some faithful R-tower. Then $R \subseteq R(V) \subseteq S$. We know, from 27.2, that $R(V)$ is not right finite. Since V is a single simple module, 13.8 shows that $R(V)$ is a minimal overring of R. Thus if we choose any $x \in R(V) - R$ then $R\langle x\rangle = R(V)$ and so is not right finite

Next, say $\mathcal{Y}$ includes a complete cycle $\mathcal{C}$. Then $R \subseteq R(\mathcal{C}) \subseteq S$. We know, again from 27.2, that $R(\mathcal{C})$ is not right finite. Choose a simple module $W \in \mathcal{C}$. Then, by 24.10, $E(W)$ is uniserial of infinite length, each proper submodule is cyclic of finite length and its composition factors consist of simple modules of $\mathcal{C}$ from W upwards, cycled in order. There is a copy of $E(W)$ in R_{quo}/R with which we identify $E(W)$. Then $E(W) \subseteq R(\mathcal{C})/R$ since all its composition factors belong to $\mathcal{C}$. Pick a cyclic submodule of $E(W)$ whose composition factors include all the members of $\mathcal{C}$ and lift its generator to an element $x \in R(\mathcal{C})$. Then $R\langle x\rangle/R \supseteq (xR + R)/R$ whose composition factors include all the members of $\mathcal{C}$. Thus $R\langle x\rangle$ kills all the simple modules in $\mathcal{C}$ and so $R(\mathcal{C}) \subseteq R\langle x\rangle$. Hence $R\langle x\rangle = R(\mathcal{C})$ which is not right finite. □

The section ends with some comments about symmetry and the lack of it.

27.12. PROPOSITION. *Let S be an overring of R such that S_R is finitely generated. Then ${}_RS$ is finitely generated if and only if no simple right R-module killed by S is faithful.*

PROOF. Since S_R is finitely generated, $(S/R)_R$ has finite length [12.13] and therefore the simple right R-modules killed by S are the composition factors of $(S/R)_R$ [13.8].

Suppose that ${}_RS$ is finitely generated. Since R_{quo} is a right quotient ring of R, we can choose a common right denominator d for some finite set of generators of ${}_RS$. Then $Sd \subseteq R$ and therefore $(S/R)d = 0$. This shows that $(S/R)_R$, and hence all of its composition factors, are unfaithful.

Conversely, suppose that all composition factors of $(S/R)_R$ are unfaithful. Then $(S/R)_R$ is annihilated by a nonzero ideal of the prime Goldie ring R, and hence by a regular element d of R. Therefore $Sd \subseteq R$ and hence $S \subseteq Rd^{-1}$. Since R is a Noetherian ring, ${}_RS$ is therefore finitely generated. □

27.13. PROPOSITION. *Let S be a right integral overring of R. Then S is left integral if and only if no simple right R-module killed by S is faithful.*

PROOF. Suppose that S kills some faithful simple right R-module W, and let T be the overring of R that kills precisely W; so $T \subseteq S$. Since S is right integral over R, the ring T is right finite over R [27.10]. Therefore, by 27.12, T is not left finite over R; and yet T is a finitely generated ring extension of R since T is right finite over R. Therefore S is not left integral over R.

Conversely, suppose that S kills no faithful simple right R-module, and let $R \subseteq T \subseteq S$ where T is a finitely generated ring extension of R. Now T_R is finitely generated since S is right integral over R. Therefore, by 27.10, the set of simple right R-modules killed by T is finite. Since it is a subset of those killed by S, it consists only of unfaithful modules; and so, by 27.12, T is left finite over R, as desired. □

28. Simple Modules Merged

In this section[1] we show that right integral overrings S of R merge segments of R-towers into simple S-modules, in the sense of 28.6, and are determined by the segments that they merge. This provides a detailed relationship between R-towers and S-towers and shows that this 'merging' property characterizes right integral overrings. The section applies this to characterize those HNP rings which can be obtained as multichain idealizers from Dedekind prime rings. It also shows that being a right integral overring is a transitive property.

Throughout this section we implicitly use the fact that an overring $S = R(\mathcal{Y})$ is right integral over R if and only if the set $\mathcal{Y}$ of simple R-modules is integral [27.9].

28.1. DEFINITIONS. Let $\mathcal{X}$ be a tower of simple R-modules. A *segment* of $\mathcal{X}$ is a sequence of members $X_1, \ldots, X_n$ of $\mathcal{X}$ such that each X_{i+1} $(i < n)$ is the successor of the simple module X_i, that is, $\operatorname{Ext}_R^1(X_i, X_{i+1}) \neq 0$, and such that $X_i \not\cong X_j$ whenever $i \neq j$. If $n \geq 2$ the segment is *nontrivial*. The final term X_n is called the *base* of the segment. Two segments of a tower are *contiguous* if the top of one is the successor of the base of the other.

Caution. Segments of towers are linearly ordered sequences, even if the tower is a cycle tower. So, before a complete cycle tower can be regarded as a segment, a choice of its base is required.

An *initial segment* of a faithful tower $\mathcal{X}$ — or of a cycle tower $\mathcal{X}$ whose base has been chosen — is a segment whose first term is the top element of $\mathcal{X}$.

It is sometimes convenient to view an integral set of simple modules as a collection of segments of towers. The following lemma explains how.

28.2. LEMMA.

(i) *Every integral set of simple R-modules is a unique union of a set of disjoint, non-contiguous segments of towers, each containing no entire cycle tower and no base of any faithful tower.*

(ii) *Conversely, the union of any set of segments of towers, with the above properties, is an integral set of simple modules.*

PROOF. This follows easily from the fact that towers are finite sets, and integral sets never contain the base of a faithful tower or an entire cycle tower [27.1.1]. □

28.3. DEFINITIONS. (i) Let $\mathcal{M}$ be a collection of disjoint segments of towers. Define $\mathcal{M}$ *deleted* — notation $\mathcal{M}^\delta$ — to be the set of (isomorphism types of) simple modules in the segments in $\mathcal{M}$, but excluding the base of each segment.

(ii) Let $\mathcal{Y}$ be an integral set of (isomorphism types of) simple R-modules; and view $\mathcal{Y}$ as a set of disjoint, non-contiguous segments of towers, as in 28.2. Define $\mathcal{Y}$ *extended* — notation $\mathcal{Y}^\epsilon$ — to be the set of disjoint segments of towers formed by adjoining to $\mathcal{Y}$ the successor of the base of each segment in $\mathcal{Y}$. (This adjunction makes sense: such successors exist because $\mathcal{Y}$ contains no entire cycle tower, and no base of a faithful tower; and the new segments remain disjoint because the disjoint segments in $\mathcal{Y}$ are non-contiguous.)

The next lemma uses implicitly the two ways 28.2 of viewing integral sets (as sets, and as unions of segments) and follows immediately from this.

[1]In this section R denotes an HNP ring unless the contrary is specified.

28.4. LEMMA.

(i) *If $\mathcal{M}$ is a collection of disjoint segments of R-towers then $\mathcal{M}^\delta$ is an integral set and $(\mathcal{M}^\delta)^\epsilon = \mathcal{M}$.*

(ii) *If $\mathcal{Y}$ is an integral set of simple right R-modules then $\mathcal{Y}^\epsilon$ is a collection of disjoint segments and $\mathcal{Y} = (\mathcal{Y}^\epsilon)^\delta$.* □

The following simple example may help in understanding the preceding lemma and definitions.

28.5. EXAMPLE. Let $\mathcal{C} = W_1, \dots, W_5$ be a cycle tower, and consider the integral set $\mathcal{Y} = \{W_1, W_3, W_5\}$. This is the union of two disjoint, non-contiguous segments, namely $\{W_5, W_1\}$ and $\{W_3\}$. Then $\mathcal{M} = \mathcal{Y}^\epsilon$ consists of the two disjoint (contiguous) segments $\{W_5, W_1, W_2\}$ and $\{W_3, W_4\}$ whose bases are W_2 and W_4; and $\mathcal{M}^\delta = \mathcal{Y}$.

The next definition and 28.7 show the significance of the definitions of $\mathcal{M}^\delta$ and $\mathcal{Y}^\epsilon$.

28.6. DEFINITION. Let $\mathcal{U}$ be a segment of some R-tower, and U_R the *associated uniserial module* ; that is, the unique uniserial module of finite length whose sequence of composition factors, from top to bottom, is $\mathcal{U}$ [19.2]. We say that an overring S of R *merges* $\mathcal{U}$ *(into the simple S-module U_S)* if U is naturally an S-module and, moreover, U_S is a simple module. Equivalently, we say that S *merges* U_R *into a simple module.*

Note, as mentioned in 28.1, that if a complete cycle tower is to be viewed as a segment and then merged, a base must be chosen first. This must be born in mind both in this definition and the following result, 28.7.

Let $\mathcal{M}$ be a set of disjoint segments of R-towers. We say that an overring S of R *merges precisely the segments in* $\mathcal{M}$ if:

(i) S merges each segment in $\mathcal{M}$;

(ii) every simple R-module not belonging to a segment in $\mathcal{M}$ is naturally a simple S-module; and

(iii) the S-towers are the same as those of R (and of the same type) except that each segment in $\mathcal{M}$ has been replaced by the merged simple S-module.

The first part of the next result shows that, if there is such a ring, then it is unique. So, when it exists (which, as will be shown in 28.13, is always the case), we will call it the *ring determined by merging the segments in* $\mathcal{M}$.

28.7. PROPOSITION. *Let $\mathcal{M}$ be a set of disjoint segments of R-towers and suppose that S merges precisely the segments of $\mathcal{M}$. Then:*

(i) *$S = R(\mathcal{M}^\delta)$;*

(ii) *S is a right integral overring of R.*

PROOF. (i) Consider any one of the disjoint segments $\mathcal{U}$ in $\mathcal{M}$ and let U_S be the corresponding merged simple S-module. Let W be the unique simple R-submodule of U, and so also the base of the segment $\mathcal{U}$. Now $U \otimes_R S \cong U$ via multiplication [13.2] and the image of $W \otimes_R S$ in U via multiplication is $WS = U$. Therefore $W \otimes_R S \neq 0$. However, since ${}_RS$ is flat [13.2], $W \otimes_R S \hookrightarrow U \otimes_R S$ which is S-simple; so $W \otimes_R S = U \otimes_R S$ and $(U/W) \otimes_R S = 0$. Using the flatness of ${}_RS$ again, we deduce easily that $V \otimes_R S = 0$ for each R-composition factor of U/W. The definition of merging precisely ensures that no simple R-modules outside the segments are killed by S. Thus $S = R(\mathcal{M}^\delta)$.

(ii) This is immediate from (i) together with 27.9. □

In the remainder of this section, it will be shown that $R(\mathcal{M}^\delta)$ does indeed merge precisely the segments in $\mathcal{M}$. Some preparatory results are required.

28.8. LEMMA. *Let $\mathcal{Y}$ be an integral set in R and let X, X' be simple right R-modules which do not belong to any tower containing elements of $\mathcal{Y}$. Let $S = R(\mathcal{Y})$. Then:*

(i) *X, X' are also simple right S-modules;*
(ii) $\mathrm{Hom}_S(X, X') = \mathrm{Hom}_R(X, X')$ *and* $\mathrm{Ext}^1_S(X, X') = \mathrm{Ext}^1_R(X, X')$*;*
(iii) *each of X and X' is faithful over S if and only if it is faithful over R.*

PROOF. (i) If $\mathcal{T}$ is a cycle tower such that $\mathcal{T} \cap \mathcal{Y} \neq \varnothing$, then each simple module in $\mathcal{T} \cap \mathcal{Y}$ has a successor in $\mathcal{T}$. The same is true if $\mathcal{T}$ is a faithful tower since the base of the tower is not a member of $\mathcal{Y}$. The uniqueness of successors [15.2] then shows that, for each $W \in \mathcal{Y}$, X is not a successor of W and so $\mathrm{Ext}^1_R(W, X) = 0$. This is precisely the condition under which 13.13 asserts that X is also a simple right S-module.

(ii) This now follows directly from 2.5 and 4.13.

(iii) This is shown by 12.1. □

We start the main results by considering the (lack of) effect of localization on certain towers.

28.9. COROLLARY. *Let $\mathcal{Y}$ be an integral set in R, let $S = R(\mathcal{Y})$ and let $\mathcal{T}$ be an R-tower with $\mathcal{Y} \cap \mathcal{T} = \varnothing$. Then:*

(i) *each simple R-module in $\mathcal{T}$ is also a simple S-module and each predecessor-successor pair in $\mathcal{T}$ remains a pair when viewed as S-modules;*
(ii) *if $\mathcal{T}$ is a cycle tower of R then it is also a cycle tower of S;*
(iii) *if $\mathcal{T}$ is a faithful tower of R then it is also a faithful tower of S.*

PROOF. (i) 28.8 applies to all the simple modules $X, X' \in \mathcal{T}$ and so gives this result.

(ii) In the case when $\mathcal{T}$ is a cycle tower over R, (i) establishes that the consecutive pairs over R are also pairs over S. The uniqueness of these relationships shows that $\mathcal{T}$ remains a cycle tower over S.

(iii) If $\mathcal{T}$ is a faithful tower over R, then the top module of $\mathcal{T}$ is a faithful simple R-module which, by the lemma, is also a faithful S-module. The linkages given by (i) then show that $\mathcal{T}$ must be a segment of the faithful S-tower which includes its top. It remains to consider the bottom, X say, of $\mathcal{T}$. Let U be any unfaithful simple right S-module. Then, by 18.5, U_R is uniserial of finite length and U_R is unfaithful. We know that $\mathrm{Ext}_R(X, W) = 0$ for every unfaithful simple right R-module W and so, in particular, for each composition factor W of U. Hence, by 50.7, $\mathrm{Ext}_R(X, U) = 0$. However, by 4.13(ii)(c), $\mathrm{Ext}_R(X, U) = \mathrm{Ext}_S(X, U)$. Thus we see that X does not have an unfaithful successor in S and so $\mathcal{T}$ is a complete faithful tower of S. □

In dealing with the effect of localization on segments we start with the simplest nontrivial case, where just one segment of length 2 is merged.

28.10. PROPOSITION. *Let $\mathcal{T}$ be a tower in R, let V, W be a segment of $\mathcal{T}$ and let S be the overring determined by killing the deleted integral set $\{V\}$. Then:*

(i) *S is a right finite minimal overring of R and R is a basic idealizer from S of type $U = [VW]$;*

(ii) *S is the overring merging precisely the segment V, W.*

PROOF. We note first that, because V, W is a segment, W must be unfaithful and cannot be isomorphic to V.

(i) Since S kills only one simple module, it is a minimal overring of R [13.8]. Let $A = \text{ann}(W)$; then, by 14.9, $S = O_r(A)$ and has the properties described.

(ii) We have seen from (i) that S has the simple module $U = [VW]$. Also, by 28.9, the R-towers other than $\mathcal{T}$ are also S-towers of the same type. The only simple S-modules not included in these towers are U together with the simple members of $\mathcal{T}$ other than V and W.

We know from 4.4 that U_R is uniserial of length 2. Choose any simple right S-module X other than U; then X is also a simple right R-module. We see from 16.4 that $\text{Ext}_R(X, U) \cong \text{Ext}_R(X, V)$ and $\text{Ext}_R(U, X) \cong \text{Ext}_R(W, X)$. Thus $\text{Ext}_R(X, U) \neq 0$ if and only if X is the predecessor of V in R; and likewise $\text{Ext}_R(U, X) \neq 0$ if and only if X is the successor of W in R. We also know from 4.13 that $\text{Ext}_S(X, U) = \text{Ext}_R(X, U)$ and $\text{Ext}_S(U, X) = \text{Ext}_R(U, X)$. We thus see that X is the predecessor of U in S if and only if it is the predecessor of V in R and is the unfaithful successor of U in S if and only if it is the unfaithful successor of W in R.

This completes the proof provided that the predecessor of V and the unfaithful successor of W in R are also simple S-modules. We know from 5.7 that $\text{Ext}(V, V) = 0$ and $\text{Ext}(W, W) = 0$. So there is only one situation which remains — when W is the predecessor of V and so V is also the successor of W: in other words, $\text{Ext}_R(W, V) \neq 0$ and V, W form a cycle of length 2. In that case, 5.7 tells us that $\text{Ext}_S(U, U) \cong \text{Ext}_R(W, V) \neq 0$. Thus U itself forms a cycle tower of length 1. This completes the proof. □

Next we consider the case when the segments involved all belong to a single tower. This means, of course, that the integral set involved is finite and so the overring S is right finite.

28.11. PROPOSITION. *Let $\mathcal{T}$ be a single tower of R, $\mathcal{M}$ be a set of disjoint segments of $\mathcal{T}$ and $\mathcal{Y} = \mathcal{M}^\delta$ the corresponding integral set. Let $S = R(\mathcal{Y})$. Then R is an iterated basic idealizer from S and S merges precisely the segments of $\mathcal{M}$.*

PROOF. We proceed by induction on the number of simple modules in $\mathcal{Y}$, noting that 28.10 establishes the result in case $\mathcal{Y}$ contains just one simple module. Let V be a simple module in $\mathcal{Y}$ chosen so that it has no (unfaithful) successor in $\mathcal{Y}$ and let $T = R(V)$. We know from 28.10 that V and its successor W are merged into a simple T-module U, that the towers of T are the same as those of R except that $\mathcal{T}$ is replaced by a new tower, $\mathcal{T}'$ say, which is the same as $\mathcal{T}$ except that V and W are replaced by U.

Let $\mathcal{Y}' = \mathcal{Y} - V$. Note that V was so chosen that $W \notin \mathcal{Y}$; so $U \notin \mathcal{Y}'$. By 28.10, $\mathcal{Y}'$ is a collection of simple T-modules and is a subset of $\mathcal{T}'$. We now check that $\mathcal{Y}'$ is an integral subset of $\mathcal{T}'$. Since $\mathcal{Y}'$ has one fewer member than $\mathcal{Y}$ and $\mathcal{T}'$ has one fewer member than $\mathcal{T}$, we deduce that $\mathcal{Y}' \neq \mathcal{T}'$. In particular, $\mathcal{Y}'$ cannot be a complete cycle tower. It remains to show that if $\mathcal{T}'$ is a faithful tower (in which case so is $\mathcal{T}$) then the base of $\mathcal{T}'$ is excluded from $\mathcal{Y}'$. However, the base of $\mathcal{T}'$ is the same as the base of $\mathcal{T}$ (and so is excluded from $\mathcal{Y}$ and hence from $\mathcal{Y}'$) unless the base of $\mathcal{T}$ is W; but in that case the base of $\mathcal{T}'$ is U which, as noted above, is not a member of $\mathcal{Y}'$.

Thus we see that $\mathcal{Y}'$ is indeed an integral tower set in $\mathcal{T}'$. We note, by 27.5, that $T(\mathcal{Y}') = S$. Since $\mathcal{Y}'$ has one fewer member than $\mathcal{Y}$, we can apply our induction hypothesis to T and its integral subset $\mathcal{Y}'$ of $\mathcal{T}'$. This shows that $S = T(\mathcal{Y}')$ has the claimed properties. $\square$

We note, for later use, a special case of 28.11.

28.12. COROLLARY. *Let $\mathcal{T}$ be a tower, $\mathcal{M}$ a segment of $\mathcal{T}$, $\mathcal{Y} = \mathcal{M}^\delta$ and $S = R(\mathcal{Y})$. Then R is a basic chain idealizer from S of type U_S where U_R is the unique uniserial R-module whose composition factors, from top to bottom, enumerate $\mathcal{M}$.*

PROOF. 28.11 shows that the segment $\mathcal{M}$ is merged into a single simple S-module, U say, that U is uniserial over R with composition factors precisely those of $\mathcal{M}$ and that R is an iterated idealizer from S. Since U is the only simple S-module which is not simple over R, we see that R is an iterated basic idealizer of type U from S. Hence, by 8.6, R is a basic chain idealizer from S. $\square$

We now turn to the general case which shows that merging is always possible.

28.13. THEOREM. *Let $\mathcal{M}$ be a set of disjoint segments of R and $\mathcal{Y}$ the integral set $\mathcal{Y} = \mathcal{M}^\delta$. Then $S = R(\mathcal{Y})$ merges precisely the segments of $\mathcal{M}$.*

PROOF. Consider one particular tower $\mathcal{T}$ of R and let $\mathcal{M}_1 = \mathcal{M} \cap \mathcal{T}$. Let $\mathcal{Y}_1 = \mathcal{M}_1^\delta$ and $\mathcal{Y}_2 = \mathcal{Y} - \mathcal{Y}_1$. By 27.6, we know that $R(\mathcal{Y}_1)(\mathcal{Y}_2) = R(\mathcal{Y}_2)(\mathcal{Y}_1)$. Now 28.11 tells us that $R(\mathcal{Y}_1)$ is precisely the overring which merges all the segments in $\mathcal{M}_1$ and thus provides a new $R(\mathcal{Y}_1)$-tower, $\mathcal{T}'$ say. It also says that all the other towers of R remain unchanged as towers of $R(\mathcal{Y}_1)$. The same argument demonstrates that $\mathcal{T}'$ remains unchanged when $R(\mathcal{Y}_1)$ is localized at $\mathcal{Y}_2$. Thus we see that S merges $\mathcal{M}_1$.

The same argument applies to every tower in R. Thus we see that S merges all of $\mathcal{M}$ and, by 28.9, S has no effect on any tower which does not meet $\mathcal{M}$. Hence S does indeed merge precisely the segments of $\mathcal{M}$. $\square$

The results of this section have thus shown:

28.14. THEOREM. *Given any set $\mathcal{M}$ of disjoint segments of R-towers, there is a ring S determined by merging precisely the segments in $\mathcal{M}$ and, furthermore, S is right integral over R. Conversely, every right integral overring S of R is so determined by some $\mathcal{M}$.*

PROOF. The first sentence is simply a combination of 28.13 and 28.7. The converse uses 27.9 to show S involves an integral set and then 28.4 to connect this with some $\mathcal{M}$. $\square$

28.15. COROLLARY. *Let $S = R(\mathcal{Y})$ be any integral overring of R. Then the S-towers are the same as the R-towers (and of the same type) except that each segment in $\mathcal{Y}^\epsilon$ is replaced by the merged simple module.*

PROOF. By 28.14, S is determined by merging the set of disjoint segments $\mathcal{Y}^\epsilon$. Consequently the towers are as described; and by 28.9, the types of the towers remain unaltered. $\square$

28.16. COROLLARY. *Let S be the overring of R determined by merging a disjoint set $\mathcal{M}$ of segments of (right) R-towers. Then S is right finite over R if and only if $\mathcal{M}$ is a finite set.*

PROOF. We have $S = R(\mathcal{Y})$ where $\mathcal{Y} = \mathcal{M}^\delta$ [28.13], and hence $\mathcal{Y}$ is an integral set of simple modules. Note that $\mathcal{M}$ is a finite set if and only if $\mathcal{Y}$ is a finite set. To complete the proof, recall that a right integral overring $R(\mathcal{Y})$ of R is right finite over R if and only if $\mathcal{Y}$ is a finite set [27.10]. □

We apply this to give a result mentioned at 23.8.

28.17. THEOREM. *Let R be an HNP. The following are equivalent:*

(i) *R has only finitely many nontrivial towers*
(ii) *R is an iterated basic idealizer from a Dedekind prime ring.*

PROOF. (i)⇒(ii) Suppose we merge all the nontrivial towers and obtain the overring S. By 28.15, S has no nontrivial towers; so, by 23.6, it is a Dedekind prime ring.

(ii)⇒(i) Let S be a Dedekind prime ring and R be an iterated basic idealizer from S. Being an overring means that $S = R(\mathcal{Y})$ for some set $\mathcal{Y}$; and 27.10 asserts that $\mathcal{Y}$ is a finite integral set. 28.15 then tells us that only finitely many segments have been merged. Therefore, since S has no nontrivial towers, R can have only finitely many. □

Another immediate consequence of 28.15 is the transitivity of integrality.

28.18. PROPOSITION. *Let T be a right integral overring of R and S one of T. Then S is a right integral overring of R.*

PROOF. Since T is a right integral overring of R, we know from 28.15 that its towers are the same as those of R except that the relevant segments have been merged into simple T-modules.

Now $S = T(\mathcal{Z})$ for some right integral set $\mathcal{Z}$ of simple T-modules; and then $S = R(\mathcal{Y})$ where $\mathcal{Y}$ is the set of R-composition factors of the members of $\mathcal{Z}$. Now the simple modules in $\mathcal{Z}$ are either simple R-modules or else merged segments of simple R-modules. Since $\mathcal{Z}$ is right integral, it does not include a complete cycle tower of T and hence $\mathcal{Y}$ does not include a complete cycle tower of R. Similarly, $\mathcal{Z}$ does not include a base of a faithful T-tower, and so $\mathcal{Y}$ does not include a base of a faithful R-tower. Thus $\mathcal{Y}$ is a right integral set in R and therefore S is a right integral overring of R. □

29. R-Simple Images

An interesting fact about integral overrings of R is that they are determined by the isomorphism classes of the simple R-modules that they map onto.[1] This fact will be important when we discuss canonical forms of infinitely generated projective R-modules [46.7]. We will use it in this section to determine, for any right integral overring S of R, which projective R-modules are S-modules [29.6].

The main result is the following theorem. It uses notation described in 28.6.

29.1. THEOREM. *Let $S = R(\mathcal{Y})$ be a right integral overring of R, and $\mathcal{M} = \mathcal{Y}^\epsilon$.*

(i) *The simple R-modules onto which S_R maps are those which either occur as the top element of a segment in $\mathcal{M}$ or do not belong to any segment in $\mathcal{M}$.*

[1] In this section R denotes an HNP ring unless the contrary is specified.

(ii) *If $S' = R(\mathcal{Y}')$ is any right integral overring of R with the same R-simple images as S_R, then $\mathcal{Y} = \mathcal{Y}'$ and $S = S'$.*

PROOF. (i) First we show that S maps onto the simple modules described. If W does not belong to any segment in $\mathcal{M}$, then W is already a simple S-module [28.15] and so $\operatorname{Hom}_R(S, W) \neq 0$. If W is the top simple in a segment in $\mathcal{M}$, then we know, from 28.13, that W is the top R-composition factor of a uniserial R-module U and so $U/\operatorname{rad}(U) \cong W$. Also U is a simple S-module, so there is an S-surjection (and hence an R-surjection) $S \twoheadrightarrow U$. Together, these yield the desired surjection $S_R \twoheadrightarrow W$.

Next we show that S does not map onto the other simple R-modules. Let W be one of these. So W belongs to some segment in $\mathcal{M}$ and, since W is not the top simple of that segment, W has a unique predecessor, V say. Since $\mathcal{M} = \mathcal{Y}^{\epsilon}$, we have $V \in \mathcal{Y}$; so $S \supseteq R(V) = T$, say. Therefore the integral overring T is a minimal right finite overring of R [13.8] and hence R is a basic idealizer of type $[VW]$ from T, say $R = \mathbb{I}_T(A)$ with A a generative isomaximal right ideal of T and $WA = 0$ [14.9].

Suppose, now, that $\operatorname{Hom}_R(S, W) \neq 0$. Then $W \cong S/M$ for some R-submodule of S. Since A_T is generative we have $TA = T$ and hence $SA = S$. But then $0 = WA = (S/M)A = S/M$, which is a contradiction.

(ii) This holds because every overring of R has the form $R(\mathcal{Y})$ for some unique $\mathcal{Y}$ [13.8]. □

29.2. COROLLARY. *Every right integral overring of R maps onto every faithful simple right R-module.*

PROOF. This holds because every faithful simple module is the top element of a tower. □

29.3. NOTATION. For any right R-module M we define $\mathcal{Z}(M)$ or $\mathcal{Z}_R(M)$ to be the set of (isomorphism classes of) simple R-modules which are not R-homomorphic images of M. In this section we use this mainly when $M = S$, an overring of R. Later sections apply it when M is a projective R-module.

Sometimes it is simpler to say which simple R-modules are *not* images of S, than to say which are. The next two corollaries give several instances of this, beginning with the fact that $\mathcal{Z}_R\big(R(\mathcal{Y})\big)$ is just a 'right shift' of $\mathcal{Y}$.

29.4. COROLLARY. *Let $S = R(\mathcal{Y})$ be a right integral overring of R. Then:*

(i) *$\mathcal{Z}_R(S) = \mathcal{Y}^{+}$, the result of replacing every element of $\mathcal{Y}$ by its unique successor (which exists because $\mathcal{Y}$ is an integral set).*

(ii) *S is right finite over R if and only if $\mathcal{Z}_R(S)$ is a finite set;*

PROOF. (i) This is a restatement of 29.1(i).

(ii) First recall that a right integral overring $R(\mathcal{Y})$ is right finite over R if and only if $\mathcal{Y}$ is a finite set [27.10]. Then apply (i), noting that the set $\mathcal{Y}^{+}$ is defined and has the same number of elements as $\mathcal{Y}$. □

29.5. COROLLARY. *Let $S = R(\mathcal{Y})$ and $S' = R(\mathcal{Y}')$ range over all right integral overrings of R. Then:*

(i) *$\mathcal{Z}_R(S)$ ranges over all sets of unfaithful simple R-modules containing no entire cycle tower;*

(ii) $\mathcal{Z}_R(S) = \mathcal{Z}_R(S') \Leftrightarrow S = S'$;
(iii) $S \supset S' \Leftrightarrow \mathcal{Z}(S) \supset \mathcal{Z}(S') \Leftrightarrow \mathcal{Y} \supset \mathcal{Y}'$.

PROOF. (i) Choose a right integral overring $S = R(\mathcal{Y})$. The members of $\mathcal{Z}_R(S)$ are all unfaithful, [29.2]; and since the integral set $\mathcal{Y}$ contains no entire cycle tower, the same is true of its right shift $\mathcal{Z}_R(S) = \mathcal{Y}^+$ [see 29.4].

Next, suppose that we have a set $\mathcal{X}$ of unfaithful simple R-modules containing no entire cycle tower. Note that, since every $W \in \mathcal{X}$ is unfaithful, it has a unique predecessor (not necessarily in $\mathcal{X}$) [15.2]. We conclude, using this, that the 'left shift' $\mathcal{Y} = \mathcal{X}^-$ (i.e. the result of replacing each element of $\mathcal{X}$ by its unique predecessor) exists, satisfies $\mathcal{Y}^+ = \mathcal{X}$, and contains no entire cycle tower. Thus, setting $S = R(\mathcal{Y})$ gives the desired right integral overring of R such that $\mathcal{Z}_R(\mathcal{Y}) = \mathcal{X}$.

(ii) This is a restatement of 29.1(ii).

(iii) We have $S \supset S' \Leftrightarrow \mathcal{Y} \supset \mathcal{Y}'$ by 13.8. Also $\mathcal{Y} \supset \mathcal{Y}' \Leftrightarrow \mathcal{Y}^+ \supset (\mathcal{Y}')^+$, which is equivalent to $\mathcal{Z}_R(S) \supset \mathcal{Z}_R(S')$ by 29.4. □

29.6. THEOREM. *Let S be a right integral overring of R and P_R a projective module. Then P is naturally a projective right S-module if and only if its R-simple images are among the R-simple images of S (i.e. $\mathcal{Z}_R(P) \supseteq \mathcal{Z}_R(S)$).*

PROOF. Suppose first that P_S is naturally a right S-module; so P is a direct summand of some free S-module, and hence the R-simple images of P are among those of S.

Conversely, suppose that the R-simple images of P are among those of S. We want to prove that P is naturally an S-module. Since P_R is projective, and hence flat, the natural map $P \to P \otimes_R R_{\text{quo}}$ is an injection, and hence we can view P as an R-submodule of the R_{quo}-module $P \otimes_R R_{\text{quo}} = PR_{\text{quo}}$. With respect to this identification, the product PT makes sense for every subring $T \subseteq R_{\text{quo}}$; and it is now sufficient to prove that $P = PS$.

We have $S = R(\mathcal{Y})$ for some integral set $\mathcal{Y}$ of simple right R-modules. Recall that the ring S is generated by the rings $R(V)$ such that $V \in \mathcal{Y}$ [13.8]. Therefore, to deduce that $P = PS$, it is enough to prove that $P = PR(V)$ for every $V \in \mathcal{Y}$. Fix V for the rest of this proof.

Since $V \in \mathcal{Y}$ and $\mathcal{Y}$ is an integral set of simple modules, V has a unique unfaithful successor W. Since $V \in \mathcal{Y}$, 29.4 tells us that $W \in \mathcal{Z}(S)$ and so, by hypothesis, P does not map onto W. Consequently, $PA = P$ where $A = \text{ann}_R(W)$.

Let $T = R(V)$. Then $R \subset T \subseteq S$ [13.8]. Since the 1-element set $\{V\}$ is an integral set of modules, 13.8 tells us that T is a minimal right finite overring of R; and since W is the unfaithful successor of V, 14.9 now tells us that $R = \mathbb{I}_T(A)$, a basic idealizer of type $U = [VW]$ where A is a generative right ideal of T and $A = \text{ann}_R(W)$. Therefore $PR(V) = PT = (PA)T = PA = P$. □

30. Dedekind Closures

In this section[1] we prove that every HNP ring R has at least one maximal right integral extension and this is always a Dedekind prime ring: we call it a 'Dedekind right closure' of R. We also show that Dedekind right closures: need not be right finite over R [30.5]; are unique if and only if all cycle towers are trivial [30.3]; and are Dedekind left closures if and only if all faithful towers are trivial [30.4]. The section ends by comparing Dedekind closures with classical maximal orders containing R.

[1]In this section R denotes an HNP ring unless the contrary is specified.

30.1. THEOREM. *The conditions below are equivalent for an overring S of R.*

(i) *S is a maximal right integral overring of R.*
(ii) *S is a Dedekind prime ring and is right integral over R.*
(iii) *$S = R(\mathcal{Y})$, where the set $\mathcal{Y}$ of simple right R-modules consists of all but the base element of every faithful R-tower, together with all but one element of every cycle R-tower.*

PROOF. (iii)$\Rightarrow$(i) The set $\mathcal{Y}$ listed in (iii) is maximal amongst integral sets of simple modules. Because it is integral, $R(\mathcal{Y})$ is right integral, by 27.9; and, because of the correspondence between overrings and sets $\mathcal{Y}$ given by 13.8, it is maximal.

(i)$\Rightarrow$(ii) By 23.6, S is a Dedekind prime ring if and only if all S-towers are trivial. However, if S had a nontrivial tower then it could be merged, thus producing a larger right integral overring and contradicting the maximality of S.

(ii)$\Rightarrow$(iii) Since S is right integral, we know that $S = R(\mathcal{Y})$ for some right integral set $\mathcal{Y}$. If that set was not as described in (iii), then the merging described by 28.15 would still leave some nontrivial towers. □

30.2. DEFINITION. An overring S of R is a *Dedekind right closure of R* if S satisfies the equivalent conditions in the preceding theorem.

Note that R usually has more than one Dedekind right closure, because of the non-uniqueness of the set $\mathcal{Y}$ in condition (iii) of 30.1. Moreover, despite the existence of both right and left closures, it is quite feasible that no right closure is a left closure. See 30.4(ii).

30.3. COROLLARY. *R has a unique Dedekind right closure if and only if R has no nontrivial cycle towers (equivalently, the ideals maximal amongst invertible ideals are maximal ideals).*

When the conditions hold, R also has a unique Dedekind left closure.

PROOF. The first assertion follows from 30.1 since there is no choice of elements of cycle towers to omit, when forming $S = R(\mathcal{Y})$ from R. The statement in parentheses holds by 23.6. Since this last condition is right-left symmetric, the final assertion holds. □

Despite the right-left symmetry of the above corollary, in the situation it describes, the Dedekind right closure is distinct from the Dedekind left closure except in the trivial case that R itself is a Dedekind prime ring; this is shown by part (ii) of the next result.

30.4. COROLLARY.

(i) *If R has no nontrivial faithful tower then the Dedekind right closures of R coincide with the Dedekind left closures.*
(ii) *If R has some nontrivial faithful tower then no Dedekind right closure is a Dedekind left closure.*

PROOF. (i) Note first that 21.2 shows that the transpose functor establishes a symmetry between right and left towers. So it will be enough to show that every Dedekind right closure is a Dedekind left closure.

Let $S = R(\mathcal{Y})$ be a Dedekind right closure of R. Then $\mathcal{Y}$ contains all but one element of every right R-tower [30.1(iii)]. Since all faithful right R-towers are trivial, we conclude that S kills no faithful simple right R-modules. Therefore the right integral overring S of R is also left integral [27.13] and hence S is a Dedekind left closure of R, using the left-hand version of 30.1(ii).

(ii) Let $S = R(\mathcal{Y})$ be a Dedekind right closure of R. We know that R has a nontrivial faithful tower; and the set $\mathcal{Y}$ must include its top. However, then 27.13 shows that S is not left integral over R. □

30.5. PROPOSITION. *The following statements are equivalent.*

(i) *Some Dedekind right closure of R is right finite over R.*
(ii) *Every Dedekind right closure of R is right finite over R.*
(iii) *R has only finitely many nontrivial towers of simple right modules.*
(iv) *R is a multichain idealizer from a Dedekind prime ring.*
(v) *The left-hand versions of the above.*

PROOF. Let S be a Dedekind right closure of R, and recall that the right integral overring S is right finite over R if and only if the number of segments of R-towers that are merged into simple S-modules is finite [28.16]. Since Dedekind prime rings have no nontrivial towers [23.6], we have proved (i)⇒(iii) and (iii)⇒(ii). (ii)⇒(i) is obvious; and 28.17 proves that (iii) and (iv) are equivalent.

Note finally that (iii) and its left-handed version are equivalent by the correspondence given in 21.2; so (v) follows. □

Although the Dedekind closures are not unique, they are, at least in the finite case, closely related.

30.6. PROPOSITION. *Let R have only finitely many nontrivial towers.*

(i) *Any two Dedekind right closures of R are Morita equivalent.*
(ii) *Any Dedekind right closure and any Dedekind left closure of R are Morita equivalent.*

PROOF. (i) Let S and T be two Dedekind right closures of R. Since they are right finite, there are regular elements $a, b \in R$ such that $aS \subseteq R$ and $bT \subseteq R$. Note that $bTaS \subseteq S$ and this is an essential right ideal of S because it contains the regular element ba. Now $TaS \cong bTaS$; so TaS is a right S-progenerator. On the other hand, $TaS \subseteq T$ and so TaS is a left T-progenerator. Hence T is Morita equivalent to S.

(ii) Let S be a Dedekind right closure and T a Dedekind left closure. Then S is right finite and T left finite. So there exist regular elements $a, b \in R$ such that $aS \subseteq R$ and $Tb \subseteq R$. Then one can argue, as in (i), that $TbaS$ is a left progenerator over T and a right progenerator over S. □

30.7. REMARK. We know, from the examples constructed in §25, that R can have infinitely many non-trivial towers, in which case it is not a multichain idealizer from a Dedekind prime ring [30.5]. We do not know if its Dedekind right, or left, closures are necessarily Morita equivalent. However, we can now show that it is always a (possibly infinite) intersection of basic chain idealizers.

30.8. THEOREM. *Let S be a Dedekind right closure of R. Then there are basic chain idealizers T_i from S, one per nontrivial R-tower, such that $R = \cap_i T_i$.*

PROOF. Let $\mathcal{Y}$ be the right integral set of simple modules such that $S = R(\mathcal{Y})$. Then $\mathcal{Y} = \cup \mathcal{Y}_i$ where each $\mathcal{Y}_i$ is an R-tower with one element deleted, its base in the case of a faithful tower [30.1]. Let $\mathcal{Y}'_i = \cup_{j \neq i} \mathcal{Y}_j$ and $T_i = R(\mathcal{Y}'_i)$. We know that $R = \cap_i T_i$ by considering the correspondence between overrings and subsets of simple R-modules [13.8]. By 28.13, T_i has just one nontrivial tower; and the simple T_i-modules forming that tower are just the simple R-modules making up the tower from which $\mathcal{Y}_i$ came. Moreover, 27.6 tells us that $S = T_i(\mathcal{Y}_i)$. Hence, by 28.12, T_i is a basic chain idealizer from S. □

The remainder of this section deals with (classical) maximal and hereditary D-orders in a simple algebra Σ, finite dimensional over its centre D_{quo} where D is a commutative Dedekind domain. These were briefly defined and discussed in 11.5 as examples of an HNP ring R. In particular, as mentioned there, $\Sigma = R_{\text{quo}}$ and every D-order in Σ is contained in a maximal D-order in Σ, a fact that we use implicitly, below.

We begin with a general lemma.

30.9. LEMMA. *Let T be a D-order in R_{quo}. Then every simple T-module is unfaithful.*

PROOF. T, being finitely generated over D, satisfies some polynomial identity [**McR 01**, 13.1.13]; and a primitive polynomial identity ring is simple Artinian, by Kaplansky's theorem [**McR 01**, 13.3.8]. □

30.10. THEOREM. *Let R be a hereditary D-order in R_{quo}.*

(i) *The maximal D-orders in R_{quo}, containing R, coincide with the Dedekind right closures of R which are thus right finite over R.*

(ii) *R has only finitely many nontrivial cycle towers and no faithful towers.*

PROOF. In this proof, 'maximal D-order' means 'maximal D-order in R_{quo} containing R'.

(i) Choose any maximal D-order, noting that it must have the form $S = R(\mathcal{Y})$ for some integral set $\mathcal{Y}$ of simple right R-modules [27.9]. In fact, since S is module finite over D, it is right finite over R, and hence the integral set $\mathcal{Y}$ is finite [27.10]. We claim that $\mathcal{Y}$ is a maximal integral set.

Otherwise there is an integral set $\mathcal{Y}'$ consisting of $\mathcal{Y}$ plus one more member. But then the ring $S' = R(\mathcal{Y}')$ is right finite over R and hence module finite over D; that is, S' is a D-order in R_{quo}. This contradicts the maximality of S and so proves the claim. Hence S is a Dedekind right closure of R.

Conversely, suppose that S'' is any Dedekind right closure of R. We know, from 11.5, that R is contained in a maximal D-order which, by definition, is right finite over D and hence over R. Thus R has one Dedekind right closure that is right finite over R. Therefore S'' is also right finite over R [30.5], and hence over D; i.e. S'' is a D-order. But then S'' is contained in a maximal D-order, say S, which as shown above is a Dedekind right closure of R. Hence $S'' = S$.

(ii) By (i), an arbitrary Dedekind right closure of R is right finite over R. Hence, by 30.5, R has only finitely many nontrivial towers none of which, by 30.9, are faithful towers. □

31. Serial Factor Rings

We have seen, in 15.6, that each proper factor ring of an HNP ring is an Artinian serial ring.[1] What type of serial rings can occur this way? We will show that they

[1]In this section R denotes an HNP ring unless the contrary is specified.

take a particular form which does not include all Artinian serial rings. We use the term *local*, in this next result, to mean that Δ/J is a division ring.

31.1. THEOREM.

(i) *Let Δ be an Artinian local principal (right and left) ideal ring, with radical J say. Let Λ be a subring of $M_n(\Delta)$ which contains $M_n(J)$ and which, modulo $M_n(J)$, is a block upper triangular matrix ring over the division ring Δ/J. Then Λ is a basic chain idealizer from $M_n(\Delta)$ and is an Artinian (right and left) serial ring.*

(ii) *Every indecomposable proper factor ring of an HNP ring R is a homomorphic image of such a ring Λ.*

PROOF. (i) By 9.1, $\Lambda/M_n(J)$ is a basic chain idealizer from $M_n(\Delta/J)$. It follows easily (essentially from 1.2) that Λ is a basic chain idealizer from $M_n(\Delta)$. Now $M_n(\Delta)$ is Artinian and serial. Hence, by 6.2, Λ is a right Artinian right serial ring; and the symmetry of the construction gives the left-handed properties.

(ii) Let B be any proper nonzero ideal of R such that R/B is an indecomposable ring. Then 15.6 tells us that R/B is an Artinian serial ring. So each indecomposable direct summand of R/B is uniserial of finite length and, moreover, since each consecutive pair of simple composition factors has a non-split extension, the composition factors all belong to a single tower of R. Fix one such tower $\mathcal{T}$ say. The sum of all submodules of R/B having all composition factors in $\mathcal{T}$ is evidently an invariant submodule; and R/B is the direct sum of these invariant submodules. Since R/B is an indecomposable ring, the R-composition factors of R/B belong to just one tower, $\mathcal{T}$ say.

Let S be the overring of R which merges $\mathcal{T}$ into a simple right S-module U say; and let $A = (S_R)^*$. Then 28.12 shows that R is a basic chain idealizer of type U from S and so A is the smallest member of the basic chain.

(a) Suppose that $\mathcal{T}$ is a faithful tower and so U is faithful. By 9.5, R/A is a basic chain idealizer from S/A and so, by 9.1, is a block upper triangular matrix ring over the division ring $\mathrm{End}_S(U)$. Therefore R/A has the form described in (i) with Λ being a division ring. Moreover, A annihilates each simple composition factor of R/B and A is idempotent; hence R/B is a homomorphic image of R/A.

(b) Next suppose that $\mathcal{T}$ is a cycle tower. Choose a base for the tower and let S be the overring obtained by merging $\mathcal{T}$ with that particular base. As in (a), R/A is a basic chain idealizer from S/A of type U say. In this case, U is a trivial cycle tower, e.g. by 28.13. So its annihilator, M say, is a maximal and invertible ideal; and M is contained in each of the isomaximal right ideals making up the chain concerned. Note that $M \subset R$ and then $BM \subseteq B \subseteq R \subseteq S$. The R-composition factors of each of the factors B/BM, R/B and S/R belong to $\mathcal{T}$ and so the S-composition factors of S/BM are all copies of U. It follows that some power of M annihilates R/B, say M^k. Now, using 1.2 again, R/M^k is a basic chain idealizer from S/M^k. However, by 15.4, S/M^k is an indecomposable Artinian principal ideal ring and so is a full matrix ring over a local such ring. Hence $R/M^k \cong \Lambda$, as described in (i); and R/B is a homomorphic image of this. $\square$

The following example demonstrates that the specific types of indecomposable Artinian serial ring described here are special.

31.2. EXAMPLE. Let

$$\Gamma = \begin{bmatrix} \mathbb{Z}/(4) & a\mathbb{Z}/(2) \\ b\mathbb{Z}/(2) & (\mathbb{Z}/(2))[x]/(x^2) \end{bmatrix}$$

in which addition is obvious. Define multiplication using the matrix structure in the obvious ways (noting that $\mathbb{Z}/(2)$ is a module over each of the rings on the diagonal, namely the top of each) and with $ab = 2 \in \mathbb{Z}/(4)$ and $ba = x \in (\mathbb{Z}/(2))[x]/(x^2)$. Thus $e_{11}a = a = ae_{22}$ and $e_{22}b = b = be_{11}$. Then Γ is an indecomposable Artinian serial ring but is not a homomorphic image of an HNP ring.

PROOF. It is a simple matter to check that $e_{11}\Gamma$ and $e_{22}\Gamma$ are uniserial modules of length 3, having simple tops, say W_1, W_2 respectively, with $W_1 \ncong W_2$. So Γ is right Artinian, right serial; and symmetry gives the left-handed properties.

There is a factor ring of Γ by a nilpotent ideal which is isomorphic to the 2 by 2 upper triangular matrix ring over $\mathbb{Z}/(2)$. Since the latter is visibly indecomposable as a ring, so too is Γ. To show that Γ is not the homomorphic image of any ring Λ as described in 31.1, we assume it is and, using the notation of 31.1 and of its proof, aim at a contradiction.

First we note that the two uniserial right ideals of Γ mentioned above involve precisely two isomorphism classes of simple composition factors, repeated in turn: W_1, W_2, W_1 and W_2, W_1, W_2 respectively. This shows us that the tower $\mathcal{T}$ involved in the preceding proof is a cycle tower of length 2. Thus R must be a basic idealizer from S. So Λ has precisely two diagonal blocks.

Note that Γ modulo its radical is the direct sum of two copies of $\mathbb{Z}/(2)$. This must be a factor ring of Λ. Thus Λ must have two maximal ideals with simple factor rings isomorphic to $\mathbb{Z}/(2)$. So the two diagonal blocks in Λ must both be 1 by 1. The two maximal ideals must be the annihilators of W_1 and W_2. So the two uniserial modules $e_{ii}\Gamma$ are isomorphic to the length 3 factors of the top and bottom rows of Λ. So, using 50.11(iv), $\Gamma \cong \Lambda/J(\Lambda)^3$. Thus

$$\Gamma \cong \begin{bmatrix} \Delta/J(\Delta)^2 & \Delta/J(\Delta) \\ J(\Delta)/J(\Delta)^2 & \Delta/J(\Delta)^2 \end{bmatrix} = \Gamma' \text{say}.$$

Given an idempotent element e in a ring T, $\operatorname{End}(eT) \cong eTe$. Hence, for each of the right ideals $e_{ii}\Gamma'$ of Γ', its endomorphism ring is isomorphic to $e_{ii}\Gamma' e_{ii} \cong \Delta/J(\Delta)^2$. However, the two right ideals $e_{ii}\Gamma$ of Γ have endomorphism rings $\mathbb{Z}/(4)$ and $(\mathbb{Z}/(2))[x]/(x^2)$ respectively. This contradiction establishes the result. $\square$

CHAPTER 6

Invariants for Finitely Generated Projective Modules

This chapter describes two independent invariants, Genus and Steinitz class, which are additive in direct sums and which, together, completely determine the isomorphism class of any finitely generated projective right R-module of udim ≥ 2.

The first section concerns the rank of a finitely generated projective module at an unfaithful simple module. This is then used in §33 to provide a generalization of the classical notion of 'genus', phrased in a way that avoids reference to classical localization since that is not available here. Then, after discussing direct-sum cancellation (§34), we proceed, in §35, to our generalizations of 'ideal class group' and 'Steinitz class' and the definitive result promised above.

32. Rank and Merging

In this section[1] we define the ranks of a finitely generated projective module at unfaithful simple modules and, in preparation for the following sections, we investigate how these react to merging.

32.1. Definitions. The notation modspec(R) denotes a set, the *module spectrum*, consisting of the zero R-module together with a set $\mathcal{W}$ of representatives of the isomorphism classes of unfaithful simple (right) R-modules. We note that, since R is an HNP ring, spec(R) comprises simply the zero ideal together with the nonzero maximal ideals. Thus replacing each $W \in \mathcal{W}$ by the nonzero maximal ideal $M = \operatorname{ann}_R(W)$, and the zero module by the zero ideal converts the module spectrum, modspec(R), into spec(R).

Let P_R be finitely generated projective. We define the *rank* of P at W — equivalently, at M — to be:

(32.1.1) $$\rho(P, W) = \rho(P, M) = \lambda(P/PM)$$

where λ denotes composition length. We define the *rank* of P at a tower $\mathcal{C}$ to be:

(32.1.2) $$\rho(P, \mathcal{C}) = \sum\{\rho(P, W) \mid W \in \mathcal{W} \cap \mathcal{C}\}.$$

Thus, if $\mathcal{T}$ is a faithful tower, $\rho(P, \mathcal{C})$ ignores the faithful module in $\mathcal{C}$. By slight abuse of notation, we also write (32.1.2) in the form

(32.1.3) $$\rho(P, \mathcal{C}) = \sum\{\rho(P, M) \mid M \in \mathcal{C}\}.$$

We note that the definition of the rank of P at W extends the definition of the rank of a ring T at W given in 9.2. Thus 20.5, 20.6 and 25.27 provide examples showing that the ranks of P at the various unfaithful members of a tower can be complicated.

[1] In this section R denotes an HNP ring unless the contrary is specified.

Since R/M is a simple Artinian ring, P/PM is a direct sum of $\rho(P, W)$ copies of W. It is worth noting explicitly the following consequence.

32.2. LEMMA. *Let W be a simple unfaithful R-module. Then $\rho(P, W)$ is the largest n such that P can be mapped onto $W^{(n)}$.* □

In view of 32.2, it is tempting to define $\rho(P, W)$, where W_R is faithful and simple, to be the maximum n such that P maps onto $W^{(n)}$. However, this maximum never exists for HNP rings; see 49.12 or the comment after 20.3.

32.3. DEFINITION. Let σ be a function from (isomorphism classes of) finitely generated projective R-modules to an abelian group G. We say that σ is *additive on direct sums,* if $\sigma(P \oplus Q) = \sigma(P) + \sigma(Q)$.

32.4. LEMMA. *Rank is additive on direct sums of modules.*

PROOF. Clear. □

32.5. LEMMA. *Let P_R be finitely generated projective, $\mathcal{C}$ a cycle of maximal ideals, and I the intersection of the maximal ideals in $\mathcal{C}$. Then $\rho(P, \mathcal{C}) = \lambda(P/PI)$.*

PROOF. Let $M_1, \ldots, M_n$ be the maximal ideals in $\mathcal{C}$. By the Chinese Remainder Theorem we have $R/I \cong \oplus_i R/M_i$ as both rings and R/I-modules. Since rank is additive in direct sums, the lemma follows from the natural identification $P/PI = P \otimes_R R/I$. □

32.6. DEFINITION. Let P_R be finitely generated projective. We say that P has *standard rank* at an unfaithful simple module W or at a cycle tower $\mathcal{C}$, respectively, if:

$$\rho(P, W) = \rho(R, W)\frac{\mathrm{udim}(P)}{\mathrm{udim}(R)} \quad \text{or} \quad \rho(P, \mathcal{C}) = \rho(R, \mathcal{C})\frac{\mathrm{udim}(P)}{\mathrm{udim}(R)}. \tag{32.6.1}$$

32.7. LEMMA. *For each unfaithful simple module W and each tower $\mathcal{C}$, having standard rank at W, or at $\mathcal{C}$, is preserved under direct sums. In particular, all free modules of finite rank have standard rank at every W and every $\mathcal{C}$.*

PROOF. Clear. □

32.8. THEOREM (Almost standard rank). *Let P be a finitely generated projective R-module. Then P has standard rank at W for almost all (i.e. for all but finitely many) isomorphism classes of unfaithful simple modules W. (We say, more briefly: 'P has almost standard rank'.)*

PROOF. We start by proving that if E, F are uniform right ideals of R then $\rho(E, M) = \rho(F, M)$ for almost all maximal ideals M.

By symmetry it suffices to prove $\rho(E, M) \geq \rho(F, M)$ for almost all M. By 12.4, every uniform right ideal of R is isomorphic to a submodule of every other uniform right ideal. Thus we may suppose that $E \subseteq F$. Then the R-module F/E has finite length [12.17]. Let M be any maximal ideal that does not annihilate any of the finitely many composition factors of F/E. (We are disregarding only a finite number of maximal ideals.) However, M annihilates all composition factors of F/FM. Hence $F/(E + FM) = 0$; i.e. $F = E + FM$.

Therefore there are maps

$$E/EM \twoheadrightarrow E/(E \cap FM) \cong (E + FM)/FM = F/FM,$$

and the inequality $\rho(E, M) \geq \rho(F, M)$ follows.

To make use of this, write P and R as direct sums of $a = \text{udim}(P)$ and $b = \text{udim}(R)$ uniform right ideals, respectively. Choose any maximal ideal M at which all of these uniform right ideals have the same rank, say r. Then $\rho(P, M) = ar$ and $\rho(R, M) = br$ by 32.4. Therefore, if W is the unfaithful simple module with $\text{ann}\, W = M$ then P has standard rank at W. Since this is true for almost all $W \in \mathcal{W}$, the result is proved. □

32.9. THEOREM (Cycle standard rank). *Each finitely generated projective module P_R has standard rank at every cycle tower.*

PROOF. Standard rank at $\mathcal{C}$ is preserved by direct sums [32.4]. Since every finitely generated projective R-module is isomorphic to a direct sum of uniform right ideals, it is sufficient to show that every uniform right ideal H has standard rank at every $\mathcal{C}$.

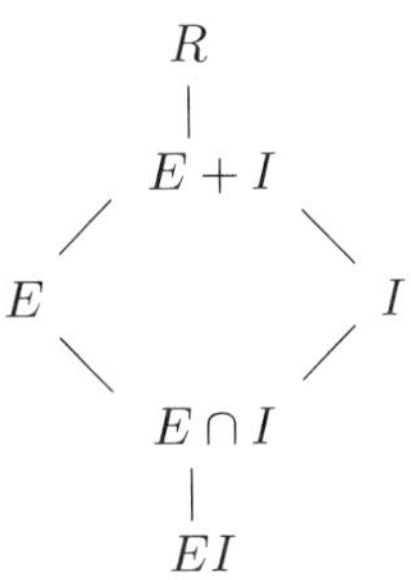

Say $\text{udim}(R) = r$; so R is a direct sum of r uniform right ideals. Since every uniform right ideal is isomorphic to a necessarily essential submodule of every other, it follows that $H^{(r)}$ is isomorphic to an essential right ideal of R. It therefore suffices to show that every essential right ideal E has standard rank at every $\mathcal{C}$.

Let I be the intersection of the maximal ideals in $\mathcal{C}$. Then I is invertible, by 22.9; and $\rho(E, \mathcal{C}) = \lambda(E/EI)$, by 32.5. In particular, $\rho(R, \mathcal{C}) = \lambda(R/I)$. Moreover, R obviously has standard rank at $\mathcal{C}$. So it is enough to show that $\lambda(E/EI) = \lambda(R/I)$.

However, since I is invertible, $\lambda(R/E) = \lambda(I/EI)$. Therefore, we see from the diagram that $\lambda\big(R/(E + I)\big) = \lambda\big((E \cap I)/EI\big)$ and hence that $\lambda(R/I) = \lambda(E/EI)$, as desired. □

32.10. COROLLARY.

(i) *Standard rank is an integer for almost all $W \in \mathcal{W}$.*
(ii) *Cycle standard rank is always an integer.*

PROOF. These are immediate consequences of 32.8 and 32.9. □

We apply these theorems to add to the description of an integral overring S of R in terms of $\mathcal{Z}_R(S)$, as defined in 29.3. Parts of the next result have counterparts in 29.5 that refer to $\mathcal{Z}_R(Q)$ where Q is any integral overring of R, and not necessarily R-projective.

32.11. LEMMA.

(i) *Let Q be a nonzero, finitely generated, projective R-module and M_R any right R-module. Then $\mathcal{Z}(M \oplus Q) \subseteq \mathcal{Z}(Q)$ and both these sets are finite and contain no cycle tower.*
(ii) *Let $\mathcal{F}$, $\mathcal{F}'$ be finite subsets of $\mathcal{W}$ which contain no cycle tower. Then there is a unique right finite overring $S(\mathcal{F})$ such that $\mathcal{Z}_R(S) = \mathcal{F}$; and $\mathcal{F}' \subset \mathcal{F} \Leftrightarrow S(\mathcal{F}') \subset S(\mathcal{F})$.*

PROOF. (i) Evidently $\mathcal{Z}(M \oplus Q) \subseteq \mathcal{Z}(Q)$ since any simple image of Q is also a simple image of $M \oplus Q$. Almost standard rank [32.8] shows that the set $\mathcal{Z}(Q)$

is finite. Cycle standard rank [32.9] shows that every cycle tower contains at least one X such that $\rho(Q, X) \neq 0$. Therefore $\mathcal{Z}(Q)$ contains no cycle tower.

(ii) This is immediate from 29.5. □

The next few results concern descending chains of three finitely generated projective modules whose composition factors reflect consecutive terms in a tower (as in (32.12.1) below). It is helpful to set up some notation.

32.12. NOTATION. Consider a nonsplit short exact sequence $W \hookrightarrow U \twoheadrightarrow V$ where W_R, V_R are simple and W is unfaithful: i.e. W is the unfaithful successor of V. Let $A = \text{ann}_R(W)$ and $S = O_r(A)$. Then 22.1(i) shows that $A = \text{ann}(W)$ is an idempotent maximal ideal of R and 14.9 shows that S is an overring of R satisfying $R = \mathbb{I}_S(A)$, a basic idealizer of type U. Moreover U_S is simple and $U_R \cong [VW]$, a uniserial R-module of length 2.

We study the existence and ranks of finitely generated projective modules P_R, P'_R, P''_R such that:

$$P \supset P' \supset P'' \quad \text{with} \quad P/P'' \cong U, \quad P/P' \cong V, \quad P'/P'' \cong W. \tag{32.12.1}$$

32.13. LEMMA. *Let $P_R \neq 0$ be finitely generated projective and X_R simple.*

(i) *If X is faithful then there exists a submodule $P' \subset P$ with $P/P' \cong X$.*

(ii) *If X is unfaithful, then there exists a submodule $P' \subset P$ with $P/P' \cong X$ if and only if $\rho(P, X) \neq 0$.*

PROOF. (i) It suffices to find a nonzero map $g\colon P \to X$, since X is simple. Choose a nonzero element $p_0 \in P$. Since P is a direct summand of a free module, there is a map $\pi\colon P \to R$ such that $\pi(p_0) \neq 0$. Since X is faithful, there exists $x_0 \in X$ such that $x_0\pi(p_0) \neq 0$. The desired g is the map $p \to x_0\pi(p)$.

(ii) This follows directly from the definitions. □

32.14. LEMMA. *Let W be the unfaithful successor of V in some R-tower.*

(i) *Let $P_R \neq 0$ be finitely generated projective and suppose that either V is faithful or $\rho(P, V) \neq 0$. Then there exist P', P'' such that* (32.12.1) *holds.*

(ii) *Let P'_R be finitely generated projective with $\rho(P', W) \neq 0$. Then there exist finitely generated projective modules P_R, P''_R such that* (32.12.1) *holds.*

PROOF. We use the notation of 32.12.

(i) The existence of a submodule P' with $P/P' \cong V$ is shown by 32.13. We now have two short exact sequences:

$$\begin{array}{ccccccccc} 0 & \to & P' & \to & P & \xrightarrow{\pi} & V & \to & 0 \\ & & & & \downarrow \beta & & \downarrow (=) & & \\ 0 & \to & W & \to & U & \xrightarrow{\alpha} & V & \to & 0 \end{array}$$

Since P is projective, there is a map $\beta : P \to U$ such that $\pi = \alpha\beta$. It follows that $\operatorname{im}\beta \not\subseteq W$ and hence $\operatorname{im}\beta = U$. So if we let $P'' = \ker(\beta)$, then $P/P'' \cong U$. Also $P'' = \ker(\beta) \subseteq \ker(\pi) = P'$. Since $P/P' \cong V$ and U_R is uniserial of length 2, then $P'/P'' \cong W$.

(ii) Consider the inclusions $P'S \supseteq P' \supseteq P'A$. Let $n = \rho(P', W) \neq 0$, so $P'/P'A \cong W^{(n)}$. Since we are working with a basic idealizer, 4.10 shows:

$$(P'S/P'A)_S \cong U^{(n)}, \quad (P'S/P')_R \cong V^{(n)} \quad \text{and} \quad (P'/P'A)_R \cong W^{(n)}$$

where the latter two isomorphisms are restrictions of the first. Thus we have an S-homomorphism $\phi\colon P'S \twoheadrightarrow U^{(n)}$ with kernel $P'A$ such that $\phi(P') = W^{(n)}$.

Let $Y = U \oplus W^{(n-1)} \subseteq U^{(n)}$ and $P = \phi^{-1}(Y)$. Then

$$P/P' \cong \phi(P)/\phi(P') = Y/W^{(n)} = (U \oplus W^{(n-1)})/W^{(n)} \cong V$$

as desired. Next let $Z = 0 \oplus W^{(n-1)} \subset Y$ and let $P'' = \phi^{-1}(Z)$. Then, as above,

$$\frac{P}{P''} \cong \frac{Y}{Z} = \frac{U \oplus W^{(n-1)}}{0 \oplus W^{(n-1)}} \cong U \quad \text{and} \quad \frac{P'}{P''} \cong \frac{W^{(n)}}{Z} = \frac{W^{(n)}}{0 \oplus W^{(n-1)}} \cong W$$

as desired. □

32.15. LEMMA. *Let W be the unfaithful successor of V in some R-tower and let $P \supset P' \supset P''$ be finitely generated projective R-modules such that $P/P'' \cong U$, $P/P' \cong V$ and $P'/P'' \cong W$, as in* (32.12.1). *Then:*

(i) $\rho(P', V) = \rho(P, V) - 1$ *if V is unfaithful;*
(ii) $\rho(P', W) = \rho(P, W) + 1$;
(iii) $\rho(P', X) = \rho(P, X)$ *for all other unfaithful simple R-modules X.*

PROOF. All undefined notation comes from 32.12.

(i) By 22.1, the nonzero maximal ideal that annihilates V, B say, is idempotent. Since $P/P' \cong V$, we have $PB \subseteq P' \subseteq P$. Hence $PB^2 = PB \subseteq P'B \subseteq PB$ and so $PB = P'B$. It follows, again since $P/P' \cong V$, that the composition length of $P'/P'B = P'/PB$ is one less than that of P/PB, as desired.

(ii) We can reason, as in (i), that since A is the nonzero, idempotent annihilator of $P'/P'' \cong W$, then $\lambda(P''/P''A) = \lambda(P''/P'A) = \lambda(P'/P'A) - 1$. In other words, if $\rho(P'', W) = k$ then $\rho(P', W) = k+1$. Thus it is sufficient to show that $\rho(P, W) = \rho(P'', W) = k$.

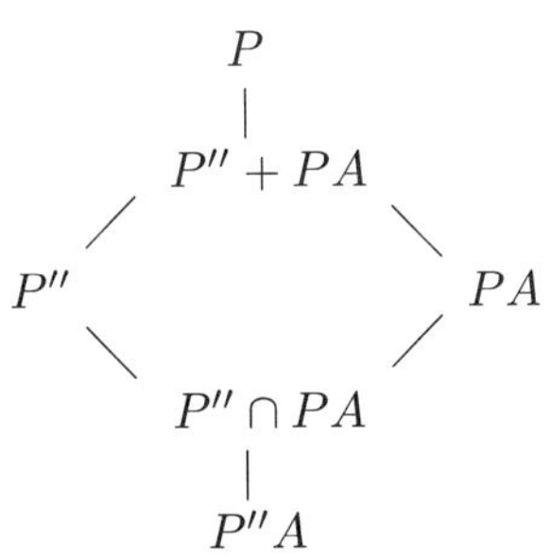

The diagram shows the submodules of P that interest us. Consider the short exact sequence

$$(W^{(k)} \cong)\ P''/P''A \hookrightarrow P/P''A \twoheadrightarrow P/P''\ (\cong U).$$

Since $\mathrm{Ext}^1_R(U, W) = 0$, by 5.8(ii), the short exact sequence splits. Hence $P/P''A \cong W^{(k)} \oplus U$. Multiplying this by A shows that $PA/P''A \cong 0 \oplus UA$. Since $R = \mathbb{I}_S(A)$ is a basic idealizer, $SA = S$. However U is an S-module, so $UA = USA = U$, which is isomorphic to P/P''. Thus $PA/P''A \cong P/P''$, and so we see from the diagram (arguing as in the last paragraph of 32.9) that $\lambda(P/PA) = \lambda(P''/P''A) = k$; that is, $\rho(P, W) = k$, as desired.

(iii) X is not the unfaithful successor of V in the tower that contains V. Therefore, by 15.1 and 15.2, $\mathrm{Ext}^1_R(V, X) = 0$. Let $N = \mathrm{ann}(X)$; then $P'/P'N \cong X^{(t)}$ where $t = \rho(P', X)$. The inclusions $P'N \subseteq P' \subset P$ show that $P/P'N$ is the middle term of some element of $\mathrm{Ext}^1_R(P/P', P'/P'N) = \mathrm{Ext}^1_R(V, X^{(t)}) = 0$. Therefore $P/P'N \cong V \oplus X^{(t)}$. Since the maximal ideal N is not the annihilator of the simple module V we have $VN = V$ and therefore

$$\frac{P}{PN} \cong \frac{P/P'N}{PN/P'N} \cong \frac{V \oplus X^{(t)}}{(V \oplus X^{(t)})N} \cong X^{(t)} \cong \frac{P'}{P'N}$$

as desired. □

It is convenient to blend facts from the preceding lemmas into a form suitable for later application.

32.16. LEMMA. *Let V_R be the predecessor of the unfaithful simple module W_R in some nontrivial tower of simple R-modules and let Q_R be nonzero, finitely generated projective.*

(i) *If $\rho(Q,W) \neq 0$, then there exists a finitely generated projective $Q'_R \supset Q$ such that $Q'/Q \cong V$, $\rho(Q',W) = \rho(Q,W) - 1$ and, if V is unfaithful, $\rho(Q',V) = \rho(Q,V) + 1$.*

(ii) *If V is faithful, or is unfaithful with $\rho(Q,V) \neq 0$, then there exists a finitely generated projective $Q' \subset Q$ such that $Q/Q' \cong V$, $\rho(Q',W) = \rho(Q,W)+1$ and, if V is unfaithful, $\rho(Q',V) = \rho(Q,V) - 1$.*

In each situation, $\rho(Q',X) = \rho(Q,X)$ for all unfaithful simple modules X_R other than V and W.

PROOF. (i) Apply 32.14 with Q in place of P', getting inclusions $Q' \supset Q \supset Q''$ such that $Q'/Q'' \cong U$, $Q'/Q \cong V$, and $Q/Q'' \cong W$ as in (32.12.1). Then apply 32.15 with P, P', P'' replaced by Q', Q, Q'' respectively.

(ii) Apply 32.14 with Q in place of P, getting inclusions $Q \supset Q' \supset Q''$ such that $Q/Q'' \cong U$, $Q/Q' \cong V$ and $Q'/Q'' \cong W$ as in (32.12.1). Then apply 32.15 with Q, Q', Q'' in place of P, P', P'' respectively. □

32.17. LEMMA. *Let S be the right finite overring of R determined by merging a segment V, W of an R-tower into a simple S-module U. Let $A = \mathrm{ann}_R(W)$, and let P be a finitely generated projective R-module.*

(i) *If Y is an S-module and $Q_R \subset PS$ with $PS/Q \cong Y$ (as R-modules) then Q is an S-submodule of PS.*

(ii) *Let $\rho_R(P,W) = t$; then $PS/PA \cong U^{(t)}$ as S-modules and $PS/P \cong V^{(t)}$ and $P/PA \cong W^{(t)}$ as R-modules.*

(iii) *The R-socle of PS/PA equals P/PA, and is isomorphic to $W^{(t)}$.*

(iv) *If U_S, or equivalently V_R, is unfaithful then $\rho_S(PS,U) = \rho_R(PS,V) = \rho_R(P,V) + \rho_R(P,W)$.*

(v) *For all unfaithful simple S-modules $X \ncong U$, $\rho_S(PS,X) = \rho_R(P,X)$.*

PROOF. Since S merges the single 2-element segment V, W, 28.10 shows that R is a basic idealizer of type U from S. Then 4.8 shows that $R = \mathbb{I}_S(A)$ and $S/A \cong U^{(n)}$ for some n.

(i), (ii) These are immediate from 4.17 and 4.10 respectively.

(iii) This follows from (ii) since the R-socle of $U^{(t)}$ is the unique R-submodule of $U^{(t)}$ that is isomorphic to $W^{(t)}$.

(iv) First we show that $\rho_S(PS,U) = \rho_R(PS,V)$. Note that $\rho_S(PS,U) = \lambda(PS/K) = a$ say, where $K = \cap\{Q \subset PS \mid PS/Q \cong U\}$. Similarly $\rho_R(PS,V) = \lambda(PS/K')$ where $K' = \cap\{Q' \subset PS \mid PS/Q' \cong V\}$. Let Q' be such that $PS/Q' \cong V\}$. By 32.14(i) applied to PS, there exists $Q'' \subset Q'$ with $PS/Q'' \cong U$. We deduce that $K \subseteq K'$. However, since $PS/K \cong U^{(a)}$, it has a submodule J/K with $PS/J \cong V^{(a)}$. So $J = K'$ and $\rho(PS,V) = a$ as claimed.

It remains to show that $\rho_R(PS,V) = \rho_R(P,V) + \rho_R(P,W)$. We know from (ii) that there is a chain of R-modules $P = P_0 \subset P_1 \subset P_2 \ldots \subset P_t = PS$ with each factor isomorphic to V where $t = \rho(P,W)$. We apply 32.15 iteratively to each pair

$P_i \subset P_{i+1}$. It asserts that $\rho(P_{i+1}, V) = \rho(P_i, V) + 1$ and $\rho(P_{i+1}, W) = \rho(P_i, W) - 1$. Thus, after the t steps involved, we get $\rho(PS, V) = \rho(P, V) + \rho(P, W)$, as required.

(v) Since S was determined by merging V, W we know [28.6] that X remains simple as an R-module. Evidently X_R is unfaithful. By 2.6(i), for every a, the S-homomorphisms of PS onto $X^{(a)}$ coincide with the R-homomorphisms of PS onto $X^{(a)}$. Therefore $\rho_S(PS, X) = \rho_R(PS, X)$ by 32.2. With the notation of (iv) above, we see from 32.15 that $\rho_R(P_{i+1}, X) = \rho_R(P_i, X)$ for each i. Hence $\rho_R(PS, X) = \rho_R(P, X)$. □

We now extend some of these results from segments of length two to segments of arbitrary length. Recall that, in any segment of a tower, all simple modules, except possibly the first, will be unfaithful.

32.18. LEMMA. *Let P be a finitely generated projective R-module with a maximal submodule P'. Suppose that P/P' is a member of a segment $\mathcal{C} = W_a, \ldots, W_b$ of some tower, and let U be the uniserial module associated with $\mathcal{C}$. Then there exist finitely generated projective R-modules P_a, P_{b+1} with $P_a \supseteq P \supset P' \supseteq P_{b+1}$ such that $P_a/P_{b+1} \cong U$.*

PROOF. By repeated use of 32.14 we obtain a chain of modules

$$P_a \supset \cdots \supset P_i \supset P_{i+1} \supset \cdots \supset P_{b+1}$$

including P and P' such that each $P_i/P_{i+1} \cong W_i$ $(a \leq i \leq b)$ and P_i/P_{i+2} is uniserial of length 2 $(a \leq i \leq b-1)$. The latter condition implies, by 16.1, that P_a/P_{b+1} is uniserial. Then 28.12 shows that $P_a/P_{b+1} \cong U$ since the composition factors of P_a/P_{b+1} enumerate $\mathcal{C}$. □

32.19. THEOREM. *Let S be an integral overring of R, P a finitely generated projective R-module, U an unfaithful simple S-module, and $W_1, \ldots, W_n$ the R-composition factors of U (and so a segment of an R-tower). Then*

$$\rho_S(PS, U) = \sum\{\rho(P, W_i) \mid 1 \leq i \leq n\}. \tag{32.19.1}$$

PROOF. (a) First suppose that $n = 1$; so U_R is also simple and unfaithful. Let $c = \rho_S(PS, U)$ and $d = \rho_R(P, U)$. We need to show that $c = d$. By 32.2, there is a surjection $P \twoheadrightarrow U^{(d)}$. Tensored with S, this gives a surjection $PS \cong P \otimes_R S \twoheadrightarrow U^{(d)}$ since $U \otimes_R S \cong U$; therefore, by 32.2 again, $c \geq d$.

Conversely, there is a surjection $PS \twoheadrightarrow U^{(c)}$ with kernel K, say. We claim that $P + K = PS$. For suppose that $PS \supset P + K$. We choose $q \in PS - (P + K)$ and note that $qR + P/P$ has finite length. (12.13 proves this for the case $P = R$; and it easily extends to a direct summand of a free R-module of finite rank.) Further, by 13.7 and 13.8, the simple R-composition factors of $(qR + P)/P$ do not include U. Hence the same is true of $(qR + P + K)/(P + K)$. However $PS/(P + K) \cong U^{(c')}$ for some $c' \leq c$. From this contradiction, we deduce that $P + K = PS$ as desired. Hence $P/(P \cap K) \cong PS/K \cong U^{(c)}$ and so $c \leq \rho_R(P, U) = d$. Thus $c = d$.

(b) Now suppose $n \geq 2$; we proceed by induction on n. Let T be the overring of R determined by merging W_{n-1} and W_n into W'_{n-1}, say. By 32.17((iv), $\rho_T(PT, W'_{n-1}) = \rho_R(P, W_{n-1}) + \rho_R(P, W_n)$. Note that each W_i with $i \geq 2$ is a simple T-module. Moreover, by part (a) of this proof, $\rho_T(PT, W_i) = \rho_R(P, W_i)$ for $i = 1, \ldots, n-2$.

Note next that $W_1, \ldots, W_{n-2}, W'_{n-1}$ are the T-composition factors of U_T. So, by induction on n, we may assume that

$$\rho_S(PS, U) = \rho_T(PT, W_1) + \cdots + \rho_T(PT, W_{n-2}) + \rho_T(PT, W'_{n-1});$$

and the result follows directly from this and the preceding two equations. □

33. Genus

This section[1] begins by defining the genus of a finitely generated projective R-module as a type of function from modspec(R) (and so implicitly from spec(R)) to the nonnegative integers. Then we give a description of those functions that occur as genera of finitely generated projective R-modules. The characterisation given is that the pair of conditions, 'almost standard rank' and 'cycle standard rank', described in §32 is necessary and sufficient.

33.1. DEFINITIONS. Let P_R be finitely generated projective. We define the *genus* of P to be the function $\Psi = \Psi(P)$ from modspec(R) to the nonnegative integers whose values are given by:

$$(33.1.1) \qquad \begin{cases} \Psi_0 = \operatorname{udim}(P) \\ \Psi_W = \rho(P, W) \quad \text{for all } W \in \mathcal{W} \end{cases}$$

and we sometimes refer to udim(P) as the *rank of P at zero.* If Q is another finitely generated module with $\Psi(Q) = \Psi(P)$, we will describe Q as being *in the genus of* P and write $Q \in \Psi(P)$.

Note. For a commutative Noetherian ring S, the genus of a finitely generated S-module P is usually defined to be the family of all finitely generated S-modules Q such that, localizing at each maximal ideal $\mathfrak{q}$ of S, $Q_\mathfrak{q} \cong P_\mathfrak{q}$. If S is a Dedekind domain and P is projective, this is easily seen to agree with the statement that $Q \in \Psi(P)$; i.e. $\Psi(Q) = \Psi(P)$.

33.2. LEMMA. *Genus is additive on direct sums of modules.*

PROOF. Clear since rank is additive [32.4]. □

33.3. COROLLARY. *If R is a Dedekind prime ring then the genus of any finitely generated projective R-module P is determined by its uniform dimension.*

PROOF. Let P_R, Q_R be finitely generated projective modules with udim(P) = udim(Q). We need to prove that $\rho(P, W) = \rho(Q, W)$ for each $W \in \mathcal{W}$. Since all towers in Dedekind prime rings are trivial [23.6], the unfaithful W is the unique member of its tower, $\mathcal{T}$ say, and $\mathcal{T}$ is a cycle tower. The desired result follows from cycle standard rank [32.9]. □

The converse, that if the genus is determined by uniform dimension, then R is a Dedekind prime ring, is clear. For if R were not a Dedekind prime ring then it would have an unfaithful simple module W in a nontrivial tower whose annihilator, M say, is a nonzero idempotent maximal ideal. Then udim(M) = udim(R) but $\Psi(M) \neq \Psi(R)$ since $\rho(M, W) = 0 \neq \rho(R, W)$.

33.4. PROPOSITION. *Let P, Q be finitely generated projective right R-modules with $\Psi(P) = \Psi(Q)$ and S be an integral extension of R. Then $\Psi(PS_S) = \Psi(QS_S)$.*

[1] In this section R denotes an HNP ring unless the contrary is specified.

PROOF. This is immediate from 32.19. □

We now aim to demonstrate what functions can occur as the genus of a finitely generated projective module.

33.5. DEFINITIONS. Let Φ be a function from $\operatorname{modspec}(R)$ to the nonnegative integers. In line with 32.6, we say Φ_W is *standard* if $\Phi_W = \rho(R, W)\cdot(\Phi_0/\operatorname{udim}(R))$. Then, motivated by 32.8 and 32.9, we define two properties of Φ. We say Φ has:

(i) *almost standard rank* if Φ_W is standard for almost all $W \in \operatorname{modspec}(R)$;
(ii) *cycle standard rank* if $\Phi_{\mathcal{C}} = \rho(R, \mathcal{C})\cdot(\Phi_0/\operatorname{udim}(R))$ for all cycle towers $\mathcal{C}$.

33.6. THEOREM (Genus Theorem). *Let Φ be a function from* $\operatorname{modspec}(R)$ *to the nonnegative integers. Then $\Phi = \Psi(P)$, the genus of some nonzero finitely generated projective R-module, if and only if $\Phi_0 > 0$ and Φ has almost standard rank and cycle standard rank.*

PROOF. The 'only if' part of the theorem is immediate from 32.8 and 32.9. The 'if' part is an immediate consequence of the following Ascent Theorem, 33.7. □

The next result is a strengthened version of the 'if' part of the Genus Theorem which involves the Dedekind closure of R, as defined in 30.2. The additional strength will become the crux of the assertion, in 35.13, that the Steinitz class and genus of P can be independently prescribed.

33.7. THEOREM (Ascent Theorem). *Let T be a Dedekind closure of R and Φ a function from* $\operatorname{modspec}(R)$ *to the nonnegative integers. Suppose that $\Phi_0 > 0$ and Φ has almost standard rank and cycle standard rank. Finally, let Y be a finitely generated projective T-module with* $\operatorname{udim}(Y) = \Phi_0$. *Then there exists a finitely generated projective R-module P such that $\Psi(P) = \Phi$ and $PT = Y$.*

PROOF. Let Q be the R-module generated by some finite set of generators of Y_T. Then Q is torsion-free and so, by 12.7, is projective. Also, $\Phi_0 = \operatorname{udim}(Q)$; and by construction $QT = Y$. By 32.8, Q has almost standard rank. Since Φ also has almost standard rank, we conclude that $\Phi_W = \rho(Q, W)$ for almost all unfaithful simple modules W. Thus Q satisfies the conditions desired except that its rank may be incorrect at finitely many W.

Let $\mathcal{C}$ be one of the finitely many R-towers containing at least one of the unfaithful simple R-modules at which some correction is needed. To prove the theorem, it is enough to show that we can change the ranks of Q arbitrarily at the elements of $\mathcal{C}$, consistent with preserving cycle standard rank if $\mathcal{C}$ is a cycle tower, without changing the rank of Q at any unfaithful simple module not in $\mathcal{C}$ and while preserving the property that $QT = Y$.

This will be achieved by making a sequence of replacements of Q by Q' which do not change any rank outside of $\mathcal{C}$ and which have $Q' \supset Q$ or $Q \supset Q'$ with Q'/Q or Q/Q', whichever is appropriate, being a simple module, W' say, killed by T; i.e. $W' \otimes_R T = 0$. Thus $Q'T = QT$.

Let $\mathcal{C} = (W_0), W_1, \ldots, W_n$, with $n \geq 1$, where the parentheses around W_0 indicate that W_0 is present if and only if $\mathcal{C}$ is a faithful tower. We repeatedly use the facts that the Dedekind right closure T of R kills all but the bottom element of every faithful R-tower, and all but one element of every cycle R-tower [30.1, 30.2]. So if $\mathcal{C}$ is a faithful tower, the simple module not killed is the base W_n of the tower;

and if $\mathcal{C}$ is a cycle tower, we can choose the numbering so that W_n is the simple not killed. Let $r_1, \ldots, r_n$ be the rank sequence of Q at $\mathcal{C}$ where $r_i = \rho(Q, W_i)$.

Choose some $i \leq n-1$. If $r_{i+1} \neq 0$, we can replace Q by a Q' such that $Q'/Q \cong W_i$, replacing the pair of ranks (r_i, r_{i+1}) by $(r_i+1, r_{i+1}-1)$ and having no other ranks altered, using 32.16(i). Moreover, since $i \neq n$, T kills W_i; and so $Q'T = QT$. Repeating this sufficiently often, we can reach the stage that $r_1 = \rho(Q, \mathcal{C})$ and $r_i = 0$ for $i > 1$.

We now consider separately two cases, starting with the case when $\mathcal{C}$ is a cycle tower; so $\rho(Q, \mathcal{C}) = \Phi_{\mathcal{C}}$. We repeatedly use 32.16(ii), each time replacing Q by Q' such that $Q/Q' \cong W_i$ $(i \leq n-1)$, replacing various pairs of ranks (r_i, r_{i+1}) by $(r_i - 1, r_{i+1} + 1)$ and thus eventually moving the appropriate ranks down to the desired positions. Since each $i \neq n$, T again kills W_i, and so each $Q'T = QT$.

Finally, consider the case of a faithful tower $\mathcal{C} = \{W_0, \ldots, W_n\}$. Faithfulness of W_0 allows us, using 32.16(i), to reduce r_1 to $r_1 - 1$ repeatedly without penalty. Thus we can reduce all these ranks to zero. Then repeated use of 32.16(ii) allows us to increase each r_i to an arbitrary value. □

We note that 34.5 shows that the module P in 33.7 is unique up to isomorphism whenever $\mathrm{udim}(P) \geq 2$.

33.8. COROLLARY. *Let $\mathcal{F}$ be a finite subset of $\mathcal{W}$ containing no complete cycle tower and, for each $W \in \mathcal{F}$, let Φ_W be a non-negative integer. Then there is a finitely generated projective right R-module H such that $\rho(H, W) = \Phi_W$ for all $W \in \mathcal{F}$.*

PROOF. We show that Φ can be extended to all of $\mathcal{W}$ so as to satisfy almost standard rank and cycle standard rank; and hence, by 33.6, to be the genus of some finitely generated projective module H_R.

First, we define $\Phi_0 = \mathrm{udim}(R) \cdot \sup\{\Phi_W \mid W \in \mathcal{F}\}$. Since this is an integer multiple of $\mathrm{udim}(R)$, it follows that the standard value of Φ_W, as defined in 33.5, is an integer for all $W \in \mathcal{W}$. This choice of Φ_0 also ensures that, for all $W \in \mathcal{F}$, Φ_W is either standard or smaller than standard.

Next, for each $W \in \mathcal{W}$ not belonging to $\mathcal{F}$ and not belonging to a cycle tower which includes a member of $\mathcal{F}$, we define Φ_W to be standard.

Finally we need to deal with those members of $\mathcal{W}$ which belong to a cycle tower which includes some member of $\mathcal{F}$. Note that, since $\mathcal{F}$ is finite, there are only finitely many such cycle towers; we choose one, $\mathcal{C}$ say. By hypothesis, there is a simple module $W_1 \in \mathcal{C}$ which does not belong to $\mathcal{F}$. For each $W \in \mathcal{C} \cap \mathcal{F}$, Φ_W is already defined. For each $W \in \mathcal{C}$ other than W_1 and those in $\mathcal{F}$, we define Φ_W to be standard. So it continues to be the case that so far all Φ_W are either standard or smaller than standard. Hence we may choose a value for Φ_{W_1} to make $\mathcal{C}$ satisfy cycle standard rank. Since this can be done for each of the finitely many cycle towers concerned, the result is proven. □

33.9. REMARK. A genus $\Psi = \Psi(P)$ involves two types of ingredients: $\Psi_0 = \mathrm{udim}(P)$ and Ψ_W for $W \in \mathcal{W}$. To what extent are these ingredients independent of each other? To answer this, we introduce the function $\Psi'(P)$ on $\mathcal{W}$ which ignores Ψ_0 and otherwise agrees with $\Psi(P)$.

33.10. THEOREM. *Let $P_R \neq 0$ be finitely generated projective.*

(i) *If R has infinitely many maximal ideals, or at least one cycle tower, then $\Psi(P)$ is determined by $\Psi'(P)$.*
(ii) *Otherwise* $\operatorname{udim}(P)$ *and* $\Psi'(P)$ *are independent invariants of the isomorphism class of P.*

PROOF. (i) If R has infinitely many maximal ideals then almost standard rank requires $\rho(P,M) = \rho(R,M)\cdot(\operatorname{udim}(P)/\operatorname{udim}(R))$ for infinitely many M, and only finitely many M for which this fails. Hence one may deduce $\operatorname{udim}(P)$ from $\Psi'(P)$.

Assume, next, that R has at least one cycle tower $\mathcal{C}$. Then cycle standard rank (32.6.1) requires that

$$\rho(P,\mathcal{C}) = \rho(R,\mathcal{C})\frac{\operatorname{udim}(P)}{\operatorname{udim}(R)}$$

which can be solved for $\operatorname{udim}(P)$, since $\rho(P,\mathcal{C})$ is determined by $\Psi'(P)$.

(ii) Now assume that R has only finitely many maximal ideals and no cycle tower. Then both of the restrictions in the Genus Theorem, almost standard rank and cycle standard rank 33.6, become vacuous. □

The next example displays a striking difference between commutative and noncommutative HNP rings.

33.11. EXAMPLE. *Suppose that R has at least one nontrivial tower and let d be any positive integer. Then R has a finitely generated projective module P with* $\operatorname{udim}(P) = d$ *such that no (finite or infinite) direct sum $P^{(\alpha)}$ of copies of P is free.*

PROOF. Choose any finitely generated projective right R-module Q such that $\operatorname{udim}(Q) = d$. Let W be any unfaithful member of a nontrivial tower, $I = \operatorname{ann}(W)$ and $P = QI$. We know [22.1] that I is an idempotent maximal ideal and so $PI = P$. Hence P is a finitely generated projective right R-module with $\operatorname{udim}(P) = \operatorname{udim}(Q)$ and $\rho(P,W) = 0$. Hence no $P^{(\alpha)}$ can be free. □

34. Cancellation

This section[1] contains the proof of the Cancellation Theorem. It starts with a result concerning the exchange of ranks between direct summands. The proof of this relies upon 49.16, the stronger form of Schanuel's Lemma.

34.1. LEMMA. *Let $P = P_1 \oplus P_2$ be a direct sum of nonzero finitely generated projective modules. Let W be an unfaithful simple module and $\mathcal{T}$ the tower containing W. Then there is a new decomposition $P \cong Q_1 \oplus Q_2$ with*

$$\rho(Q_1,W) = \rho(P_1,W) - 1 \quad \text{and} \quad \rho(Q_2,W) = \rho(P_2,W) + 1$$

and with each $\operatorname{udim} Q_i = \operatorname{udim} P_i$ *provided that:*
(i) *if $\mathcal{T}$ is a faithful tower then $\rho(P_1,W) \geq 1$;*
(ii) *if $\mathcal{T}$ is a cycle tower then $\rho(P_1,W) \geq 1$ and $\rho(P_2,W) < \rho(P_2,\mathcal{T})$.*

PROOF. First we show that P_2 maps onto some simple module V that precedes W in $\mathcal{T}$. If $\mathcal{T}$ is a faithful tower, we may take V to be the faithful top of $\mathcal{T}$, by 32.13. If $\mathcal{T}$ is a cycle tower, recall from 32.1 and 32.9 that $\rho(P,\mathcal{T}) = \Sigma\{\rho(P,W) \mid W \in \mathcal{T}\}$ and this is nonzero if $P \neq 0$. Hence the fact that $\rho(P_2,W) < \rho(P_2,\mathcal{T})$ implies that P_2 maps onto some element V of $\mathcal{T}$ other than W; and we can choose the base

[1]In this section R denotes an HNP ring unless the contrary is specified.

of $\mathcal{T}$ to make V precede W. Thus, in each case, there is such a V. Let U be the uniserial R-module associated with the segment $V, \ldots, W$ in $\mathcal{T}$.

Let $r_i = \rho(P_i, W)$. We claim that there is a finitely generated projective R-module $Q_1 \supset P_1$ and an R-homomorphism $f_1 : Q_1 \twoheadrightarrow W^{(r_1-1)} \oplus U$ such that $\operatorname{udim} Q_1 = \operatorname{udim} P_1$, $P_1 \supseteq \ker f_1$ and $f_1(P_1) = W^{(r_1-1)} \oplus W$.

We know that P_1 has a submodule, K_1 say, with $P_1/K_1 \cong W^{(r_1-1)} \oplus W$. Let $g_1 : P_1 \twoheadrightarrow W^{(r_1-1)} \oplus W$ be the canonical map and H_1, H_2 be submodules of P_1, each containing K_1, corresponding to this decomposition; so $g_1(H_1) = W^{(r_1-1)} \oplus 0$ and $g_1(H_2) = 0 \oplus W$. (Of course $H_1 + H_2 = P_1$ and $H_1 \cap H_2 = K_1$.) By 32.18, there is a finitely generated projective module, $H_3 \supset H_2$ such that $H_3/K_1 \cong U$. Let $Q_1 = P_1 + H_3$. Then $Q_1 \supset P_1$ and one can check that

$$\frac{Q_1}{K_1} = \frac{P_1 + H_3}{K_1} = \frac{H_1 + H_3}{K_1} \cong \frac{H_1}{K_1} \oplus \frac{H_3}{K_1} \cong W^{(r_1-1)} \oplus U.$$

Thus g_1 extends to an epimorphism $f_1 : Q_1 \twoheadrightarrow W^{(r_1-1)} \oplus U$ and $\ker f_1 = K_1 \subset P_1$. Also, since Q_1/P_1 has finite length, $\operatorname{udim}(Q_1) = \operatorname{udim}(P_1)$ [12.17]. This completes the proof of the claim about Q_1.

Next we show that there is an R-homomorphism $f_2 : P_2 \to W^{(r_2)} \oplus U$ and a finitely generated projective R-submodule $Q_2 \subset P_2$ with $\operatorname{udim}(Q_2) = \operatorname{udim}(P_2)$, $f_2(P_2) = W^{(r_2)} \oplus U$, $f_2(Q_2) = W^{(r_2)} \oplus W$ and $Q_2 \supseteq \ker f_2$.

Since $\rho(P_2, W) = r_2$ and $\rho(P_2, V) > 0$, there is a submodule $K_2 \subset P_2$ such that $P_2/K_2 \cong W^{(r_2)} \oplus V$. Let $g_2 : P_2 \twoheadrightarrow W^{(r_2)} \oplus V$ be the canonical map. There is an epimorphism $W^{r_2} \oplus U \twoheadrightarrow W^{(r_2)} \oplus V$. Thus g_2 lifts to a homomorphism $f_2 : P_2 \to W^{(r_2)} \oplus U$. Since the image of f_2 maps onto V it must contain the uniserial module U; so f_2 is an epimorphism. If we set Q_2 to be the inverse image in P_2 of $W^{(r_2)} \oplus W$, this has the required properties.

Let $P' = Q_1 \oplus P_2$ and $f = f_1 \oplus f_2$. By 49.16 there is an automorphism α of P' that takes $\ker f = \ker(f_1) \oplus \ker(f_2)$ onto itself and induces an automorphism β of

$$X = W^{(r_1-1)} \oplus U \oplus W^{(r_2)} \oplus U$$

that interchanges the two copies of U in X and leaves the other summands unaffected. In particular, we have $f\alpha = \beta f$. It follows that $\alpha(P_1 \oplus P_2) = Q_1 \oplus Q_2$. To see why, note that, since both $P_1 \oplus P_2$ and $Q_1 \oplus Q_2$ contain $\ker f$, it suffices to check that $\beta f(P_1 \oplus P_2) = f(Q_1 \oplus Q_2)$. However

$$f(P_1 \oplus P_2) = W^{(r_1-1)} \oplus W \oplus W^{(r_2)} \oplus U$$

and

$$f(Q_1 \oplus Q_2) = W^{(r_1-1)} \oplus U \oplus W^{(r_2)} \oplus W$$

and β interchanges the two copies of U in X; i.e. the second and fourth summands.

Finally, note that $\rho(Q_1, W) \geq r_1 - 1$ and $\rho(Q_2, W) \geq r_2 + 1$; but neither inequality can be strict because $Q_1 \oplus Q_2 \cong P_1 \oplus P_2$ and $\rho(P_1 \oplus P_2, W) = r_1 + r_2$. □

34.2. REMARK. The preceding result does not mention the ranks of the modules Q_i at other unfaithful simple modules X, although these can indeed be calculated using 32.15. In fact, modules Q_i can be chosen such that their ranks are arbitrary subject only to being consistent with almost standard rank and cycle standard rank and having $\rho(Q_1 \oplus Q_2, X) = \rho(P, X)$ for each X. This is easily seen from 33.6 and 37.1 below.

The next result concerns unique presentability, as defined in 49.7. It will be applied in the proof of the Descent Theorem.

34.3. THEOREM. *Let P be a finitely generated projective R-module such that* $\operatorname{udim} P \geq 2$, *let* $X = W^{(t)}$ *for some simple module W, and suppose there is a presentation $f\colon P \twoheadrightarrow X$. Then X is uniquely presentable by P.*

PROOF. First consider the case when W is faithful. Then given any uniform right ideal I of R there is an epimorphism of I onto X, by 32.13. Of course, P is isomorphic to the direct sum of at least two uniform right ideals. Then 49.9 shows that X is uniquely presentable by P.

Next consider the case when W is unfaithful. We write $P = P_1 \oplus \ldots \oplus P_n$ where each P_i is uniform and, by hypothesis, $n \geq 2$. Let $r_i = \rho(P_i, W)$ and $r = \sum\{r_i \mid i = 1, \ldots, n\}$. Note that the existence of f shows that $r \geq t$. By re-ordering the summands, we can arrange that the r_i come in increasing order. Further, by repeated use of 34.1, we can replace P_{n-1} and P_n by two other uniform direct summands whose ranks at W differ by no more than 1. We will suppose this to be already done.

We now consider two cases. First suppose that $r_n \leq t$. Then we can write $W^{(t)}$ in the form $W^{(t_1)} \oplus \ldots \oplus W^{(t_n)}$ with $\sum t_i = t$, $t_i \leq r_i$ for each i and $t_n = r_n$. Note that, since $t_n = r_n = \rho(P_n, W)$, we have $W^{(t_n)} \cong P_n/P_nA$ where $A = \operatorname{ann} W$. By 49.10, $W^{(t_n)}$ is uniquely presentable by P_n and then 49.8 shows that X is uniquely presentable by P.

That leaves the case when $r_n > t$ and so $r_{n-1} \geq t$. Then we can write $W^{(t)}$ in the form $W^{(t_1)} \oplus \ldots \oplus W^{(t_n)}$ with $t_{n-1} = t$ and every other $t_i = 0$. This ensures that $t_i \leq r_i$ for each i. Since 0 is uniquely presentable by P_n, we can apply 49.8 again. □

Recall that $\Psi(P)$ denotes the genus of the projective module P.

34.4. THEOREM (Descent Theorem). *Let P, Q be finitely generated projective R-modules of* $\operatorname{udim} \geq 2$, *and S be an integral overring of R. If $PS \cong QS$ and $\Psi(P) = \Psi(Q)$ then $P \cong Q$.*

PROOF. First we note that every overring T of R is a flat left and right R-module (see 13.2). Therefore (tensoring over R with) T kills an R-module X of finite length if and only if T kills all composition factors of X.

After identifying PS and QS via the given isomorphism, we may suppose that $P+Q \subseteq PS = QS$. It follows that $P, Q, P\cap Q, P+Q, PS$ all have the same uniform dimension. Hence, by 12.17, $(P+Q)/(P\cap Q)$ has finite composition length. Since $(PS/P) \otimes_R S \cong PS/PS = 0$, all composition factors of PS/P are killed by S. The same is true for all composition factors of QS/Q, and therefore the same is true for all composition factors of $(P+Q)/(P\cap Q)$. Therefore, if T is the overring which kills just these composition factors, then $R \subseteq T \subseteq S$, by 13.8. Since S is integral over R and T kills only finitely many of the simple R-modules killed by S, the ring T is a right finite overring of R, by 27.10. Since T kills all composition factors of $(P+Q)/(P\cap Q)$, T kills $(P+Q)/(P\cap Q)$, and therefore $(P\cap Q)T = (P+Q)T$, and so $PT = QT$.

This shows that, without loss of generality, we may suppose that S is a right finite overring of R. Therefore S is determined by merging a finite number of disjoint segments of R-towers (using 27.10 together with 28.16). It therefore follows that we can construct a chain of rings

$$R = S_0 \subset S_1 \subset \cdots \subset S_k = S$$

such that each ring is the overring of its predecessor determined by merging two consecutive simple modules in the tower that contains them.

We next aim to show, for each i, that if $PS_{i+1} \cong QS_{i+1}$ then $PS_i \cong QS_i$. Since we are given that $PS_k \cong QS_k$ and are aiming to prove that $PS_0 \cong QS_0$, this will complete the proof.

Note, by 33.4, that $\Psi(PS_i) = \Psi(QS_i)$. Therefore we may assume now that S is the overring of R determined by merging one segment V, W consisting of two simple R-modules of which W is necessarily unfaithful. Let U be the uniserial module associated with the segment V, W, and let $N = \operatorname{ann}(W_R)$. Since $\Psi(P) = \Psi(Q)$, we have $P/PN \cong Q/QN \cong W^{(t)}$ for some t. Hence, by 32.17(ii), $PS/PN \cong QS/QN \cong U^{(t)}$. By 34.3, the S-module presentation $0 \to PN \to PS \to U^{(t)} \to 0$ of $U^{(t)}$ by PS is unique. Yet there is a similar presentation of $U^{(t)}$ by QS; and $QS \cong PS$. Thus some isomorphism $\alpha : PS \to QS$ induces an automorphism of $U^{(t)}$, via the two presentations, and hence α induces an isomorphism $PS/PN \cong QS/QN$. However, by 32.17(iii), P and Q are, respectively, the inverse images in PS and QS of the R-socles of PS/PN and QS/QN. Hence α restricts to an isomorphism $\alpha\colon P \to Q$. □

We note one immediate consequence, an addition to the Ascent Theorem.

34.5. COROLLARY. *The finitely generated projective module P provided by* 33.7 *is unique up to isomorphism provided that its uniform dimension is at least* 2. □

The next result, at which this section has been aiming, extends to HNP rings a result known when R is a Dedekind prime ring and proved in [**McR 01**, 11.7.14(ii)] as an application of Stafford's Cancellation Theorem for Noetherian rings; see [**McR 01**, 11.7.13]. The proof uses that result.

34.6. THEOREM (Cancellation Theorem). *Let P, Q be finitely generated projective R-modules with* $\operatorname{udim} P = \operatorname{udim} Q \geq 2$ *and let X be a finitely generated projective R-module such that $P \oplus X \cong Q \oplus X$. Then $P \cong Q$.*

PROOF. Let S be a Dedekind closure of R. Now $P \oplus X \cong Q \oplus X$ implies $PS \oplus XS \cong QS \oplus XS$. Since S is a Dedekind prime ring, the result cited above shows that $PS \cong QS$. Also, since genus is an additive function to the non-negative integers and $P \oplus X \cong Q \oplus X$, we see that $\Psi(P) = \Psi(Q)$. The Descent Theorem 34.4 now shows that $P \cong Q$. □

The next example shows that this theorem does not extend to the case when $\operatorname{udim}(P) = 1$. The references in the example are to [**McR 01**].

34.7. EXAMPLE. Let A_1 be the first Weyl algebra over a field of characteristic zero. Then A_1 is a simple HNP domain [7.11.3] which has a nonzero (i.e. uniform) right ideal, P say, with $P \ncong A_1$, by [7.11.8]. However, [7.11.3, 7.11.5], any direct sum of two uniform (i.e. nonzero) right ideals is free of rank 2; so $P \oplus A_1 \cong A_1 \oplus A_1$.

We say more about this example in 35.14.

35. Ideal Class Group; Steinitz Class; Main Theorem

Every finitely generated projective R-module[1] is a direct sum of uniform right ideals [12.18]. But when are two such sums isomorphic? This question is answered by 35.13 together with 35.16. The answer involves two invariants: the genus $\Psi(P)$, which we have already studied, and the Steinitz class $\mathcal{S}(P)$ [35.12], an element of the ideal class group $G(R)$ [35.7].

Before introducing these new notions, we provide some preparatory results.

35.1. LEMMA.

(i) *Let X be an R-module, $W_1, \ldots, W_n$ non-isomorphic simple modules, and $c(1), \ldots, c(n)$ positive integers. If X can be mapped onto every $W_i^{(c(i))}$ then X can be mapped onto $\oplus_i W_i^{(c(i))}$.*

(ii) *Let U_R be a module of finite length and W_R an unfaithful simple module. Then $\rho(U,W) = \rho\big(U/\operatorname{rad}(U), W\big)$, and this equals the number of copies of W in any decomposition of $U/\operatorname{rad}(U)$ into a direct sum of simple modules.*

PROOF. (i) Let f_i be a map from X onto $W_i^{(c(i))}$ with $\ker(f_i) = K_i$ say; and let $K = \cap_i K_i$. Then $X/K \cong \oplus_i W_i^{(c(i))}$.

(ii) Recall that $\rho(U,W)$ is the largest n such that U can be mapped onto $W^{(n)}$ [32.2]. However every homomorphism from U to a semisimple module factors through the semisimple module $U/\operatorname{rad}(U)$. □

35.2. PROPOSITION. *Let P_R be finitely generated projective, let U_R have finite length, and let n be a positive integer.*

(i) *There is a surjection $P \twoheadrightarrow U$ if and only if $\rho(P,W) \geq \rho(U,W)$ for every unfaithful composition factor W of $U/\operatorname{rad}(U)$.*

(ii) *U can be generated by n elements if and only if $\rho(U,W) \leq n{\cdot}\rho(R,W)$ for every unfaithful composition factor W of $U/\operatorname{rad}(U)$.*

(iii) *If P maps onto U, then every finitely generated projective R-module in the genus of P maps onto U.*

PROOF. (i) We suppose that $\rho(P,W) \geq \rho(U,W)$ for each unfaithful composition factor W of $U/\operatorname{rad}(U)$, since the converse is trivial. Note, by 35.1(ii), that U and $U/\operatorname{rad}(U)$ have the same ranks. Moreover, since P is projective, any surjection $P \twoheadrightarrow U/\operatorname{rad}(U)$ lifts to a surjection $P \twoheadrightarrow U$. Hence we may assume that U is semisimple.

So there is a decomposition $U = (\oplus_i W_i^{c(i)}) \oplus F$, where $W_1, W_2, \ldots$ are non-isomorphic unfaithful simple modules and F is a direct sum of faithful simple modules; and, by 35.1(ii), each $c(i) = \rho(U, W_i)$. By the hypothesis on ranks, P maps onto each $W_i^{c(i)}$. Therefore, by 35.1(i), P maps onto $\oplus_i W_i^{c(i)}$ and so, because F is completely faithful, onto U [49.11].

(ii) Apply (i) with $P = R^{(n)}$.

(iii) $\rho(P,W) = \rho(P',W)$ for every P' in the genus of P and every unfaithful simple W. Now apply (i). □

35.3. LEMMA. *Let P, Q, X be nonzero finitely generated projective R-modules with $\Psi(P) = \Psi(Q) = \Psi(X)$. Then there exists $Y \in \Psi(X)$ with $P \oplus Q \cong X \oplus Y$.*

[1] In this section R denotes an HNP ring unless the contrary is specified.

PROOF. Let $d = \mathrm{udim}(Q) = \mathrm{udim}(X)$ and write each of Q and X as a direct sum of d uniform modules. Since every uniform R-module is isomorphic to a (necessarily essential) submodule of every other [12.4], we may assume that each of the d uniform direct summands of Q is an essential submodule of the corresponding summand of X. Then Q is essential in X, and therefore X/Q has finite length [12.17]. Since $\Psi(P) = \Psi(X)$, every finite-length homomorphic image of X is also a homomorphic image of P [35.2]. Therefore $X/Q \cong P/Y$ for some $Y \subseteq P$. Therefore Schanuel's lemma implies $P \oplus Q \cong X \oplus Y$, as desired. □

35.4. LEMMA. *Let P, A be nonzero finitely generated projective R-modules over a Dedekind prime ring R, with A uniform. Then $P \cong A^{(d-1)} \oplus B$ for some uniform B where $d = \mathrm{udim}(P)$.*

PROOF. Write $P = \oplus_{i=1}^{d} A_i$ with each A_i uniform. Since A, A_1, A_2 are all uniform, and hence belong to the same genus by 33.3, we have $A_1 \oplus A_2 \cong A \oplus A_2'$ for some uniform A_2', by 35.3. Similarly $A_2' \oplus A_3 \cong A \oplus A_3'$ for some A_3'. Thus $A_1 \oplus A_2 \oplus A_3 \cong A^{(2)} \oplus A_3'$. Continuing in this way completes the proof. □

We now introduce stable isomorphism and $\mathbf{K}_0(R)$, noting that, although some early definitions are valid for arbitrary rings, what ensues applies only to HNP rings.

35.5. DEFINITION. Two finitely generated projective R-modules P, Q are *stably isomorphic,* written $[P] = [Q]$, if $P \oplus X \cong Q \oplus X$ for some finitely generated projective R-module X (or, equivalently, for some free R-module X of finite rank). Thus $[P]$ denotes the *stable isomorphism class* of P.

The Cancellation Theorem 34.6 shows:

35.6. COROLLARY.

(i) *If $[P] = [Q]$ then $P \oplus X \cong Q \oplus X$ for every nonzero finitely generated projective R-module X.*

(ii) *When $\mathrm{udim}(P) \geq 2$, the stable isomorphism class $[P]$ of P coincides with the isomorphism class of P.* □

35.7. DEFINITIONS. (i) $\mathbf{K}_0(R)$ denotes the abelian group generated by the stable isomorphism classes $[P]$ of finitely generated projective modules P_R, with addition defined by $[P] + [Q] = [P \oplus Q]$. Thus every element of $\mathbf{K}_0(R)$ is a difference $[P] - [Q]$. For more information and references, see e.g. [**McR 01**, §12.1].

(ii) The genus function Ψ is additive by 33.2 and so constant on stable isomorphism classes. Hence Ψ induces a group homomorphism, which we denote by Ψ^+, from $\mathbf{K}_0(R)$ to a direct product of copies of $\mathbb{Z}$. We define $G(R) = \ker(\Psi^+)$, the set of elements of genus 0 in $\mathbf{K}_0(R)$, and call this the *ideal class group* $G(R)$. Thus

$$G(R) = \{[P] - [Q] \in \mathbf{K}_0(R) \mid \Psi(P) = \Psi(Q)\}. \tag{35.7.1}$$

[For a discussion of the relationship of this definition to earlier definitions of the ideal class group, see §36.]

(iii) *Homomorphism $\tau_{R,S}$.* If S is an overring of R, the function $P \to P \otimes_S S \to [P \otimes_R S] \in \mathbf{K}_0(S)$, for P_R finitely generated projective, is additive on direct sums, and hence induces a homomorphism $\tau = \tau_{R,S}\colon \mathbf{K}_0(R) \to \mathbf{K}_0(S)$ that occurs frequently in this section.

35.8. LEMMA. *Let S be a Dedekind right closure of R, and Q_R finitely generated projective.*

(i) *The natural map $\tau = \tau_{R,S}$ is surjective.*
(ii) *τ maps the set of stable isomorphism classes $[P]$ of modules $P \in \Psi_R(Q)$ bijectively onto the set of stable isomorphism classes $[A]$ of modules $A \in \Psi_S(Q \otimes_R S)$.*
(iii) *Every element of $G(R)$ has the form $[P] - [Q]$ for some unique $[P]$ with $P \in \Psi(Q)$.*

PROOF. (i) Since every element of $\mathbf{K}_0(S)$ has the form $[A] - [B]$ for finitely generated projective S-modules A, B, it suffices to show that every such $[A]$ is in $\operatorname{im}(\tau)$. Let P_R to be the R-module generated by some finite set of S-generators of A. Then $P \otimes_R S \cong PS = A$. Hence $\tau[P] = [A]$.

(ii) First we check the range of τ. Now $\operatorname{udim}(P) = \operatorname{udim}(Q)$, since $P \in \Psi(Q)$, and so $\operatorname{udim}(P \otimes_R S) = \operatorname{udim}(Q \otimes_R S)$. Since S is a Dedekind prime ring, this implies that $\Psi_S(P \otimes_R S) = \Psi(Q \otimes_R S)$ by 33.3. Thus $\tau[P]$ is in the desired S-genus.

Next, let A be given. By the Ascent Theorem [33.7], there is a finitely generated projective R-module P such that $PS \cong A$ and $P \in \Psi_R(Q)$. Hence τ is surjective.

Finally, suppose that $P, P' \in \Psi_R(Q)$ and $[P \otimes_R S] = [P' \otimes_R S]$. Let $A = P \otimes_R S$, $B \cong P' \otimes_R S$, and $C = U \otimes_R S$ where U is some uniform right ideal of R. Then C_S is finitely generated projective and nonzero, so 35.6 shows that $A \oplus C \cong B \oplus C$. We have $(P \oplus U) \otimes_R S \cong (P' \oplus U) \otimes_R S$ since both sides are isomorphic to $A \oplus C$. Moreover, since $\operatorname{udim}(A \oplus C) \geq 2$, the Descent Theorem 34.4 shows that $P \oplus U \cong P' \oplus U$; that is, $[P] = [P']$. Hence τ is injective.

(iii) (a) Suppose first that $R = S$. Start with an arbitrary element $[X] - [Y] \in G(R)$, necessarily with $\Psi(X) = \Psi(Y)$. By 35.4, we have $X \cong U^{(d-1)} \oplus X'$ and $Y \cong U^{(d-1)} \oplus Y'$ with U, X', Y' uniform and $d = \operatorname{udim}(X) = \operatorname{udim}(Y)$. Then, for any intcgcr $c \geq 0$, we have:

$$[X] - [Y] = [X'] - [Y'] = [U^{(c)} \oplus X'] - [U^{(c)} \oplus Y'].$$

Therefore we may assume that $\operatorname{udim}(X) = \operatorname{udim}(Y) = \operatorname{udim}(Q)$. Since R is a Dedekind prime ring, this implies $\Psi(X) = \Psi(Y) = \Psi(Q)$ [33.3], and hence $X \oplus Q \cong Y \oplus P$ for some $P \in \Psi(Q)$ [35.3]. Therefore $[X] - [Y] = [P] - [Q]$ as desired.

Any P' that works in place of P satisfies $Y \oplus P' \cong Y \oplus P$, and hence $[P'] = [P]$, proving the uniqueness of $[P]$.

(b) Suppose $R \neq S$. Start with an arbitrary $[X] - [Y] \in G(R)$. Then $\operatorname{udim}(X) = \operatorname{udim}(Y)$, and hence $\operatorname{udim}(XS) = \operatorname{udim}(YS) = \operatorname{udim}(QS)$. Since S is a Dedekind prime ring, $\Psi_S(XS) = \Psi_S(YS) = \Psi_S(QS)$, by 33.3. By (a), we have $[XS]_S - [YS]_S = [A]_S - [QS]_S$ for some $A \in \Psi_S(QS)$.

By (ii), we have $[A]_S = \tau[P]_R = [PS]$ for some $P \in \Psi_R(Q)$. The definition of equality in $\mathbf{K}_0(S)$ combined with 35.6 implies:

$$(XS \oplus QS) \oplus QS \cong (PS \oplus YS) \oplus QS.$$

Hence, by the Cancellation Theorem [34.6], $XS \oplus QS \cong PS \oplus YS$ since these modules have udim ≥ 2. Thus $\tau[X \oplus Q] = \tau[P \oplus Y]$ and so $[X \oplus Q] = [P \oplus Y]$, by (ii); equivalently, $[X] - [Y] = [P] - [Q]$, as desired.

Uniqueness of $[P]$ is proved as in Case 1. □

35.9. DEFINITION. A *base-point set* $\mathcal{B} = \mathcal{B}_R$ for R is a set consisting of exactly one module in each genus of nonzero finitely generated projective modules such that

$\mathcal{B}$ is closed under direct sums (up to isomorphism). Given P, we write $Q = \mathcal{B}(P)$ for the module $Q \in \mathcal{B}$ which is in the genus of P; i.e. $\Psi(Q) = \Psi(P)$. For all finitely generated projective $X_R, Y_R \neq 0$ we have

$$\mathcal{B}(X) \oplus \mathcal{B}(Y) \cong \mathcal{B}(X \oplus Y) \tag{35.9.1}$$

because the set $\mathcal{B}$ is closed under direct sums.

35.10. LEMMA. *Let A be a uniform right ideal of a Dedekind prime ring S. Then the set $\mathcal{B}_S = \{A^{(d)} \mid d \geq 1\}$ is a base-point set for S, and every base-point set for S has this form for some A.*

PROOF. Given A, any base-point set containing A obviously contains $\mathcal{B}_S$. Conversely, $\mathcal{B}_S$ is clearly closed under direct sums. To complete the proof, recall [33.3] that, over Dedekind prime rings, genus is determined by uniform dimension. □

35.11. LEMMA. *Let S be a Dedekind right closure of R, $\mathcal{B}_S$ be a base-point set for S (given by 35.10) and $\tau = \tau_{R,S}$. Then R has a base-point set $\mathcal{B}_R$ such that $\tau[\mathcal{B}_R] = [\mathcal{B}_S]$.*

PROOF. By 35.10, $\mathcal{B}_S = \{(A)^{(d)} \mid d \geq 1\}$ for some uniform right ideal A of S. Choose an R-genus, say $\Psi_R(Q)$. We need to define $\mathcal{B}_R(P)$ for $P \in \Psi(Q)$. By 35.8, there exists $X \in \Psi_R(Q)$, unique up to stable isomorphism, such that $\tau[X] = [A^{(d)}]$, where $d = \operatorname{udim}(Q)$. If $d = 1$, arbitrarily choose one such X to include in $\mathcal{B}_R$. If $d \geq 2$, the stable isomorphism classes of X and $A^{(d)}$ coincide with their actual isomorphism classes. Thus we choose, for $\mathcal{B}_R(Q)$, the unique X (up to isomorphism) such that $X \otimes_R S \cong A^{(d)}$.

It remains to prove that $\mathcal{B}_R$ is closed under direct sums. Any direct sum $P' \oplus P''$ of elements of $\mathcal{B}_R$ has udim ≥ 2, and hence is the unique element of its genus such that $(P' \oplus P'') \otimes_R S \cong A^{(c+d)}$ where c, d are the respective uniform dimensions of P', P''. Therefore, by the previous paragraph, $P' \oplus P'' \in \mathcal{B}_R$. □

35.12. DEFINITION. Let $\mathcal{B}$ be a base-point set for R and P a nonzero finitely generated projective R-module. We define $\mathcal{S}(P) = [P] - [\mathcal{B}(P)]$, the *Steinitz class of P with respect to $\mathcal{B}$*, and we call $\mathcal{B}$ the base-point set *associated with $\mathcal{S}$*. By (35.7.1), $\mathcal{S}(P) \in G(R)$ since $\Psi(P) = \Psi(\mathcal{B}(P))$.

Unless otherwise stated, the notation $\mathcal{S}(-)$ denotes Steinitz class with respect to some fixed (but unspecified) base-point set for R. As usual, we use the notation $\mathcal{S}_R(P)$ in place of $\mathcal{S}(P)$ if we are dealing with more than one ring.

Note that $\mathcal{S}(P)$ can change if the associated base-point set changes. For an example of this, choose any Dedekind prime ring R with at least two uniform right ideals U_0 and U_0' such that $[U_0] \neq [U_0']$. (For a noncommutative such R, one can choose the 2×2 matrix ring over any commutative Dedekind domain that is not a principal ideal domain.) By 35.10, there are base-point sets $\mathcal{B}, \mathcal{B}'$ that contain U_0, U_0' respectively. Then $\mathcal{S}(U_0) = 0$ with respect to $\mathcal{B}$ but not with respect to $\mathcal{B}'$.

35.13. THEOREM (Main Structure Theorem).

(i) *'Steinitz class' and 'genus' are independent invariants of nonzero finitely generated projective R-modules.*
(ii) *These invariants are additive in direct sums.*
(iii) *Together, they determine the stable isomorphism class of the module and, if it has uniform dimension 2 or more, its isomorphism class.*

PROOF. Let $\mathcal{B}$ be the base-point set associated with $\mathcal{S}$.

(i) We need to show that for every $\sigma \in G(R)$ and every nonzero genus ψ, there is a finitely generated projective module P_R such that $\mathcal{S}(P) = \sigma$ and $\Psi(P) = \psi$. Let $Q \in \mathcal{B}$ be the element of genus ψ. By 35.8, $\sigma = [P] - [Q]$ for some $P \in \Psi(Q)$. However, since $P \in \Psi(Q)$ we have $\mathcal{B}(P) = Q$ and so $\sigma = [P] - [\mathcal{B}(P)] = \mathcal{S}(P)$, as desired.

(ii) By 33.2, we know that genus is additive. Let $P = X \oplus Y$. We want to prove that $\mathcal{S}(P) = \mathcal{S}(X) + \mathcal{S}(Y)$. Evidently $[P] = [X] + [Y]$; and $\mathcal{B}(P) \cong \mathcal{B}(X) \oplus \mathcal{B}(Y)$, as was observed in (35.9.1). Hence $\mathcal{S}(P) = [P] - [\mathcal{B}(P)] = \mathcal{S}(X) + \mathcal{S}(Y)$.

(iii) To prove uniqueness up to stable isomorphism, suppose that $\sigma = [P] - [\mathcal{B}(P)] = [P'] - [\mathcal{B}(P')]$ with both P and P' in $\Psi(Q)$. As above, we have $\mathcal{B}(P) = \mathcal{B}(P') = Q$, and so $[P] = [P']$. Finally, if $\operatorname{udim}(P) \geq 2$ then, by 35.6, the isomorphism class of P coincides with its stable isomorphism class. □

The restriction in (iii) to the case when $\operatorname{udim}(P) \geq 2$ is necessary since the result fails with a vengeance when $\operatorname{udim}(P) = 1$.

35.14. EXAMPLE. Let A_1 be the first Weyl algebra. As noted in 34.7, A_1 is a simple noncommutative Dedekind domain; and the direct sum of any two uniform right ideals is free. The former fact tells us that all uniform right ideals share the same genus; and the latter implies that the ideal class group of A_1 is trivial. However, it is known (e.g. see [**Canning-Holland 94**]) that A_1 has infinitely many isomorphism classes of right ideals.

The next result is basically a combination of 35.13 and 33.6 in a form useful in later applications.

35.15. COROLLARY. *Let $Q_R \neq 0$ be finitely generated projective, and let Φ be any function from* $\operatorname{modspec}(R)$ *to the nonnegative integers which is equal to $\Psi(Q)$ except for finitely many $W \in \mathcal{W}$ and which satisfies cycle standard rank. Then there is a finitely generated projective P_R such that $\Psi(P) = \Phi$. Further, the Steinitz class of P can be arbitrarily prescribed.*

PROOF. According to 33.6, Φ is the genus of some finitely generated projective nonzero module if and only if it satisfies almost standard rank and cycle standard rank. In particular $\Psi(Q)$ satisfies these properties; and evidently the same is true of Φ. The final statement of the result holds because the genus and Steinitz class can be independently prescribed. □

The next corollary answers the question of when two direct sums are isomorphic, provided that at least one of the sums contains two or more nonzero terms. It follows immediately from 35.13.

35.16. COROLLARY. *Let $M = \oplus_{i=1}^m M_i$ and $N = \oplus_{i=1}^n N_i$ be finitely generated projective R-modules with* $\operatorname{udim}(M) \geq 2$. *Then*

$$M \cong N \iff \sum_i \mathcal{S}(M_i) = \sum_i \mathcal{S}(N_i) \text{ and } \sum_i \Psi(M_i) = \sum_i \Psi(N_i). \quad \square$$

In the next result, Ψ^+ is the homomorphism defined on $\mathbf{K}_0(R)$ that is induced by the genus function Ψ, as described in 35.7(ii).

35.17. COROLLARY. *$G(R)$ is a direct summand of $\mathbf{K}_0(R)$. Indeed $\mathbf{K}_0(R) \cong G(R) \oplus \operatorname{im}(\Psi^+)$.*

PROOF. Since Steinitz class is additive on direct sums [35.13], it induces a homomorphism of abelian groups $\mathcal{S}'\colon \mathbf{K}_0(R) \to G(R)$. Since every element of $G(R)$ is the Steinitz class of some module [35.13 again], the map $\mathcal{S}'$ is surjective.

To complete the proof that $G(R)$ is a direct summand of $\mathbf{K}_0(R)$, it is sufficient to verify that $\mathcal{S}'$ equals the identity on $G(R)$. Take an arbitrary $\gamma \in G(R)$. Then $\gamma = [P] - [Q]$ with $\Psi(P) = \Psi(Q)$. So

$$\mathcal{S}'(\gamma) = \mathcal{S}(P) - \mathcal{S}(Q) = [P] - [\mathcal{B}(P)] - \big([Q] - [\mathcal{B}(Q)]\big).$$

However $\mathcal{B}(P) = \mathcal{B}(Q)$ since $\Psi(P) = \Psi(Q)$; and so $\mathcal{S}'(\gamma) = [P] - [Q] = \gamma$ as required.

Finally, since $\mathcal{S}'$ is surjective and $G(R) = \ker(\Psi^+)$, every complementary summand of $G(R)$ in $\mathbf{K}_0(R)$ is isomorphic to $\operatorname{im}(\Psi^+)$. □

The complement of $G(R)$ in $\mathbf{K}_0(R)$ is easy to describe in the case when R is a Dedekind prime ring.

35.18. COROLLARY. *If R is a Dedekind prime ring then $\mathbf{K}_0(R) \cong \mathbb{Z} \oplus G(R)$.*

PROOF. Choose a uniform right ideal U in R. The genus of any finitely generated projective R-module P is determined by its uniform dimension, by 33.3. Hence $\Psi(P) = \operatorname{udim}(P) \cdot \Psi(U)$. Thus $\operatorname{im}(\Psi^+) \cong \mathbb{Z}$ and then 35.17 gives the result. □

35.19. THEOREM. *Let S be a Dedekind closure of R. Then $G(R) \cong G(S)$ via the natural map $\tau = \tau_{R,S}$.*

PROOF. *Range.* Every element of $G(R)$ has the form $[P] - [Q]$ where $\Psi_R(P) = \Psi_R(Q)$. Since the function $- \otimes_R S$ preserves udim, and since udim determines the genus in Dedekind prime rings [33.3], we see that $\Psi_S(P \otimes_R S) = \Psi_S(Q \otimes_R S)$ and therefore $\tau[P] - \tau[Q] \in G(S)$.

Bijection. Fix a uniform right ideal Q of R. Recall that every element $\gamma \in G(R)$ has the form $\gamma = [P] - [Q]$ for some unique $[P]$ with $P \in \Psi_R(Q)$ [35.8(iii)]. Similarly, every element of $G(S)$ has the form $\delta = [A] - [Q \otimes_R S]$ with $A \in \Psi_S(Q \otimes_R S)$. However, 35.8(ii) shows that τ maps $[\Psi_R(Q)]$ bijectively onto $[\Psi_S(Q \otimes_R S)]$. □

Next we show that the isomorphism in 35.19 preserves Steinitz classes provided the base-point sets match up.

35.20. COROLLARY. *Let S be a Dedekind right closure of R and let $\mathcal{B}_R, \mathcal{B}_S$ be base-point sets such that $\tau_{R,S}(\mathcal{B}_R) = \mathcal{B}_S$, as in 35.11. Then, for every finitely generated projective $P_R \neq 0$, we have $\tau_{R,S}\big(\mathcal{S}_R(P)\big) = \mathcal{S}_S(P \otimes_S S)$.*

PROOF. Let $Q = \mathcal{B}_R(P)$. The consistency hypothesis on $\mathcal{B}_R, \mathcal{B}_S$ implies that $Q \otimes_R S \in \mathcal{B}_S$. It is now routine to verify

$$\tau\big(\mathcal{S}_R(P)\big) = \tau[P] - \tau[\mathcal{B}_R(P)] = [P \otimes_R S] - [Q \otimes_R S] = \mathcal{S}_S(P \otimes_R S)$$

completing the proof. □

Before our next corollary, we prove an amusing lemma.

35.21. LEMMA. *Let $(G, +)$ be an abelian group, fix an element $\alpha_0 \in G$, and define a new addition in G by $\alpha \boxplus \beta = \alpha + \beta + \alpha_0$. Then $(G, \boxplus) \cong (G, +)$ via the bijection $\sigma(\alpha) = \alpha + \alpha_0$. In particular, $(G, \boxplus)$ is a group with zero element $-\alpha_0$.*

PROOF. We have $\sigma(\alpha \boxplus \beta) = \sigma(\alpha+\beta+\alpha_0) = \alpha+\beta+2\alpha_0 = \sigma(\alpha)+\sigma(\beta)$, showing that the bijection σ respects the two additions. Similarly $-\alpha_0 \boxplus \beta = \beta$ for all β. □

35.22. COROLLARY. *Let P_0 be a nonzero, finitely generated projective R-module. Then the set* $\operatorname{gcg}(P_0)$ *of stable isomorphism classes of modules in $\Psi(P_0)$ becomes an abelian group with zero element $[P_0]$ if we define addition $\boxplus$ in* $\operatorname{gcg}(P_0)$ *by:*

$$[X] = [P] \boxplus [Q] \quad \Leftrightarrow \quad X \oplus P_0 \cong P \oplus Q. \tag{35.22.1}$$

Moreover $\operatorname{gcg}(P_0) \cong G(R)$*; and we call* $\operatorname{gcg}(P_0)$ *the* genus class group *of P_0.*

PROOF. Let $G = G(R)$. By 35.13, for each $\alpha \in G$ there exists $P(\alpha) \in \Psi(P_0)$, unique up to stable isomorphism, such that $\mathcal{S}\big(P(\alpha)\big) = \alpha$. This puts the elements of the set G into (1,1)-correspondence with the elements of the set $\operatorname{gcg}(P_0)$. Choose $\alpha_0 \in G$ such that $[P(-\alpha_0)] = [P_0]$. By 35.16, for all $\alpha, \beta \in G$,

$$P(\alpha+\beta+\alpha_0) \oplus P(-\alpha_0) \cong P(\alpha) \oplus P(\beta).$$

We can rewrite this in the notation of 35.21:

$$P(\alpha \boxplus \beta) \oplus P_0 \cong P(\alpha) \oplus P(\beta). \tag{35.22.2}$$

Thus, defining

$$[P(\alpha)] \boxplus [P(\beta)] = [P(\alpha \boxplus \beta)] \tag{35.22.3}$$

makes the set $[\Psi(P_0)]$ into an abelian group $\operatorname{gcg}(P_0) \cong (G, \boxplus)$; and $(G, \boxplus) \cong G(R)$ via the isomorphism σ in (35.21).

Since $P(\alpha \boxplus \beta)$ is determined up to stable isomorphism by (35.22.2), this addition in the group $\operatorname{gcg}(P_0)$ can be described as in in (35.22.1). □

36. Notes on Chapters 5 and 6

The major part of these two chapters[1] comes, albeit in a refined form, from [**Levy-Robson 99a**] and [**Levy-Robson 99b**] respectively, the first main aim being to obtain the cancellation result appearing in 34.6. This is effected via a reduction to the case of a Dedekind prime ring, for which cancellation was proved in [**Levy-Robson-Stafford 94**] (or see [**McR 01**, 11.7.14]), where it is obtained as a consequence of Stafford's work on cancellation in Noetherian rings; see [**Stafford 81**] or [**McR 01**, Chapter 11].

This reduction involves the notion of a Dedekind right closure of an HNP ring R , which we discuss in §30. This originated in [**Hodges 87**], where it was shown that such a (non-unique) closure always exists. However, if, unlike the situation dealt with by [**Eisenbud-Robson 70b**], R has infinitely many idempotent maximal ideals, these Dedekind closures are not finitely generated modules.

The remainder of Chapter 5 involves a careful analysis of the behaviour of simple modules when moving between the given ring and a Dedekind closure. This was the main thrust of [**Levy-Robson 99a**], where it was tied in with the notions of integral extensions and of the merging of consecutive terms in a tower. (We note that in [**Levy-Robson 99a**], the phrase 'integral over R' generally means what is here termed 'right integral over R'.)

The description, in 31.1, of the serial rings that occur as proper homomorphic images of an HNP ring, which adds fine detail to the result [**Eisenbud-Griffith 71**]

[1] In this section R denotes an HNP ring unless the contrary is specified.

that the factor rings are serial, was first proved in [**Robson 72**]. The example [31.2] showing that these are indeed a special class of serial ring comes from [**Puninski 01**, 8.13].

Then Chapter 6 concentrates on the study of finitely generated projective modules. The idea that one could define the genus of such a module without, as in the classical case, using localization at a prime ideal (this not being available in the general case) appeared in [**Warfield 80**] and [**Stafford 81**] where they used the notion of the rank at each prime ideal, as in 33.1 here. The remainder of the arguments leading up to the Cancellation Theorem 34.6 come from [**Levy-Robson 99b**]. However, almost standard rank was demonstrated in [**Warfield 80**] and cycle standard rank in [**Goodearl-Warfield 79**]. The example [34.7] showing that cancellation fails for uniform dimension 1 comes from [**Webber 70**].

The history of the ideal class group (or genus class group) is more complex. When R is a commutative Dedekind domain, the genus class group $\mathrm{gcg}(R)$ [35.22] is what is classically called the 'ideal class group'. For the hereditary orders studied in integral representation theory — HNP rings that are module-finite over a central Dedekind domain — $\mathrm{gcg}(R)$ is called the 'locally free class group'. Following [**Hodges 87**, 5.2], our treatment shows that $\mathrm{gcg}(R) \cong G(R)$; indeed $\mathrm{gcg}(P) = G(R)$ for every nonzero finitely generated projective P_R. The isomorphism $G(R) \cong \mathrm{gcg}(R)$ is the reason that we call $G(R)$ the 'ideal class group'. The isomorphism $G(R) \cong G(S)$, where S is any Dedekind closure of R was proved in the context of integral representation theory by Jacobinski and Reiner [**Reiner 75**, 40.16], and extended to arbitrary HNP rings in [**Hodges 87**, 5.13].

The decomposition $\mathbf{K}_0(R) \cong G(R) \oplus \mathrm{im}(\Psi^+)$ in 35.17 is due to [**Hodges 87**, §5]. The Main Theorem 35.13 and 35.16 come from [**Levy-Robson 99b**, §4]; but in the present treatment, $G(R)$ is a subgroup of $\mathbf{K}_0(R)$ rather than $\mathbf{K}_0(S)$ for a Dedekind closure S of R. In the context of integral representation theory, an equivalent result, but with a considerably different statement, was proved in [**Jacobinski 71**].

CHAPTER 7

Applications of Invariants

This chapter contains a potpourri of applications of the structure theory of finitely generated projective modules P over an HNP ring R. In §37 we show how to recognize from their invariants when one is isomorphic to a proper direct summand of another. Then, in §38, we examine how many generators are needed for P, especially if P is a right ideal. §39 looks at conditions on R which ensure that every such P decomposes uniquely into a direct sum of uniform right ideals; and §40 describes special properties which occur when R is semilocal. In §41, we use the structure of projective modules to complete what we know about the structure of modules of finite length. Then, in turn, we use this in §42 to obtain what we know about simultaneous (stacked) decompositions of a projective module and a submodule. Finally, §43 surveys, without proofs, the enormous contrast between finitely generated and infinitely generated torsion modules.

37. Direct Summands

Let P_R, X_R be finitely generated projective.[1] When is P isomorphic to a direct summand of X? To make the question nontrivial, we assume that $\operatorname{udim}(P) < \operatorname{udim}(X)$. It is well-known in the commutative case that this condition is also sufficient; and this remains true for Dedekind prime rings [37.2]. But for arbitrary HNP rings a further hypothesis is needed.

Given two genera $\Psi(P), \Psi(X)$, we write $\Psi(P) \leq \Psi(X)$ if $\operatorname{udim} P \leq \operatorname{udim} X$ and $\rho(P, W) \leq \rho(X, W)$ for every $W \in \mathcal{W}$.

37.1. Theorem (Direct Summand Theorem). *Let P, X be finitely generated projective R-modules such that* $\operatorname{udim} P < \operatorname{udim} X$. *Then P is isomorphic to a direct summand of X if and only if $\Psi(P) \leq \Psi(X)$.*

Proof. We may assume that $P \neq 0$. We claim that the function $\Phi = \Psi(X) - \Psi(P)$ is the genus of some nonzero finitely generated projective R-module. By the Genus Theorem, 33.6, it suffices to show that $\Phi_0 > 0$ and Φ has almost standard rank and cycle standard rank. By the hypotheses on P and X we have $\Phi_0 > 0$ and every $\Phi_W \geq 0$. The remaining two properties hold because because the genera $\Psi(P)$ and $\Psi(X)$ satisfy these properties, and the properties are preserved by subtraction.

The Main Structure Theorem, 35.13, now shows that there is a projective module Q such that $\Psi(Q) = \Phi$ and $\mathcal{S}(Q) = \mathcal{S}(X) - \mathcal{S}(P)$. It also shows that $X \cong P \oplus Q$ since genus and Steinitz class are additive in direct sums and $\operatorname{udim}(X) \geq 2$. □

[1]In this section R denotes an HNP ring unless the contrary is specified.

37.2. COROLLARY. *Suppose that R is a Dedekind prime ring, and let P, X be finitely generated projective R-modules such that* $\operatorname{udim} P < \operatorname{udim} X$. *Then P is isomorphic to a direct summand of X.*

PROOF. Since R is a Dedekind prime ring, all R-towers are trivial [23.6]. In particular every unfaithful simple module W belongs to some trivial cycle tower, and hence is the unique element of that tower. Cycle standard rank [32.6, 32.9] therefore implies:

$$\begin{aligned} \rho(P,W) &= \operatorname{udim}(P)\cdot\rho(R,W)/\operatorname{udim}(R), \quad \text{and} \\ \rho(X,W) &= \operatorname{udim}(X)\cdot\rho(R,W)/\operatorname{udim}(R). \end{aligned} \tag{37.2.1}$$

However, $\operatorname{udim}(P) < \operatorname{udim}(X)$ and so $\rho(P,W) \leq \rho(X,W)$ for every $W \in \mathcal{W}$. The result now follows from 37.1. □

37.3. COROLLARY. *If R is a Dedekind prime ring, then every finitely generated projective module $P_R \neq 0$ satisfies $P \cong R^{(n)} \oplus A$ for some unique n and some right ideal $A \neq 0$, unique up to stable isomorphism.*

PROOF. Choose the largest n such that $n \operatorname{udim}(R) < \operatorname{udim}(P)$. Then 37.2 shows that $P \cong R^{(n)} \oplus A$ for some $A \neq 0$ with $\operatorname{udim}(A) \leq \operatorname{udim}(R)$. So A is isomorphic to a right ideal R. The uniqueness claimed is clear. □

38. Number of Generators

When is there an integer n such that all right ideals of R can be generated by n elements[1] and, when it exists, what is the smallest such n? In the commutative case the well-known bound is 1 or 2, and this remains true for Dedekind prime rings. For the full answer, see 38.3. We prepare for this by studying the number of generators of a single right ideal.

Our study of number of generators ends by comparing the number of generators required by modules sharing a common genus [38.6].

38.1. LEMMA.

(i) *Let P be a finitely generated projective R-module. Then P can be generated by n elements if and only if either $P \cong R^{(n)}$, or else* $\operatorname{udim} P < \operatorname{udim}(R^{(n)})$ *and $\Psi(P) \leq \Psi(R^{(n)})$.*

(ii) *Let P be a right ideal of R and let n be an integer with $n \geq 2$. Then P can be generated by n elements if and only if $\Psi(P) \leq \Psi(R^{(n)})$.*

PROOF. (i) P has n generators precisely when there is a surjection $R^{(n)} \twoheadrightarrow P$; and, since P is projective, this is equivalent to saying that P is isomorphic either to $R^{(n)}$ or else to a proper direct summand of $R^{(n)}$. In the latter case $\Psi(P) \leq \Psi(R^{(n)})$ and $\operatorname{udim} P < \operatorname{udim} R^{(n)}$. Conversely, if these last two condition hold, then the Direct Summand Theorem 37.1 shows that P is isomorphic to a proper direct summand of $R^{(n)}$.

(ii) Since P is a right ideal of R and $n \geq 2$, we have $\operatorname{udim} P < \operatorname{udim} R^{(n)}$ and hence $P \not\cong R^{(n)}$. Now apply (i). □

38.2. THEOREM. *Let n be an integer with $n \geq 2$. Then every right ideal P has a set of n generators if and only if:*

[1] In this section R denotes an HNP ring unless the contrary is specified.

(i) *R has no nontrivial faithful towers; and*
(ii) $\rho(R,\mathcal{C})/\rho(R,W) \leq n$ *for all cycle towers* $\mathcal{C}$ *and all* $W \in \mathcal{C}$.

PROOF. Since every right ideal of R is a direct summand of an essential right ideal of R, we need only consider the case where P denotes some essential right ideal of R. Then $\operatorname{udim}(P) = \operatorname{udim}(R)$ [12.4], and hence cycle standard rank [32.9] ensures that $\rho(P,\mathcal{C}) = \rho(R,\mathcal{C})$ for all cycle towers $\mathcal{C}$.

First suppose that (i) fails, and so R has some nontrivial faithful tower $\mathcal{C}$. Let W be any unfaithful simple module in $\mathcal{C}$. Let Φ be the function on modspec(R) which is equal to $\Psi(R_R)$ except that $\Phi_W = (n+1)\rho(R,W)$. Clearly Φ satisfies cycle standard rank and hence, by 35.15, there is a P such that $\Phi = \Psi(P)$; and P is isomorphic to an essential right ideal of R because $\operatorname{udim}(P) = \operatorname{udim}(R)$. Since, in addition, $\rho(P,W) > \rho(R^{(n)},W)$, P requires more than n generators [38.1].

Next suppose that (ii) fails and so $\rho(R,\mathcal{C}) > \rho(R^{(n)},W)$ for some $W \in \mathcal{C}$. Let Φ be the function on modspec(R) which is equal to $\Psi(R_R)$ except that $\Phi_W = \rho(P,\mathcal{C})$ and $\Phi_V = 0$ for all other $V \in \mathcal{C}$. These changes preserve cycle standard rank. Hence, by 35.15, there is a P such that $\Phi = \Psi(P)$; so $\rho(P,\mathcal{C}) = \rho(R,\mathcal{C})$. Therefore

$$\rho(P,W) = \rho(P,\mathcal{C}) = \rho(R,\mathcal{C}) > \rho(R^{(n)},W).$$

Once again, 38.1 shows that P needs more than n generators.

Conversely, suppose that conditions (i) and (ii) hold, and let P_R be essential in R. Then each unfaithful simple module W belongs to some cycle tower $\mathcal{C}$. Using (i) and (ii) together with the fact that $\rho(P,\mathcal{C}) = \rho(R,\mathcal{C})$, one gets

$$\rho(P,W) \leq \rho(P,\mathcal{C}) = \rho(R,\mathcal{C}) \leq n{\cdot}\rho(R,W).$$

Since $n \geq 2$ and P is a right ideal of R, we have $\operatorname{udim} P < \operatorname{udim}(R^{(n)})$. Hence, by 38.1, P has a set of n generators. □

38.3. COROLLARY. *Let* $s = \sup\big(\rho(R,\mathcal{C})/\rho(R,W)\big)$ *where* $\mathcal{C}$ *ranges over all cycle towers and* W *ranges over all elements of* $\mathcal{C}$.

(i) *If* $2 \leq s < \infty$ *and R has no nontrivial faithful towers then the smallest integer* $n \geq s$ *is the smallest integer such that every right ideal can be generated by n generators.*
(ii) *If* $s = 1$ *and R has no nontrivial faithful towers — equivalently, R is a Dedekind prime ring — then every right ideal can be generated by 1 or 2 elements.*
(iii) *If* $s = \infty$ *or R has a nontrivial faithful tower, then there is no finite n such that all right ideals can be generated by n elements.*
(iv) *Every right ideal is generated by n elements if and only if every left ideal is generated by n elements.*

PROOF. (i) and (iii) follow immediately from 38.2.

(ii) Having $s = 1$ is equivalent to all cycle towers being trivial. Moreover lack of nontrivial towers is equivalent to R being a Dedekind prime ring [23.6]. Suppose that R is Dedekind. Then, by 37.2, every right ideal is isomorphic to a direct summand of $R^{(2)}$, and hence can be generated by 2 elements.

(iv) Let P be any finitely generated projective right R-module, say with n generators. The Dual Basis Lemma [1.6] shows that P^* is a projective left R-module also generated by n elements. Since $P^{**} \cong P$, it follows that the minimal numbers of generators of P_R and of ${}_RP^*$ are equal. Since $\operatorname{udim}(P_R) = \operatorname{udim}({}_RP^*)$, by 17.2, P is a right ideal if and only if P^* is a left ideal (up to isomorphism). Hence the result. □

38.4. COROLLARY. *Suppose that there is no upper bound to the lengths of cycle towers of simple R-modules. Then there is no n such that every right ideal of R can be generated by n elements.*

PROOF. Let n be any integer with $n \geq 2$. By hypothesis there is a cycle tower $\mathcal{C}$ of length n', for some $n' > n$. Choose $W \in \mathcal{C}$ with the least possible value of $\rho(R, W)$; then $\rho(R, \mathcal{C}) \geq n'\rho(R, W) > n{\cdot}\rho(R, W)$. Hence, by 38.2, there are right ideals needing more than n generators. □

Even if the lengths of cycles are bounded, it can happen that there is no bound to the number of generators required for right ideals. The next example gives an extreme instance of this.

38.5. EXAMPLE. There is an HNP ring S with infinitely many cycle towers, all of length 2, no nontrivial faithful towers, and such that there is no bound to the number of generators required for the right ideals of S.

PROOF. By 25.27, one can construct an HNP ring R which has precisely one tower $\mathcal{C}_n$ for each positive integer n and such that each $\mathcal{C}_n$ is a cycle tower of length 2, say V_n, W_n, with $\rho(R, V_n) = n$ and $\rho(R, W_n) = 1$. Thus $\rho(R, \mathcal{C}_n) = n+1 > n{\cdot}\rho(R, W_n)$. Hence, by 38.2, there is no bound on the number of generators required for the right ideals of R. □

The next result compares numbers of generators of modules in the same genus. Item (i) is a familiar fact in the commutative case. Item (ii) and the supplementary statement show that the noncommutative situation is quite different. We say that a module P *requires n generators* if P can be generated by n but no fewer than n elements.

38.6. THEOREM. *Let P be a nonzero finitely generated projective R-module.*

(i) *If $\Psi(P)$ is the genus of some free R-module, say the genus of $R^{(n)}$, then either $P \cong R^{(n)}$ or P requires $n+1$ generators.*

(ii) *If $\Psi(P)$ is not the genus of a free R-module, then every P' such that $\Psi(P') = \Psi(P)$ requires the same number of generators as P.*

Moreover, if R is noncommutative but not a noncommutative Dedekind domain, then R always has projective modules P of type (ii).

PROOF. (i) Since $\Psi(P) = \Psi(R^{(n)})$, the Direct Summand Theorem 37.1 shows that P is isomorphic to a direct summand of $R^{(n+1)}$, and therefore P can be generated by $n+1$ elements. Clearly $R^{(n)}$ can be generated by n elements.

To show that these numbers of generators are minimal, recall, from 38.1(i), that an arbitrary finitely generated projective Q_R can be generated by n elements if and only if either $Q \cong R^{(n)}$ or else $\operatorname{udim} Q < \operatorname{udim}(R^{(n)})$ and $\Psi(Q) \leq \Psi(R^{(n)})$.

Since $\Psi(P) = \Psi(R^{(n)})$, we have $\operatorname{udim}(P) = \operatorname{udim}(R^{(n)})$. Therefore 38.1(i) rules out n generators for the case when $P \not\cong R^{(n)}$. A second application of 38.1(i), but with $n-1$ in place of n, shows that $R^{(n)}$ can be generated by $n-1$ elements only if $R^{(n)} \cong R^{(n-1)}$, which cannot happen since their uniform dimensions differ.

(ii) Let P, P' be as described in (ii). By symmetry, it suffices to show that if P can be generated by n elements then the same is true of P'. Suppose P can be generated by n elements. Then 38.1(i) shows that $\operatorname{udim} P < \operatorname{udim}(R^{(n)})$ and $\Psi(P) \leq \Psi(R^{(n)})$. Since $\Psi(P) = \Psi(P')$, the same inequalities are valid with P' in place of P. Therefore 37.1 implies that P' is isomorphic to a direct summand of $R^{(n)}$ and hence can be generated by n elements, as desired.

Supplementary statement. Suppose first that R is not an integral domain, and let P be any uniform right ideal of R. Then $\operatorname{udim}(R) \geq 2$ and $\operatorname{udim}(P) = 1$, which is not a multiple of $\operatorname{udim}(R)$. Therefore P is not in the genus of any free R-module.

Suppose, next, that R is a noncommutative integral domain but not a Dedekind domain. Then R has at least one nontrivial tower [23.6]. Let W be any unfaithful simple module in that tower. Then, arguing as in the proof of 38.2, there is a finitely generated projective module P with $\operatorname{udim}(P) = 1$ and $\rho(P, W) = 0$. However $0 \neq \rho(R, W)$ and so $\Psi(P)$ is not the genus of any free R-module. □

39. Unique Decompositions

We know that each finitely generated projective R-module[1] can be decomposed as a direct sum of uniform right ideals. The Main Structure Theorem for projective modules shows that if $\operatorname{udim}(P) > 1$ there can be a large number of non-isomorphic such decompositions; in fact, infinitely many if the ideal class group $G(R)$ is an infinite group.

In this section, we investigate conditions upon R which ensure the somewhat unusual circumstance that all such decompositions are essentially unique. We need some definitions to make this precise.

First, however, recall [35.5] that finitely generated projective modules P_R, Q_R are called 'stably isomorphic', with the notation $[P] = [Q]$, if $P \oplus X \cong Q \oplus X$ for some nonzero finitely generated projective X_R, and hence for all such X [by (35.6)]. It follows immediately from this, that $[P] = [Q]$ if and only if $P \oplus P \cong Q \oplus P$.

We know from 35.13 that a stable isomorphism class $[P]$ is uniquely defined by the genus and Steinitz class of P; indeed the isomorphism class of P is so determined provided $\operatorname{udim}(P) > 1$. Conversely, if two finitely generated projective modules have distinct genera or distinct Steinitz classes, then they are in distinct stable isomorphism classes (and so in distinct isomorphism classes, of course).

39.1. DEFINITIONS. By a *complete decomposition* of a finitely generated projective R-module P we mean a decomposition of the form $P = \oplus_{i=1}^{n} P_i$ where each $\operatorname{udim}(P_i) = 1$; equivalently, P_i is isomorphic to a uniform right ideal of R.

We say that P has a *stably unique* complete decomposition if the distinct stable isomorphism classes and their multiplicities in this decomposition are the same as those in any other complete decomposition of P. If, further, the isomorphism classes and their multiplicities in this decomposition of P are the same as those in any other complete decomposition of P, we say that the given complete decomposition is *isomorphically unique.*

We aim to study circumstances under which every finitely generated projective right R-module has a stably unique (resp. isomorphically unique) complete decomposition, in which case we will say that R *satisfies stable [resp. isomorphic] uniqueness.* (We should really call this right stable [resp. isomorphic] uniqueness. However, the property then holds for left R-modules too, because of the symmetry

[1] In this section R denotes an HNP ring unless the contrary is specified.

provided by the duality functor [1.6] and the fact that this functor preserves udim [17.2].)

In 39.15(i), we show that these two uniqueness properties are distinct. The following simple result shows the precise relation between them.

39.2. THEOREM. *R satisfies isomorphic uniqueness if and only if it satisfies stable uniqueness and stably isomorphic uniform right ideals are isomorphic.*

PROOF. ($\Leftarrow$) Consider two complete decompositions $\oplus_{i=1}^m P_i \cong \oplus_{i=1}^n Q_i$. Comparing uniform dimensions shows that $m = n$. Stable uniqueness shows, after suitable renumbering, that $[P_i] = [Q_i]$ for each i. Since P_i and Q_i are uniform right ideals we therefore have $P_i \cong Q_i$.

($\Rightarrow$) Evidently, stable uniqueness holds. Let U_1, U_2 be uniform right ideals with $[U_1] = [U_2]$. Then $U_1 \oplus U_1 \cong U_2 \oplus U_1$. Hence, by isomorphic uniqueness, $U_1 \cong U_2$. □

The above theorem allows us to concentrate on stable uniqueness, which is more tangibly related to properties of R than is isomorphic uniqueness. We return to isomorphic uniqueness in 39.10.

39.3. LEMMA. *Let $\mathcal{C}$ be a cycle tower. Then $\rho(R,\mathcal{C})/\operatorname{udim}(R) = \rho(A,\mathcal{C})$ for every uniform right ideal A of R. In particular, $\rho(R,\mathcal{C})/\operatorname{udim}(R)$ is a positive integer.*

PROOF. $\rho(A,\mathcal{C}) = \rho(R,\mathcal{C}) \cdot \operatorname{udim}(A)/\operatorname{udim}(R)$ since cycle standard rank holds [32.9, 32.6]. Setting $\operatorname{udim}(A) = 1$ yields the formula claimed. Also $\rho(R,\mathcal{C})/\operatorname{udim}(R)$ is positive since R has nonzero rank at every unfaithful simple module and is an integer because $\rho(A,\mathcal{C})$ is a sum of ranks. □

The next proposition lists several properties, each of which causes stable uniqueness to fail. We will show below that the list is exhaustive in the sense that, if none of these failures occur, then stable uniqueness follows. The proof makes considerable use of 35.15.

39.4. PROPOSITION. *If R has any of the following properties, then stable uniqueness fails for R:*

(i) *$G(R) \neq \{0\}$;*
(ii) *R has a nontrivial faithful tower;*
(iii) *R has at least two nontrivial cycle towers;*
(iv) *R has exactly one nontrivial cycle tower $\mathcal{C}$, and $\rho(R,\mathcal{C}) \neq \operatorname{udim}(R)$.*

PROOF. (i) Let α be any nonzero element of $G(R)$, and let A be any uniform right ideal. Since genus and Steinitz class can be independently prescribed, there are uniform right ideals $A(\alpha), A(-\alpha), A(0)$ with the same genus as A and Steinitz class respectively $\alpha, -\alpha, 0$. Then the isomorphism

$$A(\alpha) \oplus A(-\alpha) \cong A(0) \oplus A(0)$$

shows that stable uniqueness fails.

(ii) The nontrivial faithful tower contains an unfaithful simple module W. Let A be any uniform right ideal of R, and i any nonnegative integer. By 35.15, there exists a uniform right ideal $A(i)$ having the same Steinitz class as A, and the same genus as A except that the rank of $A(i)$ at W equals i. Then

$$A(0) \oplus A(2) \cong A(1) \oplus A(1)$$

since both sides have the same genus and Steinitz class and uniform dimension 2.

(iii) Let the two cycle towers be $\mathcal{C} = \{W_1, W_2, \dots\}$ and $\mathcal{C}' = \{W_1', W_2', \dots\}$; and let A be any uniform right ideal. Let $\rho = \rho(A, \mathcal{C})$ and $\rho' = \rho(A', \mathcal{C}')$. By 35.15, given any nonnegative integers x, y, x', y' such that $x+y = \rho$ and $x'+y' = \rho'$, there exists a uniform right ideal $B(x, y, x', y')$ with the same Steinitz class as A, with ranks x, y, x', y' at W_1, W_2, W_1', W_2' respectively, with rank 0 at every remaining $W_i \in \mathcal{C}$ and $W_i' \in \mathcal{C}'$, and otherwise the same ranks as A. Then the isomorphism

$$B(\rho, 0, \rho', 0) \oplus B(0, \rho, 0, \rho') \cong B(\rho, 0, 0, \rho') \oplus B(0, \rho, \rho', 0)$$

completes the proof.

(iv) Let $\mathcal{C} = \{W_1, W_2, \dots\}$, let A be any uniform right ideal and $\rho = \rho(A, \mathcal{C})$. Since $\rho(R, C) \neq \mathrm{udim}(R)$ then $\rho \geq 2$ by 39.3. By 35.15, if x, y are nonnegative integers such that $x + y = \rho$, then there is a uniform right ideal $B(x, y)$ with the same Steinitz class as A, with ranks x, y at W_1, W_2 respectively, rank 0 at every remaining $W \in \mathcal{C}$, and otherwise the same ranks as A. Then the isomorphism

$$B(\rho, 0) \oplus B(0, \rho) \cong B(\rho - 1, 1) \oplus B(1, \rho - 1)$$

completes the proof. □

39.5. THEOREM. *The following are equivalent:*

(i) *R satisfies stable uniqueness;*

(ii) $G(R) = 0$, *and either R is a Dedekind prime ring or else R has exactly one nontrivial tower $\mathcal{C}$ which is a cycle tower with* $\rho(R, \mathcal{C}) = \mathrm{udim}(R)$.

PROOF. (i)⇒(ii) This is given by 39.4.

(ii)⇒(i) First suppose that R is a Dedekind prime ring. Since $G(R) = \{0\}$, the stable isomorphism class of every finitely generated projective P_R is determined by the genus $\Psi(P)$; and, by [33.3], $\Psi(P)$ is determined by $\mathrm{udim}(P)$. Hence, if we choose a uniform right ideal A and a complete decomposition $P = \oplus_{i=1}^k P_i$, where $k = \mathrm{udim}(P)$, then each $[P_i] = [A]$. Hence stable uniqueness holds.

Next consider the alternative case, when R has the one nontrivial cycle tower $\mathcal{C}$ and $\rho(R, \mathcal{C}) = \mathrm{udim}(R)$. Let $\mathcal{C} = \{W_1, W_2, \dots, W_t\}$ and $1 \leq i \leq t$. Since R maps onto W_i and R is a direct sum of uniform right ideals, some uniform right ideal, A_i say, maps onto W_i; hence $\rho(A_i, W_i) \neq 0$. Since $\rho(R, \mathcal{C}) = \mathrm{udim}(R)$, we see from 39.3 that $\rho(A_i, \mathcal{C}) = 1$; so $\rho(A_i, W_i) = 1$ and $\rho(A_i, W_j) = 0$ for $j \neq i$.

We claim that for every uniform right ideal B, we have $[B] = [A_i]$ for some unique i. For B must map onto some W_i; that is, $\rho(B, W_i) \geq 1$. In the same way as with A_i, we see that $\rho(B, W_i) = 1$ and $\rho(B, W_j) = 0$ for each $j \neq i$. Since all towers other than $\mathcal{C}$ are trivial, we see that $\Psi(B) = \Psi(A_i)$. Since $G(R) = 0$, this implies $[B] = [A_i]$, as claimed.

Now consider any decomposition $P \cong \oplus_k B_k$ where each B_k is projective with $\mathrm{udim}(B_k) = 1$. The preceding paragraph shows that the number of summands with $[B_k] = [A_i]$ equals $\rho(P, W_i)$. Thus stable uniqueness holds □

We now aim to explain the explicit structure and construction of HNP rings satisfying stable uniqueness. Our results form a trichotomy, the disjoint situations being: when R is an integral domain; when R is a Dedekind prime ring but not an integral domain; and when R is neither a Dedekind prime ring nor an integral domain. We proceed to the first situation.

39.6. THEOREM. *If D is a hereditary Noetherian domain, the following statements and their left-handed versions are equivalent.*

(i) *D has the stable uniqueness property.*
(ii) *$G(D) = \{0\}$ and D is a noncommutative Dedekind domain.*
(iii) *All uniform right ideals are stably isomorphic to each other.*
(iv) *All finitely generated projective right D-modules of* udim ≥ 2 *are free.*
(v) *$M_n(D)$ is a principal right ideal ring for all $n \geq 2$.*
(vi) *$M_n(D)$ is a principal right ideal ring for some $n \geq 2$.*

PROOF. Recall from 39.1 that property (i) is a symmetric condition. So it is enough to prove the equivalence of the conditions as listed.

(i)⇒(ii) By 39.5, we need only show that D cannot have a nontrivial cycle tower $\mathcal{C}$. Otherwise, since D is an integral domain, 39.5 would require $\rho(D, \mathcal{C}) = \operatorname{udim}(D) = 1$. But since each member of $\mathcal{C}$ makes a contribution of at least 1 to $\rho(D, \mathcal{C})$, this is impossible.

(ii)⇒(iii) Since D is a Dedekind prime ring, all uniform right ideals are in the same genus [33.3]. Since they all have the same Steinitz class, (iii) holds [35.13(i)].

(iii)⇒(iv) Consider $P = \oplus_{i=1}^{t} P_i$ with each $\operatorname{udim}(P_i) = 1$ and $t \geq 2$. Since each P_i is isomorphic to a uniform right ideal and D is a domain, (iii) gives $[P_i] = [D]$. Therefore $[P] = [D^{(t)}]$. Since $t \geq 2$, this implies that $P \cong D^{(t)}$ [34.6].

(iv)⇒(v) Let $R = M_n(D)$, this being an HNP ring since D is. By (iv) and Morita equivalence, all finitely generated projective R-modules P with $\operatorname{udim}(P) = n$ are isomorphic to each other (since $n \geq 2$), and hence isomorphic to R_R.

Let H be a right ideal of R with $\operatorname{udim}(H) = h$ say. Then $h \leq \operatorname{udim}(R) = n$. Therefore, if K is any direct sum of $n - h$ uniform right ideals of R we have $H \oplus K \cong R$. Thus H is a homomorphic image of R_R, and hence a principal right ideal.

(v)⇒(vi) Trivial.

(vi)⇒(i) Let $R = M_n(D)$ for some $n \geq 2$ such that R is a principal right ideal ring (and is an HNP ring since D is). We first show that R is its own Dedekind right closure. To see this, suppose that S is a right finite overring of R. Then S_R is isomorphic to some essential right ideal of R and so is principal; say $S = sR$. Note that s is a unit of S and $s^2 \in S = sR$; hence $s \in R$ and then $S = R$, as required. So R is a Dedekind prime ring.

Since R is a principal right ideal ring, it has just one isomorphism class of essential right ideals. Since Steinitz class and genus are independent invariants [35.13(i)], $G(R)$ must be trivial. Therefore 39.5 gives (i). □

We proceed to the second situation in our trichotomy.

39.7. THEOREM. *Suppose that R is not an integral domain. The following are equivalent.*

(i) *R is a Dedekind prime ring and has the stable uniqueness property.*
(ii) *$R \cong M_n(D)$ where $n = \operatorname{udim}(R) > 1$ and D is a noncommutative Dedekind domain which has the stable uniqueness property [and so is as described in 39.6].*
(iii) *R is a principal (right and left) ideal ring.*

PROOF. (i)⇒(ii) Given (i), 39.5 shows that $G(R) = \{0\}$ (so all uniform right ideals have the same Steinitz class) and 33.3, asserts that all uniform right ideals share the same genus. Hence, by 35.13, they share the same stable isomorphism class.

Decompose R as a direct sum of uniform right ideals, $R_R = \oplus_{i=1}^n A_i$, and let A be any uniform right ideal. Then every $[A_i] = [A]$, and hence $[\oplus_i A_i] = [A^{(n)}]$. Since $n \geq 2$ this implies $\oplus_i A_i \cong A^{(n)}$ and therefore $R \cong A^{(n)}$. This provides a ring isomorphism $R \cong M_n(D)$ where $D = \operatorname{End}(A_R)$ which, by [**McR 01**, 3.3.5], is an integral domain. Of course, R and D are Morita equivalent and so D is a noncommutative Dedekind domain. Also all uniform projective D-modules are stably isomorphic to each other, since this is a Morita invariant property of R. So D is as described in (ii).

(ii)⇒(iii) Clear from 39.6, using the left-right symmetry it provides.

(iii)⇒(i) Let M be a nonzero maximal ideal; so $M = mR$, where m is regular because M is an essential right ideal. Because of this regularity, the relation $mRq \subseteq mR$ with $q \in R_{\text{quo}}$ implies that $q \in R$; that is, $O_r(M) = R$. This, in turn shows that M is invertible [14.3]. Hence R has no idempotent maximal ideals and thus, by 23.6, R is a Dedekind prime ring.

Say $\operatorname{udim}(R) = n$ where $n > 1$ since R is not an integral domain. Any direct sum of n uniform right ideals is isomorphic to an essential right ideal which is cyclic and so isomorphic to R_R. Hence, if we choose any three uniform right ideals, U_1, U_2, U_3, then $U_1 \oplus U_3^{(n-1)} \cong U_2 \oplus U_3^{(n-1)}$ and so $[U_1] = [U_2]$. Hence all uniform right ideals are stably isomorphic and so R must satisfy stable uniqueness. □

We now turn to the final situation in our trichotomy. Here R has an explicit construction from a Dedekind prime ring; indeed, by 39.7, from a matrix ring over a principal ideal domain. The construction involves basic chain idealizers, as described in Section 8.

39.8. THEOREM. *The following conditions on R are equivalent.*

(i) *R has the stable uniqueness property and is neither a Dedekind prime ring nor an integral domain.*
(ii) *R is a basic chain idealizer, of type U say, from a Dedekind prime ring S such that S has the stable uniqueness property, U is unfaithful and $\rho(S, U) = \operatorname{udim}(S)$.*

PROOF. (i)⇒(ii) By 39.5, R has trivial ideal class group and precisely one nontrivial tower $\mathcal{C}$, that being a cycle tower with $\rho(R, \mathcal{C}) = \operatorname{udim}(R)$. Let S be the ring determined by merging the nontrivial tower $\mathcal{C}$ into a simple module, U_S say, using 28.15. From there, we also see that U_S is unfaithful (because $\mathcal{C}$ is a cycle tower) and that S has no nontrivial towers, this making S a Dedekind prime ring. Moreover, 28.12 shows that R is a basic chain idealizer from S of type U.

Further, S has trivial ideal class group (since S and R share the same ideal class group by 35.19). Thus, by 39.5, S has the stable uniqueness property.

Note that U is obtained by merging $\mathcal{C}$ into the trivial cycle tower whose only member is U [28.15]. Because merging preserves cycle rank [32.19], we have $\rho(S, U) = \rho(R, \mathcal{C})$. We already know that $\rho(R, \mathcal{C}) = \operatorname{udim}(R)$; and $\operatorname{udim}(R) = \operatorname{udim}(S)$ because $R_{\text{quo}} = S_{\text{quo}}$. Therefore $\rho(S, U) = \operatorname{udim}(S)$, as desired.

(ii)⇒(i) By 35.19, $G(S) = G(R)$; and $G(S) = 0$ by 39.5. So $G(R) = 0$. Being a chain idealizer from S implies that R has precisely one nontrivial tower; and, since U is unfaithful, that tower is a cycle tower [28.9]. Because merging preserves cycle rank [32.19], $\rho(S,U) = \rho(R,\mathcal{C})$; and $\mathrm{udim}(S) = \mathrm{udim}(R)$. Hence, reasoning similar to that above yields $\rho(R,\mathcal{C}) = \mathrm{udim}(R)$. Statement (i) now follows from 39.5. □

39.9. EXAMPLE. For each integer $n \geq 2$, there is an HNP ring R with $\mathrm{udim}(R) = n$ which satisfies stable uniqueness, and has a nontrivial cycle tower.

PROOF. Choose a commutative principal ideal domain D with a maximal ideal $\mathfrak{m}$. Then let $S = M_n(D)$, a Dedekind prime ring with $\mathrm{udim}(S) = n$. Morita equivalence shows that S inherits the isomorphic uniqueness property from D. Moreover, $M = M_n(\mathfrak{m})$ is a maximal ideal of S, and $(S/M)_S \cong U^{(n)}$ where U is an unfaithful simple S-module. Since $n \geq 2$, some maximal right ideal A properly contains M. Therefore, by 39.8, the basic idealizer $R = \mathbb{I}_S(A)$ (which has type U) satisfies stable uniqueness. □

We now turn to the more special case of isomorphic uniqueness. As was shown in 39.2, this occurs precisely when an HNP ring with stable uniqueness has the additional property that stably isomorphic uniform right ideals are isomorphic. We will work through the trichotomy again.

39.10. LEMMA. *If D is a hereditary Noetherian domain, the following statements are equivalent.*

(i) *D has the isomorphic uniqueness property.*
(ii) *D is a principal ideal domain.*

PROOF. (i)⇒(ii) By 39.6, D is a noncommutative Dedekind domain; and by 39.2, all its nonzero right ideals are isomorphic.

(ii)⇒(i) Clear. □

39.11. THEOREM. *Let R be a Dedekind prime ring which is not an integral domain. The following are equivalent:*

(i) *R has the isomorphic uniqueness property.*
(ii) *$R \cong M_n(D)$ where $n = \mathrm{udim}(R) > 1$ and D is a principal ideal domain.*

PROOF. (i)⇒(ii) Here stable uniqueness for R implies that $R \cong M_n(D)$ with $n > 1$ and some noncommutative Dedekind domain D [39.7]. Morita equivalence then shows that D inherits isomorphic uniqueness from R. Finally, 39.10 shows that D is a principal ideal domain.

(ii)⇒(i) Morita equivalence gives this. □

What remains is the non-Dedekind case.

39.12. LEMMA. *Let T be a right finite overring of R, and let P_R, Q_R be finitely generated projective modules.*

(i) *If $[P] = [Q]$ then $[PT] = [QT]$ as T-modules. (As usual, we make the identification $PT = P \otimes_R T$ here.)*
(ii) *If R satisfies stable uniqueness, then so does T.*

PROOF. (i) $P \oplus X \cong Q \oplus X$ implies that $PT \oplus XT \cong QT \oplus XT$ as T-modules.

(ii) Suppose $\oplus_{i=1}^n A_i \cong \oplus_{i=1}^n B_i$ for uniform right ideals A_i, B_i of T. Since T is right finite over R, this is also an isomorphism of finitely generated projective R-modules. Hence, by stable R-uniqueness and after suitable renumbering, each $[A_i] = [B_i]$ as R-modules. Therefore $A_i \oplus X \cong B_i \oplus X$ as R-modules for some finitely generated projective X_R. Tensoring with T over R and remembering that each A_i and B_i is a T-module then demonstrates that $A_i \oplus XT \cong B_i \oplus XT$ as T-modules; that is $[A_i] = [B_i]$ as T-modules. □

The next lemma provides the crux of our passage from stable to isomorphic uniqueness.

39.13. LEMMA. *Let $R = \mathbb{I}_T(A)$, a basic idealizer of type $U = [VW]$. If T satisfies isomorphic uniqueness and R satisfies stable uniqueness, then R satisfies isomorphic uniqueness.*

PROOF. Since R satisfies stable uniqueness, it suffices to show that stably isomorphic uniform right ideals of R are actually isomorphic [39.2]. Since R satisfies stable uniqueness and V, W is a segment of a nontrivial R-tower, say $\mathcal{C}$, we see that $\mathcal{C}$ is a cycle tower.

Now fix uniform right ideals I, I' of R such that $[I] = [I']$. Then $[IT] = [I'T]$ as T-modules [39.12]; and so the isomorphic uniqueness property for T yields a T-module isomorphism ϕ: $IT \cong I'T$. To complete the proof it suffices to show that $\phi(I) = I'$.

Since I is an R-submodule of the T-module IT, we have $I/IA \cong_R W^{(t)}$ and $IT/IA \cong_T U^{(t)}$ for some integer $t \geq 0$ [4.10]. Note that $t = \rho(I, W)$. Also I and I' have the same genus since $[I] = [I']$; and so a second application of 4.10 yields an analogous pair of isomorphisms for I', with the same value of t. In summary, we have the following R- and T-isomorphisms:

$$I/IA \cong_R W^{(t)} \cong_R I'/I'A \quad \text{and} \quad IT/IA \cong_T U^{(t)} \cong_T I'T/I'A. \tag{39.13.1}$$

Since R satisfies stable uniqueness $\rho(R, \mathcal{C}) = \mathrm{udim}(R)$ [39.5] and so $\rho(I, \mathcal{C}) = 1$ for every uniform right ideal I of R [39.3]. Therefore $t = \rho(I, W) \leq 1$, so there are two cases to consider.

Case 1: $t = \rho(I, W) = 0$; that is, $I = IA$. Hence I is a right T-module and so $IT = I$. The same holds for I'. Therefore $\phi(I) = \phi(IT) = I'T = I'$, as desired.

Case 2: $t = \rho(I, W) = 1$. Since $\rho(I, \mathcal{C}) = 1$ then $\rho(I, V) = 0$. However, $\rho_T(IT, U) = \rho_R(I, V) + \rho_R(I, W)$ [32.19], and so $\rho_T(IT, U) = 1$. Thus $IT/[IT\,\mathrm{ann}_T(U)] \cong U_T$. However, $IT/IA \cong U_T$, by (39.13.1); and hence $IT \supset IA \supseteq IT\,\mathrm{ann}_T(U)$. We deduce that $IT\,\mathrm{ann}_T(U) = IA$.

The same is true for I', of course. However, the isomorphism $\phi : IT \to I'T$ restricts to an isomorphism $IT\,\mathrm{ann}_T(U) \to I'T\,\mathrm{ann}_T(U)$. Thus $\phi(IA) = \phi(I'A)$. Since $(IT/IA)_R$ is uniserial of length 2, with unique simple submodule I/IA, we see that $\phi(I) = \phi(I')$ as required. □

We can now prove our main theorem on isomorphic uniqueness for HNP rings which are not Dedekind.

39.14. THEOREM. *The following conditions on R are equivalent.*

(i) *R has the isomorphic uniqueness property and is not a Dedekind prime ring.*
(ii) *R is a basic chain idealizer, of type U say, from a ring $S \cong M_n(D)$ such that D is a principal ideal domain, U_S is unfaithful and $\rho(S, U) = \mathrm{udim}(S) = n > 1$.*

PROOF. (i)$\Rightarrow$(ii) Since R satisfies stable uniqueness, R is a basic chain idealizer from a Dedekind prime ring S which has the stable uniqueness property [39.8]. Right finiteness of S over R implies that any complete decomposition of a finitely generated projective right S-module P is also a complete decomposition as a right R-module. The decomposition is therefore isomorphically unique over R and hence over S. Now apply 39.11.

(ii)⇒(i) We know that S satisfies isomorphic uniqueness [39.11], and the basic chain idealizer R satisfies stable uniqueness [39.8]. Every multichain idealizer is an iterated basic idealizer [8.6]. Therefore we have a chain of rings $S = T_0 \supset T_1 \supset \ldots \supset T_m = R$ where each T_{i+1} is a basic idealizer from T_i. Since R satisfies stable uniqueness, so does every T_i [39.12]. A simple induction, using 39.13, now carries the isomorphic uniqueness property from $S = T_0$ to $R = T_m$. □

39.15. REMARKS. (i) The first Weyl algebra $A_1(F)$, where F is any field of characteristic zero, is a simple hereditary Noetherian domain that is not a principal ideal domain, and such that all finitely generated projective $A_1(F)$-modules of udim ≥ 2 are free [11.8]. Hence, all right ideals of this ring are stably free. Thus isomorphic and stable uniqueness are actually distinct properties.

(ii) In the case of isomorphic uniqueness, the ring D in 39.7(ii) above is unique up to isomorphism. For suppose that $R \cong M_n(D) \cong M_n(C)$. The top row of each of $M_n(D)$ and $M_n(C)$ are uniform right ideals of R so are isomorphic. Hence they have isomorphic endomorphism rings; i.e. $C \cong D$.

However, in the case of stable uniqueness this is not so. To see why, let $D = A_1(F)$, as in (i). In [**Smith 81**], it is shown that D has a uniform right ideal A such that $\mathrm{End}_D(A) \ncong D$. Since all projective modules of rank 2 are free, $A \oplus A \cong D \oplus D$ and so $M_2(\mathrm{End}_D(A)) \cong M_2(D)$.

40. Semilocal HNP Rings

Among noncommutative ring theorists, a ring S is usually called *semilocal* if the ring $S/J(S)$ is semisimple Artinian, where $J(S)$ is the Jacobson radical.[1] The definition of (noncommutative) 'local ring' is less standard, sometimes meaning that $S/J(S)$ is simple Artinian, and sometimes, as in §31, meaning that $S/J(S)$ is a division ring. In this section, where both types of ring occur, we avoid using this term, except for the commutative case where the two definitions coincide.

40.1. PROPOSITION. *The following conditions are equivalent for (an HNP ring) R.*

(i) *R is semilocal.*
(ii) *R has no faithful towers, and has only finitely many cycle towers.*
(iii) *$J(R) \neq 0$.*

PROOF. (i)⇒(ii) Recall that $J(R)$ is the intersection of the annihilators of the simple (right) R-modules. Therefore every simple R-module is a simple $J(R)$-module. But since $R/J(R)$ is Artinian, it has only finitely many simple modules (up to isomorphism). Therefore the same is true of R, and hence R has only finitely many towers. Note that $J(R) \neq 0$ since, by our nontriviality convention, HNP rings are not Artinian. Hence every simple module W_R is unfaithful.

(ii)⇒(iii) By (ii), R has only finitely many simple modules, all unfaithful. Since R is a prime ring, the product A of the annihilators of these finitely many simple modules, taken in some order, is nonzero and annihilates all simple R-module. Therefore $0 \neq A \subseteq J(R)$.

---

[1]In this section R denotes an HNP ring unless the contrary is specified.

(iii)$\Rightarrow$(i) We only need to prove that the ring $R/J(R)$ is Artinian which is so since R/A is Artinian for every nonzero ideal A of R. □

The next theorem summarizes what special decomposition properties hold in the semilocal case. First we prove a well-known fact.

40.2. LEMMA. *Noetherian projective modules P, Q over any ring S are isomorphic if and only if $P/PJ(S) \cong Q/QJ(S)$.*

PROOF. For the nontrivial direction, suppose that $P/PJ(S) \cong Q/QJ(S)$ and choose such an isomorphism h, as shown in the next diagram

$$\begin{array}{ccc} P & & Q \\ \downarrow & & \downarrow \\ P/PJ(S) & \xrightarrow{h} & Q/QJ(S) \end{array}$$

where the downward arrows denote the natural surjections. Since P is projective, h can be lifted to a map $f\colon P \to Q$ making the diagram commute; and since Q is finitely generated, Nakayama's Lemma implies that f is a surjection. Similarly, there is a surjection $g\colon Q \to P$. Since P is Noetherian, the surjection $gf\colon P \to P$ is necessarily an isomorphism. Hence f is an isomorphism. □

40.3. THEOREM. *Suppose that R is semilocal. Then:*

(i) *Stable isomorphism of finitely generated projective R-modules implies isomorphism.*
(ii) $\Psi(P) = \Psi(Q) \Rightarrow P \cong Q$ *for finitely generated projective R-modules P, Q.*
(iii) $G(R) = \{0\}$.
(iv) *If R is a Dedekind prime ring, then $R \cong M_n(D)$ where $n = \mathrm{udim}(R)$ and D is a semilocal PID.*

PROOF. Let $J = J(R)$.

(i) If $[P] = [Q]$, then $P \oplus R \cong Q \oplus R$. Reducing modulo J and recalling that the Krull-Schmidt theorem holds for modules of finite length shows that $P/PJ \cong Q/QJ$. Hence $P \cong Q$, by 40.2.

(ii) Let P, Q share the same genus. Since the ring R/J is semisimple Artinian, it follows that $P/PJ \cong Q/QJ$. Thus $P \cong Q$ by 40.2.

(iii) This follows from (ii) by the Main Structure Theorem 35.13.

(iv) Let A, B be uniform right ideals. Since R is a Dedekind prime ring, $\Psi(A) = \Psi(B)$, by 37.2, and hence $A \cong B$, by (ii). Since every finitely projective R-module is isomorphic to a direct sum of uniform right ideals, it follows that all finitely generated projective R-modules have isomorphically unique complete decompositions. Therefore 39.7 shows that $R \cong M_n(D)$ for some PID R; and D is semilocal because R is. □

One simple, faithful module. There is a tempting, purely noncommutative, notion of 'local': namely having a unique (isomorphism class) of simple module. This would include a type of ring which is excluded by the definition of 'semilocal', namely a ring with exactly one simple faithful module and no other simple modules. The following (intriguingly incomplete) result, which we give without proof, comes from [**Levy-Robson 99b**, Theorem 5.8].

40.4. THEOREM. *Suppose that R has exactly one simple module W and W is faithful. Then $G(R)$ is a finite cyclic group.* □

Such a ring R is easily constructed, as a localization of any HNP ring that has a faithful simple module.

Since R has exactly one tower, and that tower is faithful and trivial, the transpose functor shows that the same is true of left R-towers [21.2]. Therefore R has a unique, necessarily faithful simple left module. Also, R can have no nonzero ideal. Hence R is simple and so is a Dedekind prime ring.

Obviously R has exactly one genus of (right) modules for each uniform dimension and hence 40.4 implies: *For every nonzero uniform dimension, there exist exactly $|G(R)|$ stable isomorphism classes of finitely generated projective modules.*

What makes this information incomplete is that (a) we do not know whether or not $G(R)$ must be trivial; and (b) we do not know whether stable isomorphism implies isomorphism, over this ring. [See Problem 54.6.]

41. Modules of Finite Length

Our first result[1] shows that a module of finite length splits into two parts, and the subsequent results show that the two parts are quite different from each other.

41.1. THEOREM. *Every R-module M of finite length has a unique decomposition $M = F \oplus C$ where all composition factors of F belong to faithful towers and all composition factors of C belong to cycle towers.*

PROOF. Recall, from 19.3, that every simple module belongs to some unique tower. Hence any decomposition of the form $M = F \oplus C$ is unique.

We prove existence of the decomposition by induction on the length $\lambda(M)$ of M, the assertion being clear if $\lambda(M) = 0$. Suppose now that $M \neq 0$. Then M has a simple submodule W, and this yields a short exact sequence

(41.1.1) $$0 \to W \to M \to M' \to 0\,.$$

By induction there is a decomposition $M' = F' \oplus C'$ where the composition factors F' and C' belong to faithful towers and cycle towers respectively. Let the full inverse images of F' and C' in M be F'' and C'' respectively. Then

(41.1.2) $$M = F'' + C'' \qquad \text{with} \quad F'' \cap C'' = W\,.$$

Recall that $\operatorname{Ext}^1_R(X,Y) = 0 = \operatorname{Ext}^1_R(Y,X)$ whenever X belongs to a faithful tower and Y belongs to a cycle tower, this being an immediate consequence of the unique order of succession within towers [19.1].

First, suppose that W is an element of a faithful tower. Then $\operatorname{Ext}^1_R(Y,W) = 0$ for every composition factor Y of $C''/W \cong C'$ because all composition factors of C' belong to cycle towers. This implies that $\operatorname{Ext}^1_R(C''/W, W) = 0$ [50.7]; and so the short exact sequence $W \hookrightarrow C'' \twoheadrightarrow C''/W$ splits. Say $C'' = W \oplus C'_1$ with $C'_1 \cong C'$. Then (41.1.2) implies that $M \cong F'' \oplus C'$, as desired.

The case when W is an element of a cycle tower is dealt with similarly, with the roles of F and C reversed. □

We now study each of the two types of modules of finite length.

41.2. THEOREM.

---

[1] In this section R denotes an HNP ring unless the contrary is specified.

(i) *Let C be an indecomposable R-module of finite length whose composition factors belong to cycle towers. Then C is uniserial (and hence cyclic) and its composition factors all belong to a single cycle tower $\mathcal{T}$.*
(ii) *For every simple R-module W belonging to a cycle tower and every positive integer n there is, up to isomorphism, exactly one uniserial module U of length n with bottom (alternatively, top) composition factor W.*

Caution. Since the ordering in cycle towers is cyclic, rather than linear, the sequence of composition factors of C is not necessarily a segment of $\mathcal{T}$. It might, for example, consist of several repetitions of $\mathcal{T}$.

PROOF. (i) We may assume that $C \neq 0$. Since all composition factors of C are unfaithful and R is a prime ring, C itself is unfaithful. Let $A = \operatorname{ann}(C)$. Then the proper homomorphic image R/A of R is an Artinian serial ring [15.6], and hence all of its modules are direct sums of uniserial modules of finite length [50.19]. The remaining assertion of (i) is a consequence of unique order of succession within towers [19.1].

(ii) Suppose that we want W to be the top composition factor of U. Because of the unique, cyclic order of succession within cycle towers, there is only one possibility for the sequence of composition factors of the module U we are trying to build. Let the sequence (from top to bottom) be $W = W_1, \ldots, W_n$. The result is now an immediate consequence of 16.5. A similar argument covers the case when W is to be the bottom composition factor. □

We can now prove the main result on modules whose composition factors belong to faithful towers.

41.3. THEOREM. *Let F be an R-module of finite length whose composition factors belong to faithful towers. Then:*
(i) *F is a homomorphic image of a uniform right ideal of R; and*
(ii) *F is a submodule of a cyclic R-module of finite length.*

PROOF. (i) We need only find a uniform right ideal H such that $\rho(H, W) \geq \rho(F, W)$ for each of the finite set of unfaithful composition factors W of $F/\operatorname{rad}(F)$, for then 35.2 asserts that H can be mapped onto F. Let the distinct unfaithful composition factors of $F/\operatorname{rad}(F)$ be $W_1, \ldots, W_n$ and let H' be any uniform right ideal of R. By 35.15, there is a uniform right ideal H having the same ranks as H' except at $W_1, \ldots, W_n$ and having $\rho(H, W_i) = \rho(F, W_i)$ for each i since the changes do not involve any cycle towers and so cycle standard rank is preserved.

(ii) We have $F \cong H/K$ for some K, by part (i), and $K \neq 0$ because F has finite length but H does not. Let A be any cyclic, uniform right ideal. Since every uniform right ideal is isomorphic to a submodule of every other [12.4], we may assume that $H \subseteq A$. But then $F \cong H/K$ is contained in A/K, which is cyclic; and A/K has finite length because all nonzero submodules of A are essential in A. □

Remark. An extreme form of behaviour related to that described in 41.3(ii) is that, over any ring which is not right Artinian ring, any right module of finite length whose composition factors are all faithful is cyclic [49.12].

We return to our HNP ring R. Recall that R is said to have 'enough invertible ideals' if every nonzero ideal contains an invertible ideal [23.1]; and this is equivalent to saying that every faithful R-tower is trivial [23.2]. This class of rings includes all classical hereditary orders (HNP rings that are finitely generated modules over a central Dedekind domain). For such rings, our previous results take the following simple form.

41.4. COROLLARY. *Suppose that every faithful R-tower is trivial. Then every R-module of finite length is a direct sum of:*

(i) *uniserial, and hence cyclic, modules that are unfaithful; and*
(ii) *one cyclic module (not necessarily indecomposable) whose composition factors are all faithful.*

PROOF. Let M be the given R-module. By 41.1, there is a decomposition $M = F \oplus C$ where the composition factors of F and C belong, respectively, to faithful and cycle towers. By 41.2, C is a direct sum of uniserial modules whose composition factors belong to cycle towers, and hence are unfaithful. Since R is a prime ring, unfaithfulness of the composition factors implies unfaithfulness of C itself.

Now consider F. Its composition factors belong to faithful towers, and these composition factors are faithful since all faithful R-towers are trivial. Then 49.12 shows that F is cyclic. □

One of the surprises in this subject is that, for an arbitrary HNP ring R, indecomposable R-modules of finite length can be very far from cyclic and very far from uniserial as the next few examples show.

41.5. EXAMPLE. Let m be a positive integer. Then there is a hereditary Noetherian domain R and an indecomposable non-uniserial R-module M of finite length requiring precisely m generators. Furthermore:

(i) $\operatorname{soc}(M) = \operatorname{rad}(M) = V$ for some simple R-module V;
(ii) $M/V \cong W^{(m)}$ where W is unfaithful and simple.

PROOF. Begin with the simple hereditary Noetherian domain $S = B_1(F)$, where F has characteristic 0 [11.7]. Choose any maximal right ideal A and let $U_S = S/A$ and $R = \mathbb{I}_S(A)$. Thus R is a basic idealizer of type $U = [VW]_R$ for some V, W where W is unfaithful with annihilator A. Also V, W is a faithful R-tower and $\rho(R, W) = 1$ [20.5].

Let $E = E(U_S)$ and recall that, for this ring, $\operatorname{soc}_S(E/U)$ consists of the direct sum of infinitely many copies of each isomorphism type of simple S-module [24.12]. Therefore there is a submodule $X_S \subset E$ such that $X/U \cong U^{(m)}$. So we can choose Y_R such that $U \subset Y \subset X$ and $Y/U \cong W^{(m)}$. Now $W \subset U$; we define $M = Y/W$.

Since V, W is a tower of length 2, the uniserial R-module U_R is the unique length-2 submodule of E_R [24.11]. Hence $V = U/W$ is the unique simple submodule of M; so $V = \operatorname{soc}(M)$ and thus M is indecomposable. Moreover, since M/V is semisimple, we deduce that $V = \operatorname{rad}(M)$. Finally, since $\rho(R, W) = 1$ then $M/\operatorname{rad}(M) \cong W^{(m)}$ needs m generators [35.2] and hence so too does M. □

The next example complements the previous one by showing that indecomposable R-modules of finite length need not have simple socles.

41.6. EXAMPLE. Let m be a positive integer. Then there is an HNP ring R which has an indecomposable cyclic module N_R of finite length such that $\operatorname{soc}(N)$ has length m and $N/\operatorname{soc}(N)$ is simple.

PROOF. By the version of 41.5 for left modules, there is an HNP ring R and a left module M of finite length, with socle a simple module V such that $Z = M/V$ is semisimple with length m. Apply the transpose functor [17.1] to the nonsplit short exact sequence $V \hookrightarrow M \twoheadrightarrow Z$. Then 17.7 asserts that $Z^{\mathrm{tr}} \hookrightarrow M^{\mathrm{tr}} \twoheadrightarrow V^{\mathrm{tr}}$ is a nonsplit short exact sequence of right R-modules of finite length; that V^{tr} is simple; and that Z^{tr} is semisimple of length m. We deduce that $Z^{\mathrm{tr}} = \mathrm{soc}(M^{\mathrm{tr}})$. Also, using the lattice anti-isomorphism given by 17.7, we deduce that M^{tr} is indecomposable and that $Z^{\mathrm{tr}} = \mathrm{rad}(M^{\mathrm{tr}})$ which is, therefore, the unique maximal submodule of M^{tr}. Hence M^{tr} is cyclic. □

42. Stacked Decompositions

This section[1] studies the existence of stacked decompositions over an arbitrary HNP ring R, together with the uniqueness and nonuniqueness properties of these decompositions. These theorems have an extensive history, outlined in §44. Before beginning, we note that the Lifting and Straightening Theorem 49.6 plays an extensive role in this section.

42.1. THEOREM. *Let N be a submodule of a finitely generated projective R-module P. Then there exist 'stacked decompositions':*

(42.1.1) $$P = \oplus_i P_i \qquad N = \oplus_i (P_i \cap N)$$

with each P_i uniform.

PROOF. Since every projective R-module is a direct sum of uniform projective modules, 42.2 reduces the proof to the case that N is essential in P. Then 42.3 proves that case, in sharpened form. □

42.2. THEOREM. *Let N be a submodule of a finitely generated projective R-module P. Then there is a decomposition $P = P' \oplus P''$ such that N is an essential submodule of P'' (equivalently, P''/N has finite length).*

The direct summands P' and P'' are unique up to isomorphism.

PROOF. There is a decomposition $P/N = Q \oplus T$ where Q is projective and T has finite length [12.18]. Consider the composite map $P \overset{\nu}{\twoheadrightarrow} P/N \overset{\pi}{\twoheadrightarrow} Q$, where π is coordinate projection in the direct sum $Q \oplus T$. Since Q is projective, $P'' = \ker(\pi\nu)$ is a direct summand of P. Say $P = P' \oplus P''$. Then $N \subseteq P''$ and hence $P''/N \cong \nu(P'')/\nu(N) = \nu(P'') = \ker(\pi) = T$. Since T has finite length, N is essential in P'' [12.14].

Uniqueness. Since N is essential in P'', P''/N is the torsion submodule [12.5] of P/N, and so P'' is a uniquely determined submodule of P (not just unique up to isomorphism). Hence its complement P' is unique up to isomorphism. □

Recall that $\mathcal{S}(P)$ and $\Psi(P)$ denote the Steinitz class [35.12] and genus [33.1.1] of the finitely generated projective R-module P. The next proof makes use, several times, of the Main Structure Theorem [35.13].

42.3. THEOREM. *Let N be an essential submodule of a finitely generated projective R-module $P = Q_1 \oplus \ldots \oplus Q_n$ with each* $\mathrm{udim}(Q_i) = 1$.

(i) *There exist stacked decompositions of P and N, as in* (42.1.1), *with each* $\mathcal{S}(P_i) = \mathcal{S}(Q_i)$.

[1] In this section R denotes an HNP ring unless the contrary is specified.

(ii) *If all faithful R-towers are trivial, the decompositions in* (i) *can be chosen with each* $P_i \cong Q_i$.

Remark. We show in 42.4 that (ii) above does not apply to arbitrary HNP rings.

PROOF. We may assume that $n \geq 2$ since otherwise there is nothing to prove. Let $\nu\colon P \twoheadrightarrow T = P/N$ be the natural surjection. There is a decomposition $T = F \oplus C$ where all composition factors of F belong to faithful towers and those of C to cycle towers) [41.1].

The proof involves consideration of complete decompositions of P (i.e. as a direct sum of n uniform modules) and of C as a direct sum of n submodules.

We will call a complete decomposition

(42.3.1) $$P = P_1 \oplus \ldots \oplus P_n; \quad \text{with} \quad \mathcal{S}(P_i) = \mathcal{S}(Q_i) \quad \text{for all} \quad i$$

a *good decomposition* of P. (Further, in the case when all faithful R-towers are trivial, we also require that $P_i \cong Q_i$.)

The proof of the theorem relies on two main steps. Step 1 demonstrates that there is a good decomposition of P such that there is a surjection $P_1 \twoheadrightarrow F$.

Then Step 2 finds a decomposition of C and surjections as shown next:

(42.3.2) $$C = U_1 \oplus \ldots \oplus U_n; \quad P_1 \twoheadrightarrow F \oplus U_1; \quad P_i \twoheadrightarrow U_i \quad \text{for all} \quad i > 1.$$

It then follows, by Lifting and Straightening [49.6], that the summands P_i can be isomorphically replaced in (42.3.1), to achieve $\nu(P_1) = F \oplus U_1$, and $\nu(P_i) = U_i$ for every $i > 1$. Hence $N = \ker(\nu) = \oplus_i (P_i \cap N)$, as desired.

The following fact from 35.2 will be needed in several places below.

(42.3.3) Let P_R be finitely generated projective and let U_R have finite length. Then there is a surjection $P \twoheadrightarrow U$ if and only if $\rho(P, W) \geq \rho(U, W)$ for every unfaithful composition factor W of $U/\operatorname{rad}(U)$.

Step 1: There is a good decomposition of P *such that* P_1 *maps onto* F.

First we consider the special case described in (ii); i.e. when all faithful towers are trivial. In this case, of course, F has no unfaithful composition factors; so 42.3.3 applies to show that the given decomposition is as required.

We now turn to the general case of Step 1 and start with a preliminary step.

Step 0: P *has a uniform direct summand* P_1 *that can be mapped onto* F *and such that* $\mathcal{S}(P_1) = \mathcal{S}(Q_1)$.

Since all composition factors of F belong to faithful towers, there is an R-module surjection $\pi\colon H \twoheadrightarrow F$ for some uniform right ideal H of R [41.3]. Also, $P/N \cong F \oplus C$, so there is a surjection $g\colon P \twoheadrightarrow F$. Since P is projective, g can be factored through π yielding $g\colon P \xrightarrow{\alpha} H \twoheadrightarrow F$. Let $H' = \alpha(P)$, this being a uniform right ideal of R. Then $P \cong H' \oplus \ker(\alpha)$ since H' is projective and, since P maps onto F, so does H'.

Since H' can be mapped onto F, every module in the genus $\Psi(H')$ of H' can be mapped onto F [35.2]. However, the genus and Steinitz class of finitely generated projective R-modules can be independently prescribed [35.13]; so there exists $P_1 \in \Psi(H)$ such that $\mathcal{S}(P_1) = \mathcal{S}(Q_1)$ and P_1 maps onto F. To complete the proof of Step 0, it remains to show that P_1 is isomorphic to a direct summand of P. Since $\Psi(P_1) = \Psi(H')$ and H' is a direct summand of P, we have $\Psi(P_1) = \Psi(H') \leq \Psi(P)$.

This, together with the fact that $\operatorname{udim}(P_1) = 1 < n = \operatorname{udim}(P)$ implies, by the Direct Summand Theorem [37.1], that P_1 is isomorphic to a direct summand of P, as desired.

We now turn to the proof of Step 1 itself. We have just shown that there is a decomposition $P \cong P_1 \oplus X$ with P_1 as in Step 0. Write X as a direct sum of $n-1$ uniform modules: $X \cong \oplus_{i=2}^n H_i$. By 35.13, Steinitz class and genus can be independently prescribed and are additive; hence there exist uniform projective R-modules P_i for each $i > 1$ such that each $\mathcal{S}(P_i) = \mathcal{S}(Q_i)$ and $\Psi(P_i) = \Psi(H_i)$. Further, since $n > 1$, 35.13 implies that $\oplus_{i=1}^n P_i \cong P_1 \oplus (\oplus_{i=2}^n H_i) \cong P$, completing Step 1.

Step 2: There is a decomposition $C = \oplus_{i=1}^n U_i$ such that P_1 maps onto $F \oplus U_1$ and P_i maps onto U_i for $i > 1$.

Since all composition factors of C belong to cycle towers, we have a decomposition $C = \oplus_j V_j$, in which each V_j is uniserial with unfaithful composition factors [41.2]. Each U_i will be the direct sum of a subset of these uniserial modules. Note that $\rho(V_j, W) = 1$ if W is the top composition factor of V_j, and $\rho(V_j, W) = 0$ otherwise.

By (42.3.3), the surjections needed in Step 2 exist if

$$\text{(42.3.4)} \qquad \rho(P_1, W) \geq \rho(F, W) + \rho(U_1, W) \quad \text{and} \quad \rho(P_i, W) \geq \rho(U_i, W) \ (i > 1)$$

for each unfaithful composition factor W of $(F \oplus C)/\operatorname{rad}(F \oplus C)$. Note that (42.3.4) can be verified one W at a time.

We now allot the V_j to form each U_i iteratively as follows. Collect as many V_j as possible to form U_1 whilst keeping $\rho(U_1, W) \leq \rho(P_1, W)$ for each composition factor W of $C/\operatorname{rad}(C)$. Then, with any remaining V_j, repeat the process for $U_2, \ldots, U_n$. Since $\rho(P, W) \geq \rho(C, W)$ for all such W, this process exhausts the supply of V_j. Note that $U_i = 0$ if no V_j are allotted to it.

To verify that (42.3.4) holds, fix W. If W belongs to a faithful tower, then the first inequality becomes $\rho(P_1, W) \geq \rho(F, W)$, which holds since P maps onto F; and the second inequality merely becomes $\rho(P_i, W) \geq 0$. On the other hand, if W belongs to a cycle tower then $\rho(F, W) = 0$; and what remains of the two inequalities holds because of the way we included the V_j in the various U_i.

The two steps being completed, the result follows from the Lifting and Straightening Theorem as indicated earlier. □

42.4. Example. There exist finitely generated projective modules $N \subset P = Q_1 \oplus Q_2$ over a suitable hereditary Noetherian domain R, such that N is essential in P, each Q_i is uniform, and there exist no stacked decompositions

$$\text{(42.4.1)} \qquad P = P_1 \oplus P_2, \quad N = (P_1 \cap N) \oplus (P_2 \cap N)$$

in which each $P_i \cong Q_i$.

Proof. Let R be a hereditary Noetherian domain with an indecomposable module M of finite length such that M requires two generators. For example, R, M can be as in 41.5. Let $P = Q_1 \oplus Q_2$ with each $Q_i \cong R$. Since M is 2-generated, P maps onto M; say $M \cong P/N$. Then N is essential in P since M has finite length [12.17].

On the other hand, in any stacked decompositions we have $M \cong P/N \cong P_1/(P_1 \cap N) \oplus P_2/(P_2 \cap N)$. Therefore the indecomposability of M implies that

$P_i/(P_i \cap N) \cong M$ for $i = 1$ or $i = 2$. But neither P_i can map onto M since $P_i \cong R$ and M requires two generators. □

43. Infinitely Generated Torsion Modules (short survey)

This short section[1] surveys the relationship between modules of finite length and their counterparts for infinitely generated torsion modules. Essentially all of the results are negative: Krull-Schmidt fails, as does direct-sum cancellation. Moreover, there exist super-decomposable modules (modules with no indecomposable direct summands). The proofs of these examples all make use of wild representation type, and the examples themselves were suggested by earlier results of Brenner and Ringel.

We begin with the unique positive result in this section. An old, well-known theorem of Prüfer states that every abelian group whose elements have bounded orders is the direct sum of copies of cyclic groups of prime power orders (and therefore its decompositions satisfy the Krull-Schmidt theorem). Recall that a bounded HNP R [18.1] has the property that every module of finite length is unfaithful [18.2], and that all classical hereditary orders over a central Dedekind domain are bounded. We say a module M has finite *socle-height* if $M = soc^n(M)$ for some n.

43.1. PROPOSITION. *Let R be a bounded HNP ring. Then every R-module of finite socle-height is a direct sum of cyclic uniserial modules of finite length, and hence its decompositions satisfy the Krull-Schmidt theorem.*

PROOF. See [**Klingler-Levy 95**, 3.4]. □

43.2. EXAMPLE. Let $R = A_1(F)$ where F has characteristic 0. Then there exist (non-finitely-generated), indecomposable R-modules $M \ncong M'$ of socle-height 2 such that $M \oplus M' \cong M \oplus M$.

PROOF. See [**Klingler-Levy 95**, 3.5]. □

43.3. EXAMPLE. Let $R = A_1(F)$ where F has characteristic 0.

(i) There exists an R-module M of socle-height 2, that has no indecomposable direct summands.

(ii) There exists an R-module M of socle-height 2 such that, for every integer $n \geq 2$, M can be written as the direct sum of n indecomposable modules.

PROOF. See [**Klingler-Levy 95**, 3.6] □

44. Notes on Chapter 7

The results[2] which appear in the first four sections, except as specified below, appeared first in [**Levy-Robson 99b**].

§37: [**Warfield 80**, Corollary 7.2] gave an earlier, different proof of 37.1, the Direct Summand Theorem; and 37.3, which describes projective modules over Dedekind prime rings, is an early result, [**Eisenbud-Robson 70a**, Theorem 2.4], motivated by the well-known commutative case.

§38: [**Stafford-Warfield 85**], in comments below Example B in the Introduction, proved that if the cycle lengths of R are unbounded, no bound n on the number of generators of right ideals and modules exists, our 38.4. [**Klingler-Levy 95**, 4.9]

---

[1] In this section R denotes an HNP ring unless the contrary is specified.

[2] In this section R denotes an HNP ring unless the contrary is specified.

gives an example of an HNP ring R that has no cycle towers, has exactly one nontrivial faithful tower, and again no n exists. On the other hand, every right ideal of a Dedekind prime ring can be generated by 2 elements [**McR 01**, 5.7.7]. It is also easy to see that, for the HNP rings that occur in integral representation theory, a bound always exists since all essential right ideals have the same rank, t say, when considered as projective modules over the underlying commutative Dedekind domain and so can be generated by $t+1$ elements over the domain and hence over the HNP ring. This bound, however, is far from sharp, and was part of the motivation to obtain the sharp bound (in terms of cycle towers) given in 38.2.

§39: This material on unique decompositions is new.

§41: The decomposition $M = F \oplus C$, where all composition factors of F belong to faithful towers and those of C belong to cycle towers [41.1] and the fact that the summand F is always a homomorphic image of a uniform right ideal (say) H [41.3] come from [**Klingler-Levy 95**, 4.1, 4.6]. The structure of the summand C as a direct sum of cyclic uniserial modules [41.2] and the structure of modules of finite length in the special case that all faithful towers are trivial [41.4] is adapted from [**Eisenbud-Robson 70b**]. [**McConnell-Robson 73**] proved that indecomposable R-modules of finite length need not be either uniserial or cyclic; and 41.5 and 41.6, which come from [**Klingler-Levy 95**, §3], make this more extreme.

§42: Let $P \supset N$ be finitely generated projective R-modules. We discuss the history of results on stacked complete decompositions:

$$P = \oplus_{i=1}^{n} P_i, \quad N = \oplus_{i=1}^{n} (P_i \cap N). \tag{44.0.1}$$

(Recall [39.1] that *complete* means that the summands P_i are uniform.)

As well as discussing the existence of stacked complete decompositions, we also discuss what we will term the 'matching question': Given an arbitrary complete decomposition $P = \oplus_{i=1}^{n} Q_i$, can we choose the stacked decompositions (44.0.1) with each $P_i \cong Q_i$; i.e. a 'matched' decomposition?

Early (commutative) history.

Related questions go back at least to [**H. J. S. Smith 1861**] and its sequel [**H. J. S. Smith 1873**] where he proved the diagonalization theorem for matrices of integers that is now known as the 'Smith Canonical Form'. This implies the existence of stacked decompositions (44.0.1) when $R = \mathbb{Z}$, the ring of integers; and this in turn implies that every finite abelian group is a direct sum of cyclic groups.

The case of a (commutative) Dedekind domain R goes back to [**Steinitz 1911**], who gave a lengthy proof that matched stacked decompositions exist and so every R-module of finite length is a direct sum of cyclic modules. (He actually obtained the full structure of finitely generated R-modules, in different terminology than is used these days.) [**Chevalley 36**, Appendix] gave a much shorter proof of the existence of stacked decompositions which did not, however, include matching. His result was strong enough to easily imply that every finitely generated R-module is a direct sum of ideals and of cyclic modules of finite length, the latter arranged with increasing annihilators — in modern terminology, with the summands arranged according to a scheme of invariant factors.

For some additional historical comments, see [**Levy-Robson 74**, 3.6].

Noncommutative HNP rings.

[**Knebusch 67**] extended the results of Chevalley about stacked decompositions and finitely generated modules to a classical maximal order.

[**Eisenbud-Robson 70a**, 3.12] proved the existence of stacked decompositions over Dedekind prime rings, again not considering the matching question. (However, unlike the previously mentioned authors, they derived the structure of modules of finite length independently of the stacked decomposition theorem, rather than as a consequence of it.) This was extended to HNP rings with enough invertible ideals (i.e. with every faithful tower trivial [23.2]) by [**Levy-Robson 74**, 3.2] and then to arbitrary HNP rings by [**Klingler-Levy 95**, 4.11]. [**Levy-Robson 74**, 3.3] also gave a positive answer to the matching question for Dedekind prime rings. New facts here are that, for arbitrary HNP rings, one cannot match arbitrary complete decompositions of P [42.4, 42.3], but one can always match the Steinitz classes that occur.

There is a related question: Can a given decomposition $P/N \cong \oplus_{i=1}^{n} X_i$ be lifted to a matched decomposition of $P = \oplus_{i=1}^{n} Q_i$? This is answered, when P/N has finite length, by the Lifting and Straightening Theorem 49.6: it can be if and only if that each Q_i can be mapped onto X_i; and, as 42.4 shows, this is not always the case.

CHAPTER 8

Infinitely Generated Projective Modules

This chapter describes the structure of infinitely generated projective modules over an HNP ring R. Throughout the chapter, *infinitely* or *countably generated* means 'not finitely generated', and *uncountably generated* means 'not countably generated'. Unlike the finitely generated case, it turns out that the isomorphism classes of infinitely generated R-modules is determined by their ranks [46.9]. Perhaps surprisingly, there is a canonical decomposition for uncountably generated projectives [46.7, 46.8], but no corresponding result for countably generated projectives, except in the special case that the Dedekind closures are finite over R [46.10].

45. Genus, Pregenus, Notation

This section[1] extends to infinitely generated projective modules results for the finitely generated case which appear in §§12, 32, 33. After introducing the basic notions mentioned in its title, we prove two theorems, [45.7 and 45.13], which together establish precisely which functions from modspec(R) to cardinal numbers can be the genus of an infinitely generated projective R-module. The section also includes a selection of illustrative examples.

We start, however, with a result about right hereditary rings.

45.1. THEOREM. *Let S be a right hereditary ring. Then every submodule M of a free module F is the direct sum of modules isomorphic to right ideals of S.*

PROOF. Choose a well-ordered basis $\{e_\alpha\}$ for F. For each α, let F_α be the submodule generated by $\{e_\beta : \beta < \alpha\}$ and F'_α that generated by $\{e_\beta : \beta \le \alpha\}$. Each element $m \in M \cap F'_\alpha$ can be written, uniquely, in the form $m = b + e_\alpha s$ with $b \in F_\alpha$ and $s \in S$. The mapping $m \to s$ maps $M \cap F'_\alpha$ to some right ideal $I_\alpha \subseteq S$ and, since S is right hereditary, this mapping splits. So

$$(45.1.1) \qquad M \cap F'_\alpha = (M \cap F_\alpha) \oplus C_\alpha$$

with $C_\alpha \cong I_\alpha$. We will show that $M = \oplus_\alpha C_\alpha$.

First, we show the sum is direct. Suppose that $c_1 + c_2 + \cdots + c_n = 0$ with $c_i \in C_{\alpha_i}$ where $\alpha_1 < \alpha_2 < \cdots < \alpha_n$. Then $c_1 + \cdots + c_{n-1} \in M \cap F_a$ and $c_n \in C_\alpha$. So (45.1.1) shows that $c_n = 0 = c_1 + \cdots + c_{n-1}$. Induction on n completes the argument.

Finally, we show the sum equals M. Assume not; then there is a least β such that there is an element $m \in M \cap F'_\beta$ with $m \notin \oplus_\alpha C_\alpha$. We can write $m = b + c$ with $b \in F_\beta$ and $c \in C_\beta$; and then $b \notin \oplus_\alpha C_\alpha$. However $b \in F'_\gamma$ for some $\gamma < \beta$ which contradicts the minimality of β □

[1]In this section R denotes an HNP ring unless the contrary is specified.

We can improve this for an HNP ring R. First we need to extend the definition of uniform dimension [12.3].

45.2. DEFINITION. Let $P_R \cong \oplus_{i\in I} P_i$, the direct sum of a (finite or) infinite cardinal number of uniform right ideals P_i of R. We call $|I|$ the *uniform dimension* of P and write $\text{udim}(P) = |I|$.

Whether I is finite or infinite, under this new definition $\text{udim}(P)$ equals the number of simple R_{quo}-module terms in the decomposition of the semisimple R_{quo}-module $P \otimes_R R_{\text{quo}} \cong \oplus_i (P_i \otimes_R R_{\text{quo}})$. Thus the new definition makes sense; and when I is finite, 12.4 shows that it agrees with the original definition.

45.3. COROLLARY. *Every infinitely generated projective module P_R has a decomposition $P = \oplus_{i\in I} P_i$ where each P_i is isomorphic to an essential right ideal of R and $|I| = \text{udim}(P)$.*

We will call such a decomposition an *essential decomposition.*

PROOF. Since R is an HNP ring, 45.1 shows that P is isomorphic to a direct sum of right ideals. By 12.7, each right ideal of R is isomorphic to a direct sum of uniform right ideals of R. Thus P is a direct sum of infinitely many uniform right ideals. These uniform right ideals can be grouped together $n = \text{udim}(R)$ at a time. This makes P a direct sum of an infinite number of essential right ideals, say $P = \oplus_{i\in I} P_i$. Then $\text{udim}(P) = |I|{\cdot}n = |I|$ since $|I|$ is infinite. □

45.4. DEFINITIONS, NOTATION. The following definitions are obvious extensions of their counterparts in the finitely generated case [32.1, 33.1]. We continue to use the notation $\mathcal{W}$ for a set of representatives of the isomorphism classes of unfaithful simple R-modules; and then $\text{modspec}(R) = \{0\} \cup \mathcal{W}$.

Let $A = \text{ann}(W)$ for some $W \in \mathcal{W}$. Since R/A is a simple Artinian ring, we can define the *rank* $\rho(P, W)$ of a projective R-module P at W to be α if $P/PA \cong W^{(\alpha)}$ where α is a cardinal number (finite or infinite). Similarly, we let $\rho(P, 0) = \text{udim}(P)$.

Then, in a similar fashion to 33.1, we define the *genus* of P to be the function $\Psi = \Psi(P)$ from $\text{modspec}(R)$ to cardinal numbers given by $\Psi_0 = \text{udim}(P)$ and $\Psi_W = \rho(P, W)$ for all $W \in \mathcal{W}$. If Q is another finitely generated module with $\Psi(Q) = \Psi(P)$, we will describe Q as being *in the genus of* P and write $Q \in \Psi(P)$.

Let Φ be any function from $\text{modspec}(R)$ to the cardinal numbers and $\Phi_{\mathcal{W}}$ be the restriction of Φ to $\mathcal{W}$. So $\text{im}(\Phi)$ and $\text{im}(\Phi_{\mathcal{W}})$ are sets of cardinals and so are well-ordered and have an order type.

If $\Phi = \Psi(P)$, we sometimes abbreviate $\text{im}(\Psi(P))$ to $\text{im}(\Psi)$ or even $\text{im}(P)$ depending on the context; and $\text{im}(\Psi_{\mathcal{W}}(P))$ may be abbreviated to $\text{im}(\Psi_{\mathcal{W}})$ or $\text{im}_{\mathcal{W}}(P)$. A cardinal belonging to $\text{im}(\Psi_{\mathcal{W}}(P))$, i.e. a rank of the form $\rho(P, W)$ for some $W \in \mathcal{W}$, is called a $\mathcal{W}$*-rank* of P.

The Genus Theorem 33.6 shows that, for finitely generated projective modules, any function from $\text{modspec}(R)$ to the set of non-negative integers can occur as a genus, provided only that the conditions of almost standard rank [32.8] and cycle standard rank [32.9] are satisfied. This will be used in this section, often in the following special case.

45.5. COROLLARY. *Let Φ be a function from* $\text{modspec}(R)$ *to the set of non-negative integers such that $\Phi_0 = \text{udim}(R)$. Then $\Phi = \Psi(H)$ for some essential right ideal H if and only if the following two conditions hold.*

(i) (Almost standard rank) $\Phi_W = \rho(R, W)$ *for almost all* $W \in \mathcal{W}$.
(ii) (Cycle standard rank) $\sum_{W \in \mathcal{C}} \Phi_W = \rho(R, \mathcal{C})$ *for every cycle tower* $\mathcal{C}$. □

We now describe what will turn out to be the appropriate conditions on a function from modspec(R) to cardinal numbers for it to be the genus of an infinitely generated projective R-module.

45.6. DEFINITION. Let Φ be a function from modspec(R) to cardinal numbers. We call Φ a *pregenus* (for R) if it satisfies the following three conditions.

(i) Φ_0 is infinite, and $\Phi_W \le \Phi_0$ for all $W \in \mathcal{W}$.
(ii) For every cardinal number $\alpha < \Phi_0$, there are only finitely many $W \in \mathcal{W}$ such that $\Phi_W \le \rho(R^{(\alpha)}, W)$.
(iii) Each cycle tower $\mathcal{C}$ contains at least one $W \in \mathcal{W}$ such that $\Phi_W = \Phi_0$.

We make two comments about (ii). Firstly, if $\mathcal{W}$ is a finite set, then (ii) is redundant; and secondly, in practice, we will use two differing interpretations of (ii) depending upon whether or not α is infinite. When α is infinite, $\rho(R^{(\alpha)}, W) = \alpha$; and then (ii) says:

(ii)$^\infty$ For every infinite cardinal number $\alpha < \Phi_0$, there are only finitely many $W \in \mathcal{W}$ such that $\Phi_W \le \alpha$;

and when $\alpha = n$ is finite, $\rho(R^{(n)}, W) = n\rho(R, W)$ and (ii) says:

(ii)^{fin} For all natural numbers n, there are only finitely many $W \in \mathcal{W}$ with $\Phi_W \le n{\cdot}\rho(R, W)$.

We note an immediate consequence of (ii)^{fin} and (ii)$^\infty$:

(iv) For all $\alpha < \Phi_0$, whether infinite or finite, there are only finitely many $W \in \mathcal{W}$ with $\Phi_W \le \alpha$.

The next theorem shows that these conditions are satisfied by the genus of every infinitely generated projective module.

45.7. THEOREM. *Let P be an infinitely generated projective R-module. Then $\Psi = \Psi(P)$ is a pregenus.*

PROOF. We check the three conditions in the definition of pregenus. Let $P = \oplus_{i \in I} P_i$ be an essential decomposition.

(i) As observed in 45.3, $\Psi_0 = \text{udim}(P) = |I|$, which is infinite by hypothesis. Now consider any $W \in \mathcal{W}$. Since each P_i is finitely generated, we have $\rho(P_i, W) < \aleph_0$ and therefore

$$\rho(P, W) = \sum_{i \in I} \rho(P_i, W) \le |I| \cdot \aleph_0 = |I| = \Psi_0.$$

(ii) Let $\alpha < \Psi_0$. First, suppose that α is infinite and, contrary to (ii)$^\infty$, there are infinitely many $W \in \mathcal{W}$ such that $\rho(P, W) \le \alpha$. Choose a countably infinite subset $\{W_1, W_2, \dots\}$ of these. Let $I(W_j) = \{i \in I \mid \rho(P_i, W_j) \ne 0\}$. Then

$$|I(W_j)| \le \sum_{i \in I(W_j)} \rho(P_i, W_j) = \rho(P, W_j) \le \alpha.$$

Let $I' = \bigcup I(W_j)$. Since the number of sets $I(W_j)$ is $\aleph_0$ we have $|I'| \le \aleph_0{\cdot}\alpha = \alpha < |I|$. Hence there exists $i \in I - I'$; and then $\rho(P_i, W_j) = 0$ for all W_j, contradicting almost standard rank of the essential right ideal P_i [45.5].

Next suppose that $\alpha = n$ is finite. Choose any $n+1$ of the essential right ideals P_i, calling them $P_1, \dots, P_{n+1}$. For almost all $W \in \mathcal{W}$, each of $P_1, \dots, P_{n+1}$ has the standard $\mathcal{W}$-rank, $\rho(P_i, W) = \rho(R, W)$. Thus, after discarding the finite number of nonstandard W, all remaining W satisfy:

$$\rho(P, W) \geq \rho(P_1 \oplus \dots \oplus P_{n+1}, W) = (n+1)\cdot\rho(R, W) > n\cdot\rho(R, W)$$

which establishes (ii)^{fin}.

(iii) Let $\mathcal{C}$ be a cycle tower. For each P_i, cycle standard rank [45.5] requires that $\sum_{W\in\mathcal{C}} \rho(P_i, W) = \rho(R, \mathcal{C})$. Summing over i and remembering that $\rho(R, \mathcal{C})$ is finite and nonzero, we see that

$$\sum_{W\in\mathcal{C}} \rho(P, W) = |I|\cdot\rho(R, \mathcal{C}) = |I| = |\Psi_0|.$$

Since $\mathcal{C}$ is a finite set, at least one term $\rho(P, W)$ must therefore equal $|\Psi_0|$, as desired. □

45.8. Remark. There are some useful, further consequences of the conditions on a pregenus which concern the order type of $\mathrm{im}(\Phi)$ and $\mathrm{im}_{\mathcal{W}}(\Phi)$. Since the genus $\Psi(P)$ of any infinitely generated projective module P is a pregenus, these consequences also apply to $\Psi(P)$. This fact will be used henceforth without special comment.

We use the notation ω to denote the order type of the natural numbers.

45.9. Lemma. *Let Φ be a pregenus.*

(i) *If there is a $W \in \mathcal{W}$ with $\Phi_W = \Phi_0$ then the order types of $\mathrm{im}(\Phi)$ and $\mathrm{im}_{\mathcal{W}}(\Phi)$ are equal. Otherwise the order type of $\mathrm{im}(\Phi)$ is 1 more than the order type of $\mathrm{im}_{\mathcal{W}}(\Phi)$.*

(ii) *The order type of $\mathrm{im}(\Phi)$ is either finite or is $\omega + 1$. Consequently, the order type of $\mathrm{im}_{\mathcal{W}}(\Phi)$ is either finite or ω or $\omega + 1$.*

(iii) *If the order type of $\mathrm{im}_{\mathcal{W}}(\Phi)$ is finite and $\mathcal{W}$ is an infinite set, then $\max(\mathrm{im}_{\mathcal{W}}(\Phi)) = \Phi_0$.*

(iv) *If the order type of $\mathrm{im}_{\mathcal{W}}(\Phi)$ is ω or $\omega + 1$, then $\sup(\Phi_W | W \in \mathcal{W}, \Phi_W < \Phi_0) = \Phi_0$.*

(v) *If Φ_0 is uncountable, then $\mathrm{im}(\Phi)$ contains only finitely many, finite cardinals.*

(vi) *The set $\{W \in \mathcal{W} \mid \Phi_W < \Phi_0\}$ is countable.*

Proof. (i) $\mathrm{im}(\Phi)$ has only one possible additional member, namely Φ_0; and $\Phi_0 \geq \Phi_W$ for all $W \in \mathcal{W}$.

(ii) Let $\alpha < \Phi_0$. By 45.6(iv), there are only finitely many $W \in \mathcal{W}$ with $\Phi_W \leq \alpha$ and hence only finitely many cardinals $\beta \in \mathrm{im}(\Phi)$ with $\beta \leq \alpha < \Phi_0$. The only well-ordered sets which satisfy this condition on $\mathrm{im}(\Phi)$ are finite sets, ω and $\omega + 1$. However, $\mathrm{im}(\Phi)$ has a final element, namely Φ_0; so Φ must either have finite order or order type $\omega + 1$.

(iii) Let $\beta = \max(\mathrm{im}_{\mathcal{W}}(\Phi))$ and suppose $\beta < \Phi_0$. By 45.6(iv), there are only finitely many $W \in \mathcal{W}$ with $\Phi_W \leq \beta$, thus contradicting the hypothesis that $\mathcal{W}$ is an infinite set.

(iv) Note first that, since Φ_W takes infinitely many values, $\mathcal{W}$ must be an infinite set. Let $\beta = \sup(\Phi_W | W \in \mathcal{W}, \Phi_W < \Phi_0)$. If $\beta < \Phi_0$ then, by 45.6(iv), there are only finitely many $W \in \mathcal{W}$ with $\Phi_W \leq \beta$ and so $\mathcal{W}$ is finite, a contradiction.

(v) Since in this case $\aleph_0 < \Phi_0$, condition (i) of 45.6 implies that $\Phi_W < \aleph_0$ for only finitely many $W \in \mathcal{W}$.

(vi) By (ii), there are only countably many $\alpha \in \operatorname{im}(\Phi)$ and, for each such α, only finitely many $W \in \mathcal{W}$ with $\Phi_W = \alpha$, by 45.6(iv). □

We now aim towards proving that, for every pregenus, there is a projective module whose genus takes that form. Note that (v) above suggests that we might need to allow for different behaviours in the countable and uncountable cases. The next few results provide two different constructions of infinitely generated projective modules, the first of which is aimed specifically at the countable case.

45.10. LEMMA. *Let $\mathcal{W}$ be the disjoint union of two subsets $\mathcal{W}_{fin}$ and $\mathcal{W}_\infty$. For each positive integer n and each $W \in \mathcal{W}_{fin}$, let $q(W)$ be a non-negative integer and $r(W)$ be an integer with $0 \le r(W) < n$. Let P_n be an essential right ideal such that:*

$$\begin{cases} \text{for each } W \in \mathcal{W}_{fin},\ \rho(P_n, W) = \begin{cases} \rho(R,W) & \text{if } n \le q(W) \\ r(W) & \text{if } n = q(W)+1 \\ 0 & \text{if } n > q(W)+1 \end{cases} \\ \text{for each } W \in \mathcal{W}_\infty,\ \rho(P_n, W) \ne 0. \end{cases} \tag{45.10.1}$$

Let $P = \oplus_{n=1}^{\infty} P_n$. Then P is a countably generated projective module and its genus $\Psi(P)$ is given as follows:

$\Psi_W = \Psi_0 = \aleph_0$ for each $W \in \mathcal{W}_\infty$; and

$\Psi_W = q(W) \cdot \rho(R, W) + r(W)$ for each $W \in \mathcal{W}_{fin}$.

PROOF. $\Psi_W = \Psi_0 = \aleph_0$ for each $W \in \mathcal{W}_\infty$ since $\rho(P_n, W)$ is nonzero in each of the infinitely many terms of the sum. For $W \in \mathcal{W}_{\text{fin}}$, the first two lines of 45.10.1 contribute $q(W) \cdot \rho(R, W) + r(W)$ to the sum and the third line gives 0. □

45.11. REMARK. Recall from 29.3 that, for a right integral overring S of R, $\mathcal{Z}_R(S)$ denotes the set of isomorphism classes of simple R-modules which are not images of S_R. By 29.4, $\mathcal{Z}_R(S)$ is finite if and only if S is a right finite overring of R; and 29.5 shows that $\mathcal{Z}_R(S)$ uniquely specifies S and that if S' is another right integral overring of R then $\mathcal{Z}_R(S) \subset \mathcal{Z}_R(S') \Leftrightarrow S \subset S'$.

45.12. LEMMA. *Let*

$$P = H \oplus S_1^{(\alpha_1)} \oplus S_2^{(\alpha_2)} \oplus \dots \tag{45.12.1}$$

where H_R is finitely generated projective, $S_1, S_2, \dots$ is a nonempty, finite or infinite sequence of right finite overrings of R with $S_n \subset S_{n+1}$ for each $n \ge 1$, and $\alpha_1 < \alpha_2 < \dots$ are infinite cardinal numbers. For each n, let $\mathcal{F}_n = \mathcal{Z}_R(S_n)$, so that $\mathcal{F}_1 \subset \mathcal{F}_2 \subset \dots$.

Then $\rho(P, W)$, for $W \in \mathcal{W}$, is given by exactly one of the following statements.

(i) *If $W \in \mathcal{F}_1$ then $\rho(P, W) = \rho(H, W)$ and so is finite.*

(ii) *If $W \in \mathcal{F}_{n+1} - \mathcal{F}_n$ with $n \ge 1$ then $\rho(P, W) = \alpha_n$.*

(iii) *Suppose $W \in \mathcal{W} - \cup_n \mathcal{F}_n$.*

(a) *If (45.12.1) contains only finitely many terms and $S_m^{(\alpha_m)}$ is the last term, then $\rho(P, W)$ equals α_m, and is the largest $\mathcal{W}$-rank of P.*
If, further, $\mathcal{W}$ is an infinite set or R has at least one cycle tower, then $\alpha_m = \operatorname{udim}(P)$.

(b) *If (45.12.1) contains infinitely many terms then $\rho(P, W) = \operatorname{udim}(P)$.*

PROOF. Since each $\rho(S_n, W)$ is finite and each α_n is infinite, the definition of $\mathcal{F}_n$ shows that $\rho(S_n^{(\alpha_n)}, W) = 0$ if $W \in \mathcal{F}_n$ and $\rho(S_n^{(\alpha_n)}, W) = \alpha_n$ if $W \notin \mathcal{F}_n$.

(i) Since $W \in \mathcal{F}_1$ we have $W \in \mathcal{F}_i$ for all $i > 1$ and therefore $\rho(P, W) = \rho(H, W)$, which is finite since H is finitely generated.

(ii) $\rho(P, W) = \rho(H, W) + \alpha_1 + \alpha_2 + \cdots + \alpha_n = \alpha_n$ where the final equality holds because each α_i is infinite and $\alpha_i < \alpha_{i+1}$.

(iii)(a) As in (ii), $\rho(P, W) = \rho(H, W) + \alpha_1 + \alpha_2 + \cdots + \alpha_m = \alpha_m$, which is maximal since the only other possible $\mathcal{W}$-ranks of P are those given in statements (i) and (ii).

If $\mathcal{W}$ is an infinite set or R has at least one cycle tower, then 45.9 states that $\operatorname{udim}(P)$ is the largest $\mathcal{W}$-rank of P.

(iii)(b) In this case $\rho(P, W) = \rho(H, W) + \sum_{n=1}^{\infty} \alpha_n = \sup_n \alpha_n$; and $\sup_n \alpha_n = \operatorname{udim}(P)$ by 45.9. □

The next theorem is the converse to 45.7 and is the main result of this section.

45.13. THEOREM. *Every pregenus is the genus of some infinitely generated projective right R-module.*

PROOF. Let Φ be a pregenus for R. We give two different arguments which, together, cover all cases.

(i) *First, we deal with the case when Φ_0 is countable; i.e. $\Phi_0 = \aleph_0$.*

By 45.6(i), every Φ_W is a nonnegative integer or $\aleph_0$. Let $\mathcal{W}_{\text{fin}} = \{W \in \mathcal{W} \mid \Phi_W < \aleph_0\}$ and $\mathcal{W}_\infty = \{W \in \mathcal{W} \mid \Phi_W = \aleph_0\}$.

For every $W \in \mathcal{W}_{\text{fin}}$, the division algorithm yields an integral expression $\Phi_W = q(W) \cdot \rho(R, W) + r(W)$ with $q(W) \geq 0$ and $0 \leq r(W) < \rho(R, W)$. If we can find, for each n, an essential right ideal P_n satisfying the four conditions of 45.10.1 then 45.10 tells us that Φ is the genus of $P = \oplus_{n=1}^{\infty} P_n$.

Note that, if one simply set $P_n = R$, then the first and last of the conditions on P_n would automatically be satisfied. So the only problems arise from those W in the set $\mathcal{W}_n = \{W \in \mathcal{W}_{\text{fin}} \mid \Phi_W < n \cdot \rho(R, W)\}$; for these W, the ranks need to be smaller. Note that $W \in \mathcal{W}_n \Leftrightarrow q(W) < n$ and that, by 45.6(ii)^{fin}, $\mathcal{W}_n$ is a finite set.

We see from 45.5 that if W belongs to a faithful tower, reducing the rank at W does not violate the conditions. So there is an essential right ideal having the same ranks as R except that the rank at W has the smaller value sought. Suppose however that W belongs to a cycle tower. Note that, by 45.6(iii), there is a $W' \in \mathcal{W}_\infty$ which belongs to that same cycle tower. So, using 45.5 again, there is an essential right ideal having the desired smaller rank at W, a correspondingly larger rank at W', and all other ranks unchanged. In this way, we see that the ranks at the finitely many members of $\mathcal{W}_n$ can all be organized as required; and the essential right ideal thus obtained is the P_n required.

(ii) *Next we deal with the case when $\{n \in \operatorname{im}(\Phi) \mid n < \aleph_0\}$ is finite. This, by 45.9(v), includes the case when Φ_0 is uncountable.*

It follows from (iii) and (iv) of the definition of a pregenus that the set $\mathcal{F} = \{W \in \mathcal{W} \mid \Phi_W < \aleph_0\}$ is a finite set and that $\mathcal{F}$ contains no complete cycle tower. Therefore, by 33.8, there is a finitely generated projective module H such that $\rho(H, W) = \Phi_W$ for each $W \in \mathcal{F}$.

Let $\alpha_1 < \alpha_2 < \dots$ be the infinite cardinals occurring in $\mathrm{im}_{\mathcal{W}}(\Phi)$. Note, using 45.6(ii)$^\infty$ and (iii), that for each i,

$$\mathcal{F}_i = \{W \in \mathcal{W} \mid \Phi_W < \alpha_i\} = \{W \in \mathcal{W} \mid \Phi_W \leq \alpha_{i-1}\}$$

is a finite set which contains no complete cycle tower. Let S_i be the right finite overring which has rank 0 precisely for the $W \in \mathcal{F}_i$ and let $P = H \oplus \Sigma_{i=1}^\infty S_i^{(\alpha_i)}$. Then 45.12 shows that $\Phi(P) = \Psi$. □

Part (ii) of 45.9 describes various possibilities for the pair of order types of $\mathrm{im}(\Phi)$ and $\mathrm{im}_{\mathcal{W}}(\Phi)$. The proof of the theorem above shows that, if a particular pair of order types $\mathrm{im}(\Phi)$ and $\mathrm{im}_{\mathcal{W}}(\Phi)$ exists for some pregenus Φ over R, then Φ is the genus of some P. What is not yet clear is whether, or when, such pairs of order types can exist. The answer, as the next result shows, depends upon the ring R itself; but for each feasible pair, there is indeed a ring R for which it occurs. We need to distinguish between the countable and uncountable cases. The argument makes use of §25 which demonstrates that any specified configuration of towers will occur for some HNP ring R. We write Ψ for $\Psi(P)$ throughout the next result.

45.14. COROLLARY. *In each of the following parts, (a) and (b) are equivalent.*

(i) (a) *There is both a countably generated and an uncountably generated projective R-module P with $\mathrm{im}(\Psi)$ of order type $n+1$ and $\mathrm{im}_{\mathcal{W}}(\Psi)$ of order type $n+1$.*

(b) *$|\mathcal{W}| > n$ and $|\mathcal{W}^-| \geq n$, where $\mathcal{W}^-$ denotes $\mathcal{W}$ after the removal of one simple module from each cycle tower.*

(ii) (a) *There is both a countably generated and an uncountably generated projective R-module P with $\mathrm{im}(\Psi)$ of order type $n+1$ and $\mathrm{im}_{\mathcal{W}}(\Psi)$ of order type n.*

(b) *R has no cycle towers and $\aleph_0 > |\mathcal{W}| \geq n$.*

(iii) (a) *There is both a countably generated and an uncountably generated projective R-module P with $\mathrm{im}(\Psi)$ of order type $\omega+1$ and $\mathrm{im}_{\mathcal{W}}(\Psi)$ of order type $\omega+1$.*

(b) *R has infinitely many nontrivial towers.*

In both cases described in (a), it can be arranged that all but finitely many members of $\mathrm{im}_{\mathcal{W}}(\Psi)$ are infinite; and, in the countably generated case only, it can be arranged that all but one of the cardinals in $\mathrm{im}_{\mathcal{W}}(\Psi)$ are finite.

(iv) (a) *There is both a countably generated and an uncountably generated projective R-module P with $\mathrm{im}(\Psi)$ of order type $\omega+1$ and $\mathrm{im}_{\mathcal{W}}(\Psi)$ of order type ω.*

(b) *R has no cycle towers and countably many nontrivial faithful towers.*

In both cases described in (a), it can be arranged that all but finitely many members of $\mathrm{im}_{\mathcal{W}}(\Psi)$ are infinite; and, in the countably generated case only, it can be arranged that all the cardinals in $\mathrm{im}_{\mathcal{W}}(\Psi)$ are finite.

PROOF. (i) Suppose that $|\mathcal{W}^-| \geq n$ and choose $W_1, W_2, \dots, W_n \in \mathcal{W}^-$. It is clear from (ii) of the proof of the theorem that, since we have avoided choosing at least one simple module in each cycle tower, we can construct a countably or an uncountably generated P so that $\rho(P, W_i) = i$ for each i and $\rho(P, W) = \rho(P, 0)$ for all other $W \in \mathcal{W}$.

Conversely, given such a P, we need to have n members of $\mathcal{W}$ having distinct ranks different from the uniform rank of P. Since these simple modules cannot include the whole of any cycle tower, $|\mathcal{W}^-| \geq n$. There also needs to be at least one $W \in \mathcal{W}$ with $\rho(P, W) = \rho(P, 0)$; so $|\mathcal{W}| > n$.

(ii) Given the conditions on R, partition $\mathcal{W}$ into n nonempty sets $\mathcal{S}_i$ say, with $i = 1, \dots, n$; and for each $W \in \mathcal{S}_i$, we can arrange in the construction in (ii) of the proof of the theorem, that the rank of P at W_i is i. That exhausts the members of $\mathcal{W}$; yet $\rho(P, 0)$ is any infinite cardinal.

Conversely, suppose that there is an infinitely generated projective R-module P with $\operatorname{im}(\Psi)$ of order type $n + 1$ and $\operatorname{im}_{\mathcal{W}}(\Psi)$ of order type n. Evidently there can be no cycle tower, by (iii) of the definition of a pregenus. We know that for every $W \in \mathcal{W}$, $\rho(P, W) < \rho(P, 0)$, that there are only n smaller values that this rank takes, and that there are only finitely many $W \in \mathcal{W}$ having any particular rank in P less than $\rho(P, 0)$. Hence $\mathcal{W}$ is as described.

(iii) First suppose R has the towers described. For each $i \geq 0$, we choose $W_i \in \mathcal{W}$, each from a different nontrivial tower. We also choose an ascending chain of cardinals α_i for $i \geq 1$ which are either all finite or finitely many are finite and the remainder infinite. Depending upon which of these alternatives is chosen, parts (i) or (ii) of the proof of the theorem describe how to find a module P having $\rho(P, W_i) = \alpha_i$, as required, for each i; and for each other $W \in \mathcal{W}$ not already dealt with (including W_0), $\rho(P, W) = \rho(P, 0) = \alpha > \alpha_i$ for each i.

Conversely, if such a P exists, then there are infinitely many simple modules $W_i \in \mathcal{W}$ each having a different rank in P. These must all belong to nontrivial towers and must exclude at least one member of each cycle tower. Since each tower is finite, R must have infinitely many nontrivial towers.

(iv) The hypothesis on the towers in R ensures that $\mathcal{W}$ is countable. List its members as $W_1, W_2, \dots$. Choose increasing values α_i desired for each $\rho(P, W_i)$ and pick $\alpha > \alpha_i$ to be $\rho(P, 0)$. Once again, depending upon whether the α_i are all finite or not, we use the construction in part(i) or part (ii) of the proof of the theorem to yield P as desired. Note that, since all the W have been dealt with, then $\alpha \notin \operatorname{im}_{\mathcal{W}}$.

Conversely, if such a P exists, then the set of $W \in \mathcal{W}$ having rank in P less than $\rho(P, 0)$ is countable and must equal $\mathcal{W}$. So there are no cycle towers and only countably many nontrivial faithful towers. □

45.15. REMARK. We note that the symmetry on towers given by 21.2 shows that the hypotheses on R are equivalent to their left-handed versions. Hence, in each case, there is also a left module with similar properties.

We note a consequence of 45.14 (iii).

45.16. COROLLARY.

(a) *The following conditions are equivalent for R and a Dedekind right closure S.*
   (i) *R has a countably generated projective module P_R with infinitely many distinct finite ranks.*
   (ii) *R has infinitely many nontrivial towers.*
   (iii) *S is not right finite over R.*

(b) *No such P_R occurs if R is a finitely generated module over its centre.*

PROOF. (a) The equivalence of the first three conditions is a combination of 45.14(iii) with 30.5.

(b) This holds by 30.10. □

46. Main Structure Theorems

Throughout this section P denotes an infinitely generated projective R-module[1] which therefore has an essential decomposition $P = \oplus\{P_i \mid i \in I\}$ with $|I|$ infinite.

As usual, $\Psi(P)$ denotes the genus of P which, by 45.7 is a pregenus. We proceed directly to our first main result which will be extended to all infinitely generated P in 46.9.

46.1. THEOREM. *If P is countably generated, Q_R is projective, and $\Psi(P) = \Psi(Q)$ then $P \cong Q$.*

PROOF. Consider essential decompositions of both P and Q: $P = P_1 \oplus P_2 \oplus \dots$ and $Q = Q_1 \oplus Q_2 \oplus \dots$. We claim:

(46.1.1) for every initial segment $A = P_1 \oplus P_2 \oplus \dots \oplus P_m$ of the above decomposition of P, there is an initial segment $B = Q_1 \oplus Q_2 \oplus \dots \oplus Q_n$ of the decomposition of Q, with $n > m$, such that A is isomorphic to a proper direct summand of B.

The Direct Summand Theorem 37.1 tells us that, since $\operatorname{udim}(A) < \operatorname{udim}(B)$, this is so provided that $\Psi_W(A) \leq \Psi_W(B)$ for all $W \in \mathcal{W}$. Let $B' = \oplus_{i=1}^{m+1} Q_i$. By almost standard rank [see 32.8], $\rho(A, W) = \rho(B', W)$ for almost all $W \in \mathcal{W}$. Let W be one of the finitely many exceptions. Since $\Psi(P) = \Psi(Q)$, we can choose $n(W) > m$ such that

$$\sum_{i=1}^{n(W)} \rho(Q_i, W) \geq \sum_{i=1}^{m} \rho(P_i, W).$$

Taking $n = \max(n(W))$ as W varies across the finite number of exceptions (or $m+1$ if there are no exceptions), establishes the claim.

Now fix any initial segment A_1 of the decomposition of P, containing (say) $m(1)$ terms. By (46.1.1) there is an initial segment of B_1 of the decomposition of Q containing $n(1) > m(1)$ terms such that the first isomorphism in (46.1.2) holds for some $X_1 \neq 0$. Reversing the roles of P and Q, we find an initial segment A_2 of the decomposition of Q, containing $m(2) > n(1)$ terms, such that the second isomorphism in (46.1.2) holds for some $Y_1 \neq 0$.

(46.1.2) $$A_1 \oplus X_1 \cong B_1, \qquad B_1 \oplus Y_1 \cong A_2.$$

Continuing in this fashion, we obtain an infinite sequence of initial segments $A_1, B_1, A_2, B_2, \dots$ each containing strictly more terms than the previous one, such that

(46.1.3) $$A_2 \oplus X_2 \cong B_2, \qquad B_2 \oplus Y_2 \cong A_3;$$

and so on.

For $j > i$ (and by slight abuse of notation) define $A_j - A_i = \oplus_{m(i)<k\leq m(j)} P_k$, the direct sum of all P_k that appear in A_j but not in A_i; and define $B_j - B_i$ analogously. The two isomorphisms in (46.1.2) yield:

$$A_1 \oplus (A_2 - A_1) \cong A_2 \cong A_1 \oplus X_1 \oplus Y_1.$$

We have $\operatorname{udim}(X_1 \oplus Y_1) \geq 2$ since both X_1 and Y_1 are nonzero. Therefore we can cancel A_1 [34.6] getting the first isomorphism in (46.1.4) below. To obtain the

[1]In this section R denotes an HNP ring unless the contrary is specified.

second isomorphism in (46.1.4), use the first isomorphism in (46.1.3) together with the second isomorphism in (46.1.2).

$$(46.1.4) \qquad A_2 - A_1 \cong X_1 \oplus Y_1, \qquad B_2 - B_1 \cong Y_1 \oplus X_2.$$

After obtaining an infinite sequence of pairs of isomorphisms analogous to (46.1.4) we show that $P \cong Q$ as follows:

$$\begin{aligned} P &\cong A_1 \oplus (A_2 - A_1) \oplus (A_3 - A_2) \oplus \dots \\ &\cong A_1 \oplus (X_1 \oplus Y_1) \oplus (X_2 \oplus Y_2) \oplus \dots \\ &\cong (A_1 \oplus X_1) \oplus (Y_1 \oplus X_2) \oplus (Y_2 \oplus X_3) \oplus \dots \\ &\cong B_1 \oplus (B_2 - B_1) \oplus (B_3 - B_2) \oplus \dots \quad \cong Q. \end{aligned}$$ □

46.2. COROLLARY. *Every infinitely generated projective module over a Dedekind prime ring is free.*

PROOF. Let P be the given projective. Then P has an essential decomposition. So we can write P as a direct sum of modules, each of which is the direct sum of countably many essential right ideals. It will be enough, therefore, to prove the result for the countable case. Let F be a free module of countably infinite rank; then $\operatorname{udim}(P) = \operatorname{udim}(F) = \aleph_0$. By 33.3 every essential right ideal of a Dedekind prime ring shares the genus of the ring. We conclude that $\Psi(P) = \Psi(F)$ and hence $P \cong F$, by 46.1. □

Next we give an extension of 32.11 to the infinitely generated case. We describe an overring $S(Q)$ associated with any nonzero projective module Q_R, finitely or infinitely generated. As in 29.3, $\mathcal{Z}(Q)$ denotes the set of simple isomorphism types which are not homomorphic images of Q; by 49.12, these are all unfaithful.

46.3. LEMMA.

(i) *Let Q be a nonzero projective R-module. Then $\mathcal{Z}(Q)$ is a finite set and contains no cycle tower.*

(ii) *There is a unique right finite overring $S(Q)$ such that $\mathcal{Z}_R(S(Q)) = \mathcal{Z}(Q)$; and if $0 \neq \mathcal{Z}(Q') \subset \mathcal{Z}(Q)$ then $S(Q') \subset S(Q)$.*

PROOF. This is simply a restatement of 32.11 once we note that any complete decomposition of Q provides a decomposition $Q = Q_1 \oplus M$ where Q_1 is a uniform right ideal. □

We now begin working toward the proof of the second main theorem, the canonical decomposition in 46.7.

46.4. LEMMA. *Suppose that $\mathcal{Z}(P_R) = \mathcal{W}$. Then:*

(i) *$\mathcal{W}$ is a finite set and R has no cycle towers;*

(ii) *$P \cong S^{(\alpha)}$, where $\alpha = \operatorname{udim}(P)$ and $S = S(P)$.*

PROOF. (i) Clear from 46.3(i).

(ii) Since $I = \operatorname{udim}(P)$ and I is a disjoint union of countably infinite sets, it is enough to prove the case when $|I| = \aleph_0$. Since both P and $S^{|I|}$ are then countably generated, we need only check that $\Psi(P) = \Psi(S^{|I|})$, by 46.1. However, this holds since all $\mathcal{W}$-ranks of both P and S are 0 and $\operatorname{udim}(P) = \operatorname{udim}(S^{|I|}) = \aleph_0$. □

46.5. LEMMA. *Suppose that $\alpha = \operatorname{udim}(P)$ is a $\mathcal{W}$-rank of P and is the unique nonzero $\mathcal{W}$-rank of P. Then $P \cong S^{(\alpha)}$ where $S = S(P)$.*

PROOF. Pick an essential decomposition $P = \oplus_{i \in I} P_i$. For a subset $J \subseteq I$, let $\rho(J, W) = \sum_{i \in J} \rho(P_i, W)$ and $\mathcal{Z}(J) = \mathcal{Z}(\oplus_{i \in J} P_i)$. The crux of this proof is to show:

(46.5.1) There is a partition $\mathcal{G}$ of I into the disjoint union of α finite sets I_ν each of which satisfies $\mathcal{Z}(I_\nu) = \mathcal{Z}(P)$.

The initial step is to prove that there is at least one such set I_ν. Choose any $i(0) \in I$. Almost standard rank [45.5] implies that $\rho(P_{i(0)}, W) = 0$ for only finitely many $W \in \mathcal{W}$ (including, of course, all $W \in \mathcal{Z}(P)$). Let $\{W_1, W_2, \ldots, W_n\} = \mathcal{Z}(P_{i(0)}) - \mathcal{Z}(P)$. For each W_j, there exists $i(j) \in I$ such that $\rho\big(P_{i(j)}, W_i\big) \neq 0$. Then $I_\nu = \{i(0), \ldots, i(n)\}$ is as claimed.

Next, consider $\mathcal{S}$, the set of all families $\mathcal{F}$ of disjoint finite subsets $I_\nu \subseteq I$ such that each each I_ν satisfies $\mathcal{Z}(I_\nu) = \mathcal{Z}(P)$. The set $\mathcal{S}$ is partially ordered by inclusion (of families of subsets); and the union of any totally ordered subset of $\mathcal{S}$ is obviously again in $\mathcal{S}$. Therefore, Zorn's Lemma can be applied to show that $\mathcal{S}$ contains a maximal element $\mathcal{F}_0$.

Our next step concerns $\mathcal{F}_0$. Let β be the number of subsets of I in the family $\mathcal{F}_0$. Evidently $\beta \leq \alpha = \operatorname{udim}(P)$; suppose that $\beta < \alpha$. Consequently, $|\cup \mathcal{F}_0| = \beta$ and hence $|I - \cup\mathcal{F}_0| = \alpha$. Therefore the module $P' = \oplus\{P_i \mid i \in I - \cup\mathcal{F}_0\}$ satisfies the same hypotheses as did P in the statement of the lemma. Therefore, by the initial step above, there is a finite subset $I_\mu \subseteq I - \cup\mathcal{F}_0$ satisfying $\mathcal{Z}(I_\mu) = \mathcal{Z}(P)$. But then the family consisting of $\mathcal{F}_0$ together with I_μ contradicts maximality of $\mathcal{F}_0$, proving that $\beta = \alpha$.

It might happen that $\cup\mathcal{F}_0 \subset I$. However, since $\mathcal{F}_0$ contains α sets and $|I - \cup\mathcal{F}_0| \leq |I| = \alpha$, we can enlarge each set in $\mathcal{F}_0$ by at most one element of $I - \cup\mathcal{F}_0$, getting a new family $\mathcal{G} \in \mathcal{S}$ satisfying the same conditions and with $\cup\mathcal{G} = I$. This family $\mathcal{G}$ is the partition of I needed to complete the proof of (46.5.1).

For each $G \in \mathcal{G}$, let $Q_G = \oplus_{\gamma \in G} P_\gamma$. Then each Q_G is a finitely generated projective R-module satisfying $\mathcal{Z}(Q_G) = \mathcal{Z}(P)$ and $P = \oplus_{G \in \mathcal{G}} Q_G$. Let X be the direct sum of any countable collection of the modules Q_G and let $S = S(P)$. Then it is readily checked that $\Psi(X) = \Psi(S^{(\aleph_0)})$ and hence, by 46.1, $X \cong S^{(\aleph_0)}$.

Now, every infinite set of cardinality α is the disjoint union of α countably infinite sets. Therefore we can rewrite $P = \oplus_{G \in \mathcal{G}} Q_G$ as the direct sum of α terms, each of which is the direct sum of $\aleph_0$ of the terms Q_G and so is isomorphic to $S^{(\aleph_0)}$. We conclude that $P \cong S^{(\alpha)}$, as desired. □

46.6. LEMMA. *Suppose that the smallest nonzero $\mathcal{W}$-rank of P is an infinite cardinal $\alpha < \operatorname{udim}(P)$. Then there is a decomposition $P = P_\alpha \oplus Q$ such that each of P_α and Q is a direct sum of the original summands P_i in the essential decomposition $P = \oplus_{i \in I} P_i$ and such that the following conditions hold:*

(i) $\alpha = \operatorname{udim}(P_\alpha)$ *and is the unique nonzero $\mathcal{W}$-rank of P_α.*
(ii) *All nonzero ranks of Q are $> \alpha$.*
(iii) $\mathcal{Z}(P) = \mathcal{Z}(P_\alpha) \subset \mathcal{Z}(Q)$.
(iv) $P_\alpha \cong S(P)^{(\alpha)}$.

PROOF. Since $\alpha < \operatorname{udim}(P)$, the set $\mathcal{X} = \{W \in \mathcal{W} \mid \rho(P, W) = \alpha\}$ is finite, by 45.6(ii)$^\infty$, and nonempty, by hypothesis. As a first approximation to our desired decomposition, let P' be the sum of those P_i such that $\rho(P_i, W) \neq 0$ for some $W \in \mathcal{X}$, and let Q' be the sum of the remaining summands P_i; then $P = P' \oplus Q'$.

Temporarily fix some $W \in \mathcal{X}$. If $\rho(P_i, W) \neq 0$ for some P_i then $\rho(P_i, W)$ is finite and nonzero; and yet $\rho(P, W) = \alpha$. Thus there must be α summands P_i of P' with $\rho(P_i, W) \neq 0$. This applies to each of the finitely many members of $\mathcal{X}$. Hence

$$(46.6.1) \qquad \operatorname{udim}(P') = \alpha; \text{ and } \rho(P', W) = \alpha \text{ for all } W \in \mathcal{X}.$$

The construction of P' and the minimality of α ensure that

$$(46.6.2) \qquad \rho(Q', W) = 0 \text{ for all } W \text{ with } \rho(P, W) \leq \alpha.$$

Also, because P' is the sum of only α summands P_i of P,

$$(46.6.3) \qquad \operatorname{udim}(Q') = \operatorname{udim}(P); \text{ and } \rho(Q', W) = \rho(P, W) \text{ if } \rho(P, W) > \alpha.$$

Let $\mathcal{Y} = \{W \in \mathcal{W} \mid \rho(P', W) < \alpha \text{ and } \rho(P, W) \geq \alpha\}$; and for each $\beta < \alpha$, let $\mathcal{Y}_\beta = \{W \in \mathcal{W} \mid \rho(P', W) \leq \beta \text{ and } \rho(P, W) \geq \alpha\}$. So $\mathcal{Y} = \cup \mathcal{Y}_\beta$. Note that $\mathcal{Y}_\beta$ is finite, by 45.6(ii)$^\infty$, since $\beta < \alpha = \operatorname{udim}(P')$.

Let $W \in \mathcal{Y}$; so $\rho(P, W) \geq \alpha$. Note that if $\rho(P, W) = \alpha$, then $W \in \mathcal{X}$ and so $\rho(P', W) = \alpha$ by (46.6.1); but this contradicts the definition of $\mathcal{Y}$. So we see that $\rho(P, W) > \alpha$ and thus $\rho(Q', W) > \alpha$ by (46.6.3). So we can identify α summands P_i belonging to Q' with $\rho(P_i, W) \neq 0$.

Consider how many summands of Q' are involved over all $W \in \mathcal{Y}$. We know that $\mathcal{Y}_\beta$ is finite for any cardinal $\beta < \alpha$; so there are only α summands involved in relation to the members of $\mathcal{Y}_\beta$. Since there are α cardinals $\beta < \alpha$, and each involves at most α summands, there are no more than $\alpha^2 = \alpha$ summands involved in total.

We remove all these summands from Q', letting Q be the sum of the remaining summands of Q'; and we let P_α be what is obtained by adding them all to P'. By construction, $\rho(P_\alpha, W) = \alpha$ for each $W \in \mathcal{Y}$. We already had $\rho(P', W) = \alpha$ for each $W \in \mathcal{X}$. Thus $\rho(P_\alpha, W) = \alpha$ for all W with $\rho(P, W) \neq 0$. So (i) holds true and also $\mathcal{Z}(P) = \mathcal{Z}(P_\alpha)$. The remainder of (iii) follows since $\mathcal{Z}(P) \subset \mathcal{Z}(Q') \subseteq \mathcal{Z}(Q)$, using (46.6.2). We have only removed α summands from Q'; so all nonzero ranks of Q' are greater than α and (ii) holds. Finally, 46.5 gives (iv). □

46.7. THEOREM (Canonical Decomposition). *Let P be an infinitely generated projective R-module with only finitely many distinct finite ranks (e.g. any uncountably generated P, by 45.9). Then P has a decomposition:*

$$(46.7.1) \qquad P \cong H \oplus S_1^{(\alpha_1)} \oplus S_2^{(\alpha_2)} \oplus \dots$$

where H_R is finitely generated, $S_1 \subset S_2 \subset \dots$ is a nonempty, strictly increasing finite or countably infinite sequence of right finite overrings of R, and $\alpha_1 < \alpha_2 < \dots$ are infinite cardinal numbers.

Remark. 46.10 shows that this canonical decomposition can fail to exist for countably generated projective modules.

PROOF. Since P has only finitely many distinct finite $\mathcal{W}$-ranks, and since, for any fixed finite rank, there are only finitely many $W \in \mathcal{W}$ that have that rank [45.6(iv)], we deduce that the set $\mathcal{X} = \{W \in \mathcal{W} \mid \rho(P, W) < \aleph_0\}$ is finite. Hence the set $\mathcal{F} = \{i \in I \mid \rho(P_i, W) \neq 0 \text{ for some } W \in \mathcal{X}\}$ is finite.

Let $H = \oplus\{P_i \mid i \in \mathcal{F}\}$ and $P' = \oplus\{P_i \mid i \notin \mathcal{F}\}$. Then H is finitely generated, P' has no nonzero finite ranks, and all infinite ranks of P' are the same as those of P (since we have removed only finitely many summands P_i from P). Since $P = H \oplus P'$ we may assume, for the rest of the proof, that P has no nonzero finite ranks.

We give an iterative proof that either terminates in a finite number of steps, yielding only finitely many terms in decomposition (46.7.1), or else never terminates, yielding infinitely many terms. We consider three cases.

Case 1: $\mathrm{udim}(P)$ is the only nonzero rank of P (equivalently, every $\mathcal{W}$-rank equals zero). In this case, 46.4 shows that $\mathcal{W}$ is finite and that $P \cong S^{(\alpha)}$, where $\alpha = \mathrm{udim}(P)$ and $S = S(P)$. The procedure terminates in this case.

Case 2: $\alpha = \mathrm{udim}(P)$ is a $\mathcal{W}$-rank, and is the unique nonzero $\mathcal{W}$-rank of P. In this case we use 46.5 to conclude that $P \cong S^{(\alpha)}$, where $S = S(P)$. Again the procedure terminates.

Case 3: $0 \neq \beta = \rho(P, W) \neq \mathrm{udim}(P)$ for some $W \in \mathcal{W}$, and hence, by 45.6, $\rho(P, W) < \mathrm{udim}(P)$. Let α_1 be the smallest possible value of β, necessarily infinite since we are assuming that P has no nonzero finite $\mathcal{W}$-ranks. Then 46.6 yields a decomposition $P = P_{\alpha_1} \oplus Q_1$ where each summand on the right-hand side is a sum of terms P_i in the essential decomposition $P = \oplus_{i \in I} P_i$. This lemma also yields the following properties with $n = 1$:

(46.7.2) (i) $\alpha_n = \mathrm{udim}(P_{\alpha_n})$ and α_n is the unique nonzero $\mathcal{W}$-rank of P_{α_n} and $P_{\alpha_n} \cong S_n^{(\alpha_n)}$.

(ii) All nonzero ranks of Q_n are greater than α_n.

(iii) $\mathcal{Z}(Q_{n-1}) = \mathcal{Z}(P_{\alpha_n}) \subset \mathcal{Z}(Q_n)$ (where $Q_0 = P$).

Now iterate the above procedure, applying Case 1, 2, or 3 as appropriate.

If Case 1 or 2 applies, we get $Q_1 \cong S_2^{(\alpha_2)}$ where $S_2 = S(Q_1)$. Moreover, $\alpha_1 < \alpha_2$, by (46.7.2)(ii). We have $P_{\alpha_1} \cong S_1^{(\alpha_1)}$ where $S_1 = S(P_{\alpha_1})$ by (46.7.2)(i); and $S_1 \subset S_2$ by (46.7.2)(iii) and 46.3. Therefore we have $P = P_{\alpha_1} \oplus Q_1 \cong S_1^{(\alpha_1)} \oplus S_2^{(\alpha_2)}$. This is the desired 2-term decomposition of P, and the procedure terminates.

Suppose that Case 3 applies. Then we get a decomposition $Q_1 = P_{\alpha_2} \oplus Q_2$ where each summand on the right is a direct sum of terms in the essential decomposition $P = \oplus_{i \in I} P_i$, and (46.7.2) holds with $n = 2$. All this yields the 3-term decomposition $P = P_{\alpha_1} \oplus P_{\alpha_2} \oplus Q_2$, and we are ready for another iteration of our procedure.

If repeated iteration of the procedure terminates with an application of Case 1 or 2, then the theorem is proved and decomposition (46.7.1) of P contains only a finite number of terms. If the procedure never terminates, then, for every positive integer n, we get a decomposition:

(46.7.3) $$P = P_{\alpha_1} \oplus P_{\alpha_2} \oplus \ldots \oplus P_{\alpha_n} \oplus Q_n$$

such that (46.7.2) holds.

Consider the direct sum $\oplus_{i=1}^{\infty} P_{\alpha_n}$. This is a direct sum of some of the original essential summands P_i of P; hence

(46.7.4) $$P = \left(\oplus_{i=1}^{\infty} P_{\alpha_n}\right) \oplus Q'$$

where Q' is the sum of the remaining P_i.

Finally, we need only show that $Q' = 0$. The procedure that leads to (46.7.3) and (46.7.4) shows that Q' is a direct summand of every Q_n and therefore $\mathcal{Z}(Q') \supseteq \mathcal{Z}(Q_n)$ for all n. On the other hand, (46.7.2)(iii) implies that $\mathcal{Z}(Q_{n-1}) \subset \mathcal{Z}(Q_n)$ for all n. Therefore $\mathcal{Z}(Q')$ is an infinite set, and so, by 46.3, $Q' = 0$, as required. □

46.8. Theorem (Uniqueness). *Let*

$$P = H \oplus S_1^{(\alpha_1)} \oplus S_2^{(\alpha_2)} \oplus \dots \tag{46.8.1}$$

where H_R is finitely generated projective, $S_1 \subset S_2 \subset \dots$ is a nonempty, strictly increasing finite or countably infinite sequence of right finite overrings of R, and $\alpha_1 < \alpha_2 < \dots$ are infinite cardinal numbers. Then a complete set of invariants for the isomorphism class of P is:

(i) *the set of ordered pairs (S_1, α_1), (S_2, α_2), ... ; and*
(ii) *the set of ordered pairs $\big(\rho(H, W), W\big)$ such that $W \in \mathcal{Z}_R(S_1)$.*

Proof. For each n, let $\mathcal{F}_n = \mathcal{Z}_R(S_n)$. Since exactly one of the situations enumerated in 45.12 applies to each $W \in \mathcal{W}$, we see that the sets $\mathcal{F}_n$ are determined by the genus $\Psi(P)$. Since $S_n = S_n(\mathcal{F}_n)$ [46.3], we conclude that the ordered pairs in (i) and (ii) are determined by $\Psi(P)$, hence by the isomorphism class of P.

To see that the ordered pairs in (i) and (ii) determine the isomorphism class of P, consider an analogous decomposition $Q = K \oplus T_1^{(\beta_1)} \oplus T_2^{(\beta_2)} \oplus \dots$. Then (i) immediately shows that each $S_i = T_i$ and $\alpha_i = \beta_i$. To complete the proof, we will show that $H \oplus S_1^{(\alpha_1)} \cong K \oplus S_1^{(\alpha_1)}$.

Since $H \oplus S_1^{(\alpha_1)} \cong H \oplus S_1^{(\aleph_0)} \oplus S_1^{(\alpha_1)}$ and the analogous isomorphism holds, with K in place of H, it is enough to show that $H \oplus S_1^{(\aleph_0)} \cong K \oplus S_1^{\aleph_0}$. However, these are both countably generated, so their isomorphism classes are determined by their genera [46.1]. Therefore it is sufficient to check that $\rho(H \oplus S_1^{(\aleph_0)}, W) = \rho(K \oplus S_1^{(\aleph_0)}, W)$ for all $W \in \mathcal{W}$. If $\rho(S_1, W) \neq 0$ this holds since both sides equal $\aleph_0$; and if $\rho(S_1, W) = 0$ it holds by (ii). □

46.9. Theorem. *If P and Q are infinitely generated projective R-modules and $\Psi(P) = \Psi(Q)$ then $P \cong Q$.*

Proof. We have already proved this if P is countably generated [46.1]. Suppose, therefore, that P is uncountably generated. Then Q is also uncountably generated since $\operatorname{udim}(P) = \operatorname{udim}(Q)$.

Therefore P and Q have canonical decompositions, as described in 46.7; and 46.8 gives two types of invariants that determine the isomorphism class of these canonical decompositions. Since these invariants are determined by $\Psi(P)$ and $\Psi(Q)$, we have $P \cong Q$. □

Which HNP rings have the property that all infinitely generated projective modules have canonical decompositions? The short answer is: all HNP rings that were known before the appearance of [**Stafford-Warfield 84**] [see §26]. In more detail, we have:

46.10. Proposition. *The following conditions are equivalent for R.*

(i) *Every infinitely generated projective right R-module has a canonical decomposition, as in* (46.7.1).
(ii) *Some Dedekind right closure of R is right finite over R.*
(iii) *R has only finitely many nontrivial towers of simple right R-modules.*
(iv) *The left-hand versions of the above.*

These conditions are satisfied if R is a classical order over a Dedekind domain.

PROOF. 30.5 shows that (ii) is left-right symmetric and that (ii)$\Leftrightarrow$(iii). So it will be enough to prove (i)$\Leftrightarrow$(iii). Since every uncountably generated projective R-module has a canonical decomposition [46.7], we need only consider countably generated projective right R-modules.

(iii)$\Rightarrow$(i) Let P_R be countably generated. We need to prove that P has only finitely many distinct finite ranks. In proving this, we may disregard the finite number of $W \in \mathcal{W}$ that belong to the finitely many nontrivial R-towers. Any other W must belong to a trivial cycle tower; but then, by condition (iii) of the definition of a pregenus, $\rho(P, W) = \aleph_0$.

(i)$\Rightarrow$(iii) Suppose that condition (iii) fails for R; then, by 45.16, there is a countably generated projective P_R with infinitely many finite ranks. But this is not true of the module P in 46.7.

Supplementary statement. By 30.10, such an R satisfies (ii) and (iii). □

47. Direct Sums

Infinitely generated projective R-modules[1] have nontrivial direct-sum relations, and these relations are particularly simple to describe. They are all consequences of the following simple result, which is used repeatedly in this section without explicit mention.

47.1. THEOREM. *Let P_R be infinitely generated projective and let $\{P_i \mid i \in I\}$ be a family of projective R-modules. Then* $P \cong \oplus_{i\in I} P_i \quad \Leftrightarrow \quad \Psi(P) = \sum_{i\in I} \Psi(P_i)$.

PROOF. By 46.9, infinitely generated projective R-modules are isomorphic if and only if they have the same genus. The theorem therefore follows since ranks are additive in direct sums. □

47.2. COROLLARY. *Let P be any infinitely generated projective R-module. Then* $P \cong P \oplus P \Leftrightarrow$ *P has no nonzero finite ranks.*

PROOF. To apply 47.1, we need to check that $\Psi(P) = 2\Psi(P)$ if and only if P has no nonzero finite ranks. The 'if' statement is obvious. Conversely, if P has nonzero finite ranks, and α is the smallest of these, then α is not a rank of $P \oplus P$. □

47.3. COROLLARY. *Let P_R be an infinitely generated projective module. If either P is uncountably generated or some Dedekind right closure of R is right finite over R, then P is isomorphic to a proper direct summand of itself.*

PROOF. In both situations, 46.7 shows that P has a canonical decomposition $P = H \oplus Q$ where H is finitely generated and Q has no nonzero finite ranks. Therefore $Q \cong Q \oplus Q$, by 47.2, and so $P \cong P \oplus Q$, as desired. □

The next example shows the necessity of the restrictions in 47.3.

47.4. EXAMPLE. For a suitable R, there is a countably generated projective P_R that is not isomorphic to a proper direct summand of itself.

PROOF. Let R be any HNP ring with infinitely many nontrivial towers. Then, by 45.14(iii), R has a countably generated projective P_R with infinitely many distinct finite $\mathcal{W}$-ranks.

[1] In this section R denotes an HNP ring unless the contrary is specified.

Suppose some X_R satisfies $P \cong X \oplus P$. Then $\Psi(P) = \Psi(X) + \Psi(P)$. This implies that $\rho(X, W) = 0$ for the infinitely many $W \in \mathcal{W}$ such that $\rho(P, W)$ is finite and nonzero. Therefore $X = 0$, by 46.3(i). □

Unfortunately, the canonical decompositions in 46.7 are not well-behaved in direct sums: given the decompositions of two modules, the canonical decomposition of their direct sum may not be immediately obvious. The next example illustrates this in a particularly simple fashion.

47.5. EXAMPLE. Let S be a right finite overring of R such that $S \neq R$, and let $\alpha \geq \beta$ be infinite cardinal numbers. The decompositions $P_R = R^{(\alpha)}$ and $Q_R = S^{(\beta)}$ are canonical. What is the canonical decomposition of $P \oplus Q$?

Solution. First note that, since $\alpha \geq \beta$, the decomposition $R^{(\alpha)} \oplus S^{(\beta)}$ is not canonical. Note also, from 46.3, that $\mathcal{Z}_R(R)$ is empty and $\mathcal{Z}_R(S)$ is not. Since all ranks of R and S are finite and α, β are infinite, then, for $W \in \mathcal{W}$:

$$\rho\big(R^{(\alpha)} \oplus S^{(\beta)}, W\big) = \begin{cases} \alpha & \text{if } \rho(S, W) = 0; \\ \alpha + \beta = \alpha & \text{if } \rho(S, W) \neq 0. \end{cases}$$

Also $\operatorname{udim}\big(R^{(\alpha)} \oplus S^{(\beta)}\big) = \alpha + \beta = \alpha$. Therefore $\Psi(P \oplus Q) = \Psi(P)$ and hence $R^{(\alpha)} \oplus S^{(\beta)} \cong R^{(\alpha)}$, which is a canonical decomposition.

For an assortment of other results of these types, see [**Levy-Robson 00**, §5].

48. Notes on Chapter 8

[**Kaplansky 52**, Lemma 3, Theorems 2, 4] proved that, over a commutative Dedekind domain, every projective module is a direct sum of ideals and every infinitely generated projective module is free. He did this without mentioning the term 'projective', which was not yet in use.

The first part of this was generalized in [**Cartan-Eilenberg 56**, Chapter I, 5.3] which showed, as in our 45.1, that every projective module over a right hereditary ring is a direct sum of right ideals.

[**Bass 63**] made some headway generalizing the second part, showing that, if R is a noncommutative Noetherian ring and P is an infinitely generated projective R-module then P is free if and only if P is *uniformly big* — that is, P/PM requires the same infinite cardinal number of generators as does P for each maximal ideal M. In the case of HNP rings, this is a special case of the fact that the isomorphism class of every infinitely generated projective is determined by its udim and ranks at maximal ideals; our 46.9.

The results in this chapter appeared first (even for classical hereditary orders) in [**Levy-Robson 00**]; but the proofs here are more self-contained and there are some improvements of both results and proofs. They show that, when the 'uniformly big' hypothesis is deleted, infinitely generated projective modules P over an HNP ring R can be quite far from free. Indeed, 33.11 provides a simple but striking example. It shows that if R is any HNP ring with a nontrivial tower, then there is a uniform right ideal P such that at least one rank of P is zero. Hence every infinite direct sum of copies of P has that rank zero and so is not free.

This R and P have another interesting property: For cardinal numbers α, β, with β infinite, $P^{(\alpha)} \oplus R^{(\beta)}$ is free $\Leftrightarrow$ $\alpha \leq \beta$. This holds because $P^{(\alpha)} \oplus R^{(\beta)}$ is uniformly big if and only if $\alpha \leq \beta$. This shows that 'stably free', if one were to define it in the infinitely generated case, would be quite different from in the finitely generated case. It also shows that the complicated analysis in §45 of what combinations of ranks constitute a genus cannot be avoided.

On the other hand, the Canonical Decomposition Theorem and its associated Uniqueness Theorem [46.7, 46.8] provide an interesting twist. If P is an uncountably generated projective R-module for some HNP ring R then, although P may not be free, it is the direct sum of uniquely determined free modules over uniquely determined overrings of R, together with a (non-unique) finitely generated R-module.

APPENDIX

Related Topics

This appendix covers various topics which are needed in the book but which otherwise have have nothing to do with idealizers or HNP rings, and hence would distract from the main flow of ideas. Some are relatively complete accounts and others are brief summaries. It ends with a list of some open problems.

49. Projective Modules and their Images

This section is a combination of three loosely-linked topics concerning an arbitrary ring S and a projective module P_S.

(i) Lifting and straightening. Given a surjection $P \twoheadrightarrow U$, with U of finite length, which direct-sum decompositions of U can be lifted to a decomposition of P? And, when this is possible, under what circumstances can the resulting decomposition of P be chosen to match a given decomposition, up to isomorphism ('straightening'). The answers to these question give some insight into the question of when a presentation of U by P is unique (up to an appropriate equivalence).

(ii) Irrelevance of faithful composition factors. In deciding whether a given P not of finite length can be mapped onto a given U of finite length, one may disregard all faithful composition factors of U.

(iii) Stronger Schanuel's Lemma. We prove a strengthened version of the strong form of Schanuel's Lemma by lifting an interchange of two isomorphic direct summands of a homomorphic image of some P to an automorphism of P itself.

Lifting and straightening; unique presentability

49.1. Lemma. *Let $M = X \oplus Y$ be a right S-module and $\theta \in \mathrm{Hom}(X, Y)$. Then:*

(i) *$1 + \theta$ is an automorphism of M;*
(ii) *$X \cong (1 + \theta)X$;*
(iii) *$M = (1 + \theta)X \oplus Y$.*

Proof. $1 + \theta$ is an automorphism since its inverse is $1 - \theta$. The remaining statements are equally straightforward. □

The above lemma will be used many times in what follows, without explicit reference, when we say that X can be *isomorphically replaced* by $(1 + \theta)X$ in the direct sum $M = X \oplus Y$. To ease notation, we then relabel $(1 + \theta)X$ as X before proceeding.

49.2. Lemma (Erasing). *Consider an S-module surjection $f\colon M = P \oplus N \twoheadrightarrow U$ where P is projective and $f(N) = U$. Then P can be isomorphically replaced to arrange that $f(P) = 0$.*

PROOF. Since P is projective, there is a map θ making the diagram

$$\begin{array}{ccc} & & P \\ & {\scriptstyle -\theta}\swarrow & \downarrow{\scriptstyle f} \\ N & \xrightarrow{f} & U \end{array}$$

commute. Then $f(1+\theta)P = 0$ and so $(1+\theta)P$ is the desired replacement for P. □

49.3. LEMMA (Building). *Let $f\colon M = P \oplus N \twoheadrightarrow U$ be an S-module surjection where P is projective and U has finite length. Suppose, in addition, that there is an S-module surjection $g\colon P \twoheadrightarrow U$. Then P can be isomorphically replaced so that $f(P) = U$.*

PROOF. Let $\nu\colon U \twoheadrightarrow \bar{U} = U/\operatorname{rad}(U)$ be the natural homomorphism. To check that $f(P') = U$ for a submodule P' of M (e.g. our replacement for P), it is enough to check that $\nu f(P') = \bar{U}$. Therefore we may assume that U is semisimple. In that case, we can write $U = V \oplus W$ where $V = f(P)$ and $W \subseteq f(N)$.

Next we show that there is a surjection

$$\phi\colon P \twoheadrightarrow W \text{ such that } \phi\big(P \cap \ker(f)\big) = W.$$

We start with the given surjection $g\colon P \twoheadrightarrow U$. Note that $\ker(f) \cap \ker(g) \subseteq P$. Let $\nu\colon P \twoheadrightarrow \bar{P} = P/\big(\ker(f) \cap \ker(g)\big)$ be the natural map. Let $\bar{f}\colon \bar{P} \twoheadrightarrow V$ and $\bar{g}\colon \bar{P} \twoheadrightarrow V \oplus W$ be the induced maps such that $\bar{f}\nu = f$ and $\bar{g}\nu = g$. Note that $\bar{P}$ is semisimple since that is true of both $P/(P \cap \ker(f))$ and $P/\ker(g)$. Hence there exist modules V', V'', W'' with $V \cong V' \cong V''$, $W \cong W''$ such that $\bar{P} = \ker(\bar{f}) \oplus V' = \ker(\bar{g}) \oplus V'' \oplus W''$. The Krull-Schmidt Theorem ensures now that there is a surjection $\ker(\bar{f}) \twoheadrightarrow W$; and this can be extended to a surjection $\bar{\phi}\colon \bar{P} = \ker(\bar{f}) \oplus V' \twoheadrightarrow W$. Then $\phi = \bar{\phi}\nu$ is as desired.

Projectivity of P yields θ such that the diagram

$$\begin{array}{ccc} & & P \\ & {\scriptstyle \theta}\swarrow & \downarrow{\scriptstyle \phi} \\ f^{-1}(W) \cap N & \xrightarrow{f} & W \end{array}$$

commutes. Hence, using the fact that $\phi(P \cap \ker(f)) = W$,

$$f(1+\theta)(P) \supseteq f(1+\theta)(P \cap \ker(f)) = f\theta(P \cap \ker(f)) = \phi(P \cap \ker(f)) = W.$$

Therefore, using the facts that $f(1+\theta)(P) \supseteq W$, $f\theta(P) = W$ and $f(P) = V$,

$$f(1+\theta)(P) = f(1+\theta)(P) + W = f(P) + W = V + W = U.$$

Hence $(1+\theta)P$ is the desired replacement for P. □

49.4. DEFINITION. Maps $f_i : M_i \to U_i$ for $i = 1, 2$ are said to be *equivalent* precisely when there are isomorphisms δ and ϵ such that the diagram

$$\begin{array}{ccc} M_1 & \xrightarrow{f_1} & U_1 \\ \downarrow{\scriptstyle \delta} & & \downarrow{\scriptstyle \epsilon} \\ M_2 & \xrightarrow{f_2} & U_2 \end{array}$$

commutes.

For our applications, we need to do more than simply isomorphically replace summands: we must arrange, as in the next lemma, that the resulting map from a specified summand is equivalent to a specified map.

49.5. LEMMA (Straightening). *Let $f\colon P = P_1 \oplus P_2 \twoheadrightarrow U$ be an S-module surjection where P is projective and U has finite length. Suppose in addition that, for $i = 1, 2$, there is an S-module surjection $g_i : P_i \twoheadrightarrow U$. Then P_1 and then P_2 can be isomorphically replaced so that $f(P_1) = U$, $f(P_2) = 0$ and the restriction of f to P_1 is equivalent to g_1.*

PROOF. The Building Lemma 49.3 allows us to isomorphically replace P_2 so that $f(P_2) = U$. Then we can use the Erasing Lemma 49.2 to arrange that $f(P_1) = 0$. The fact that P_1 is projective means that there is a θ making the first diagram below commute.

$$\begin{array}{ccc} & & P_1 \\ & {\scriptstyle\theta}\swarrow & \downarrow{\scriptstyle g_1} \\ P_2 & \xrightarrow{f} & U \end{array} \qquad\qquad \begin{array}{ccc} & & P_1 \\ & {\scriptstyle 1+\theta}\swarrow & \downarrow{\scriptstyle g_1} \\ (1+\theta)P_1 & \xrightarrow{f} & U \end{array}$$

Then, since $f(P_1) = 0$, the second diagram also commutes. So we can isomorphically replace P_1 by $(1+\theta)P_1$; and then the restriction of f to the new P_1 is equivalent to g_1. Finally, we can use the Erasing Lemma 49.2 again to arrange that $f(P_2) = 0$. □

We are now in a position to prove the theorem at which we have been aiming.

49.6. THEOREM (Lifting and Straightening). *Consider an S-module surjection*

$$f\colon P = P_1 \oplus \ldots \oplus P_n \twoheadrightarrow U = U_1 \oplus \ldots \oplus U_n$$

where P is projective and U has finite length.

(i) *Suppose that, for each i, there is an S-module surjection $g_i\colon P_i \twoheadrightarrow U_i$. Then the P_i can be isomorphically replaced to arrange that $f(P_i) = U_i$ for each i.*
(ii) *Suppose also that, for all $i < n$, there is an S-module surjection $P_n \twoheadrightarrow U_i$. Then it can be arranged, in addition, that the restricted map $f\colon P_i \twoheadrightarrow U_i$ is equivalent to g_i for each $i < n$.*

PROOF. The proof of (ii) follows exactly the same path as that of (i) except that some small extra details are required. We will indicate these in the course of the proof of (i).

We first consider the case $n = 2$. So we have $f\colon P = P_1 \oplus P_2 \twoheadrightarrow U_1 \oplus U_2$, g_1, g_2 as described (and, in case (ii), a surjection $h\colon P_2 \twoheadrightarrow U_1$). Let $\pi_i\colon U \twoheadrightarrow U_i$ be the projection map; then $\pi_1 f\colon P_1 \oplus P_2 \twoheadrightarrow U_1$. We can use the Building Lemma 49.3 to arrange, by isomorphic replacement, that $\pi_1 f(P_1) = U_1$ and then the Erasing Lemma 49.2 to get $\pi_1 f(P_2) = 0$. (In case (ii), we simply apply the Straightening Lemma 49.5 to give these together with the property that $\pi_1 f\colon P_1 \twoheadrightarrow U_1$ is equivalent to g_1.)

Now $\pi_1 f(P_2) = 0$ means that $f(P_2) \subseteq U_2$; and so $f^{-1}(U_2) \supseteq P_2$. Since P_2 is a direct summand of $P_1 \oplus P_2$, it follows that $f^{-1}(U_2) = P_2 \oplus L$ for some $L \subseteq P_1$. Since g_2 exists, we can apply the Building Lemma 49.3 to the surjection $f\colon P_2 \oplus L \twoheadrightarrow U_2$. Thus there exists $\theta \in \mathrm{Hom}(P_2, L)$ which we may view as an element in $\mathrm{Hom}(P_2, P_1)$, and which provides an isomorphic replacement of P_2 such that $f(P_2) = U_2$.

So we now have $f(P_2) = U_2$ and still have $\pi_1 f(P_1) = U_1$ (and, in case (ii), the latter map is equivalent to g_1). We need to isomorphically replace P_1 so that these statements remain true with the π_1 terms removed. First we use the projectivity of P_1 to provide ψ making the first diagram below commute.

$$\begin{array}{ccc} & & P_1 \\ & {\scriptstyle -\psi}\swarrow & \downarrow{\scriptstyle \pi_2 f} \\ P_2 & \xrightarrow{f} & U_2 \end{array} \qquad\qquad \begin{array}{ccc} & & P_1 \\ & {\scriptstyle 1+\psi}\swarrow & \downarrow{\scriptstyle \pi_1 f} \\ (1+\psi)P_1 & \xrightarrow{f} & U_1 \end{array}$$

Since $\pi_2 = 1 - \pi_1$, the second diagram commutes. This provides the desired isomorphic replacement $(1+\psi)P_1$ for P_1.

The extension of this to the case $n > 2$ is an easy induction exercise. □

49.7. DEFINITION. A module U_S is said to be *uniquely presentable* by a module P_S if there is an S-module surjection $f\colon P \twoheadrightarrow U$ such that every S-module surjection from P to U is equivalent to f. As a trivial example, 0 is always uniquely presentable.

49.8. COROLLARY. *Consider an S-module surjection*

$$f\colon P = P_1 \oplus \ldots \oplus P_n \twoheadrightarrow U = U_1 \oplus \ldots \oplus U_n$$

where P is projective and U has finite length. Suppose that, for each i, there is an S-module surjection $P_i \twoheadrightarrow U_i$ and an S-module surjection $P_n \twoheadrightarrow U_i$. Suppose also that U_n is uniquely presentable by P_n. Then U is uniquely presentable by P.

PROOF. 49.6 allows us to assume that each $f(P_i) = U_i$. Let $g\colon P \twoheadrightarrow U$ be any other surjection. Then a second application of 49.6 gives us a decomposition $P = P_1' \oplus \ldots P_n'$ with the properties:

(i) for each i, $P_i' \cong P_i$ and $g(P_i') = U_i$; and

(ii) for each $i < n$, $g\colon P_i' \twoheadrightarrow U_i$ is equivalent to $f\colon P_i \twoheadrightarrow U_i$.

The final hypothesis ensures that $g\colon P_n' \twoheadrightarrow U_n$ is equivalent to $f\colon P_n \twoheadrightarrow U_n$. Thus f and g are equivalent. □

If we set $n = 2$ and $U_2 = 0$, we obtain:

49.9. COROLLARY. *Let P_1 and P_2 be projective modules each having a module surjection onto U, a module of finite length. Then U is uniquely presentable by $P_1 \oplus P_2$.* □

There is another useful example of unique presentability.

49.10. PROPOSITION. *Let A be an ideal of a ring S and H be a Noetherian S-module. Then H/HA is uniquely presentable by H.*

PROOF. Suppose that $H/HA \cong H/K$ and let f and g in the next diagram be the natural surjections. Now $K \subseteq HA$ since $(H/K)A = 0$. Therefore there is a surjection γ as shown making the diagram commute.

$$\begin{array}{ccc} & & H \\ & {\scriptstyle f}\swarrow & \downarrow{\scriptstyle g} \\ H/HA & \xrightarrow{\gamma} & H/K \end{array}$$

Let δ be any isomorphism from H/K to H/HA. Then $\delta\gamma$ maps the Noetherian module H/HA onto itself and hence is a monomorphism. Hence γ is an isomorphism and f and g are equivalent. □

The material in this subsection comes from [**Levy-Robson 74**, §1].

Irrelevance of faithful composition factors

49.11. LEMMA. *Let U_S be finitely generated over some ring S and suppose that $U/\operatorname{rad}(U)$ has finite length (and hence is semisimple). Write*

$$U/\operatorname{rad}(U) = Y \oplus F \tag{49.11.1}$$

where all composition factors of F are faithful. If a projective S-module P not of finite length maps onto Y, then P maps onto U.

PROOF. It suffices to show that P maps onto $Y \oplus F$, for then projectivity of P and finite generation of U show that any surjection $P \twoheadrightarrow Y \oplus F$ can be lifted to a surjection $P \twoheadrightarrow U$. Thus we may assume that $P = Y \oplus F$. Furthermore, after writing F as a direct sum of simple modules, we may assume inductively that F is simple. Let $\psi\colon P \twoheadrightarrow Y$ be given. Since Y, but not P, has finite length, we have $\ker(\psi) \neq 0$. Choose $0 \neq p_0 \in \ker(\psi)$. Since P is projective, there exists a map $\pi\colon P \to S$ such that $\pi(p_0) \neq 0$. Since F is faithful, there exists $x_0 \in F$ such that $x_0\pi(p_0) \neq 0$. Define $g\colon P \to U$ by $g(p) = \psi(p) + x_0\pi(p)$. Then $\operatorname{im}(g)$ contains the nonzero element $g(p_0) = x_0\pi(p_0)$ of the simple module F and hence contains F. Since ψ maps P onto Y, therefore g maps P onto $U = Y \oplus F$ as desired. □

49.12. COROLLARY. *Let P_S be a projective module not of finite length over a ring S that has a module U_S of finite length whose composition factors are all faithful. Then P maps onto U. If S is not simple Artinian, then U is cyclic.*

PROOF. We may suppose that $U \neq 0$. Taking $Y = 0$ in 49.11 shows that P maps onto U.

If S is not simple Artinian, then it cannot be right Artinian (because it has a faithful simple module). Therefore S_S does not have finite length, and hence maps onto U; i.e. U is cyclic. □

A special case of 49.12 appeared as 20.3 with an independent proof.

The material in this subsection is adapted from [**Levy-Robson 74**, 1.11].

Stronger Schanuel

49.13. DEFINITION. Suppose that $M = M_1 \oplus \ldots \oplus M_n$ is a direct sum of right S-modules. A *transvection* of this direct sum is defined to be a map of the form $E_{i,j}(\theta) = 1 + \theta$ where $\theta \in \operatorname{Hom}(M_i, M_j)$ for some $i \neq j$. By 49.1, every such transvection is an automorphism of M.

49.14. LEMMA. *Suppose that $M_i = M_j$ for some i, j. Let $\langle i, j\rangle$ denote the automorphism of $M_1 \oplus \ldots \oplus M_n$ which is the identity on all the summands other than M_i and M_j, and which multiplies the ith entry by -1 and then interchanges it with the jth entry. Then $\langle i, j\rangle = E_{ij}(1)E_{ji}(-1)E_{ij}(1)$.*

PROOF. This is easily checked. □

49.15. LEMMA. *For each $i \in \{1, \ldots, n\}$, let P_i be a projective S-module and U_i be a homomorphic image of P_i. Each transvection of $U_1 \oplus \ldots \oplus U_n$ lifts to a transvection of $P_1 \oplus \ldots \oplus P_n$.*

PROOF. Let $\theta' \in \mathrm{Hom}(U_i, U_j)$. Since P_j maps onto U_j and P_i is projective, there is a map $\theta : P_i \to P_j$ such that the diagram

$$\begin{array}{ccc} P_i & \xrightarrow{\theta} & P_j \\ \downarrow & & \downarrow \\ U_i & \xrightarrow{\theta'} & U_j \end{array} \tag{49.15.1}$$

commutes. Then $1 + \theta'$ is the image of $1 + \theta$. □

We can now prove the main result of this subsection (needed in 34.1).

49.16. LEMMA (Stronger Schanuel). *Consider the following short exact sequences of R-modules in which P_1, P_2 are projective and $U_1 \cong U_2$:*

$$0 \to K_1 \to P_1 \to U_1 \oplus V_1 \to 0$$

$$0 \to K_2 \to P_2 \to U_2 \oplus V_2 \to 0$$

We identify the U_i. Then there is an automorphism θ of $P_1 \oplus P_2$ such that $\theta(K_1 \oplus K_2) = K_1 \oplus K_2$, and the effect of θ on

$$(P_1 \oplus P_2)/(K_1 \oplus K_2) \cong U_1 \oplus V_1 \oplus U_2 \oplus V_2$$

is to map (u_1, v_1, u_2, v_2) to $(u_2, v_1, -u_1, v_2)$. Thus it interchanges U_1 and U_2 while acting as the identity on $V_1 \oplus V_2$.

PROOF. $P_1/K_1 \oplus P_2/K_2 \cong U_1 \oplus V_1 \oplus U_2 \oplus V_2$. We view the latter as a direct sum of four summands. By 49.14, $\langle 1, 3\rangle = E_{13}(1)E_{31}(-1)E_{13}(1)$. However, each of these three factors can also be viewed as a transvection of the direct sum of two terms $P_1/K_1 \oplus P_2/K_2$. Hence, by 49.15, each of these transvections lifts to an automorphism of $P_1 \oplus P_2$ which acts as an automorphism on $K_1 \oplus K_2$. □

In the case when V_1 and V_2 are both zero, this becomes what is sometimes called the *strong form* of Schanuel's lemma.

The material in this subsection is adapted from [**Levy-Robson-Stafford 94**].

50. Artinian Serial Rings

Recall that we call a module *uniserial* if its submodules are linearly ordered by inclusion; and we call a ring *right serial* if, when viewed as a right module over itself, it is a direct sum of uniserial modules, and *serial* if it is both right and left serial. Artinian serial rings are also called *generalized uniserial rings*, terminology that we do not use here.

This section gives a simple and succinct account of the basic theory of one- and two-sided Artinian serial rings, including several characterizations [50.12] and Nakayama's Theorem [50.19] that every module is a direct sum of uniserial modules [**Nakayama 41**, Theorem 17]. Non-Artinian serial rings have been studied in e.g. [**Warfield 75**], [**Puninski 01**]; but we do not deal with that here.

We spend some time on right serial rings, thereby demonstrating that two-sided hypotheses are needed in order to get the main results. Our theory is organized so as to apply easily to idealizer rings and HNP rings; and this leads to some possibly novel results. See [**Anderson and Fuller 92**, §32] for additional results.

Before turning to Artinian serial rings, we first introduce some definitions and give some rather general (and well-known) results about extensions of one simple module by another.

50.1. DEFINITIONS. Recall that $J(\Lambda)$ denotes the Jacobson radical of any ring Λ and, if M is a Λ-module, $\operatorname{top}(M) = M/MJ$. If $\operatorname{top}(M)$ is a simple module, we sometimes call it the *top composition factor of* M.

We define $\operatorname{soc}^n(M)$ inductively: $\operatorname{soc}^0(M) = 0$ and, for $n > 1$, $\operatorname{soc}^n(M)$ is the submodule of M such that $\operatorname{soc}^n(M)/\operatorname{soc}^{n-1}(M) = \operatorname{soc}\big(M/\operatorname{soc}^{n-1}(M)\big)$. We call $\operatorname{soc}^n(M)$ the nth *socle* of M. Then, for each $n \geq 1$, the nth *socle layer* of M is $\operatorname{soc}_n(M) = \operatorname{soc}^n(M)/\operatorname{soc}^{n-1}(M)$. Thus $\operatorname{soc}^1(M) = \operatorname{soc}_1(M) = \operatorname{soc}(M)$, the socle of M.

The facts in the next lemma will be used frequently, without explicit reference.

50.2. LEMMA. *Let e be a primitive idempotent element of a ring Λ such that $J = J(\Lambda)$ is nilpotent and Λ/J is Artinian.*

(i) $\operatorname{top}(e\Lambda)$ *is a simple module; and all simple (right) Λ-modules arise in this way.*
(ii) *If P is a projective Λ-module with* $\operatorname{top}(P) \cong \operatorname{top}(e\Lambda)$ *then* $P \cong e\Lambda$.

PROOF. (i) This is an easy consequence of the fact that idempotents can be lifted modulo nil ideals. See, e.g. [**Anderson and Fuller 92**, 27.1], [**Jacobson 80**, p.405].

(ii) Since $\operatorname{top}(e\Lambda) \cong \operatorname{top}(P)$, there is a surjection $e\Lambda \to \operatorname{top}(P)$. Since $e\Lambda$ is projective, this lifts to a map $f\colon e\Lambda \to P$ which, by Nakayama's Lemma, must be surjective; and since P is projective, $\ker(f)$ is a direct summand of the indecomposable module $e\Lambda$. Hence $\ker(f) = 0$; and so f is an isomorphism. □

50.3. LEMMA. *Let Λ be a ring and let $J = J(\Lambda)$. Let V, W be simple right Λ-modules.*

(i) *Let P be the projective Λ-module such that* $\operatorname{top}(P) \cong V$. *Suppose that $PJ/PJ^2 \cong W^{(n)} \oplus X$ for some integer n. Then* $\dim(\operatorname{Ext}^1(V,W)) = n + \dim(\operatorname{Hom}(X,W))$ *where the dimensions are as a left vector spaces over* $\operatorname{End}(W)$.
(ii) *Let H be the injective envelope of W. Suppose that* $\operatorname{soc}_2(H) \cong V^{(k)} \oplus X$ *for some integer k. Then* $\dim(\operatorname{Ext}^1(V,W)) = k + \dim(\operatorname{Hom}(V,X))$ *where the dimensions are as right vector spaces over* $\operatorname{End}(V)$.

PROOF. (i) Apply $\operatorname{Hom}(-,W)$ to the short exact sequence $PJ \hookrightarrow P \twoheadrightarrow V$, getting the exact sequence of left $\operatorname{End}(W)$-modules

$$\dots \to \operatorname{Hom}_\Lambda(P,W) \xrightarrow{f} \operatorname{Hom}_\Lambda(PJ,W) \xrightarrow{g} \operatorname{Ext}^1_\Lambda(V,W) \xrightarrow{h} \operatorname{Ext}^1_\Lambda(P,W) \to \dots .$$

Since P is projective, $\operatorname{Ext}^1_\Lambda(P,W) = 0$ and so $h = 0$. Since W is simple, $WJ = 0$. Therefore every element of $\operatorname{Hom}(P,W)$ takes PJ to zero and so $f = 0$. Hence g is an isomorphism. Note also that $\operatorname{Hom}(PJ,W) \cong \operatorname{Hom}(PJ/PJ^2,W)$ since $WJ = 0$. Thus

$$\begin{aligned}\operatorname{Ext}(V,W) &\cong \operatorname{Hom}(PJ/PJ^2,W) \cong \operatorname{Hom}(W^{(n)} \oplus X, W)\\ &\cong \operatorname{End}(W)^{(n)} \oplus \operatorname{Hom}(X,W)\end{aligned}$$

as desired.

(ii) Apply $\operatorname{Hom}(V,-)$ to the short exact sequence $W \hookrightarrow H \twoheadrightarrow H/W$, getting the exact sequence of $\operatorname{End}(V)$-modules

$$\dots \to \operatorname{Hom}(V,H) \xrightarrow{f} \operatorname{Hom}(V,H/W) \xrightarrow{g} \operatorname{Ext}^1(V,W) \xrightarrow{h} \operatorname{Ext}^1(V,H) \to \dots .$$

Since H is injective, $\mathrm{Ext}^1(V,H) = 0$ and so $h = 0$. Any nonzero map from V to H must map V to the socle of H, i.e. to W; hence $f = 0$. Therefore g is an isomorphism. Of course, any homomorphism from V to H/W maps V to the socle of H/W. So

$$\begin{aligned}\mathrm{Ext}^1(V,W) &\cong \mathrm{Hom}(V, \mathrm{soc}(H/W)) \cong \mathrm{Hom}(V, V^{(k)} \oplus X)\\ &\cong \mathrm{End}(V)^{(k)} \oplus \mathrm{Hom}(V,X). \end{aligned} \qquad \square$$

Recall, from 15.1, that if $\mathrm{Ext}^1_\Lambda(V,W) \neq 0$, for simple modules V, W over some ring Λ, we say that W is a *successor* of V and V is a *predecessor* of W. We now refine this definition.

50.4. DEFINITIONS. Let V, W be simple modules over a ring Λ with W a successor of V. Let $J = J(\Lambda)$.

(a) If there is a Λ-module M with $\mathrm{top}(M) \cong V$ and $MJ \cong W^{(n)}$ for some integer $n \geq 1$, we will say that W is *a successor of V of multiplicity $\geq n$*. If there is a largest such n, we call it the *multiplicity*.

(b) If there is an essential extension M of W with $M/W \cong V^{(k)}$ for some integer $k \geq 1$, we will say that V is *a predecessor of W of multiplicity $\geq k$*. If there is a largest such k, we call it the *multiplicity* .

50.5. PROPOSITION. *Let W be a successor of a simple module V over a ring Λ, let $J = J(\Lambda)$, and suppose that $J = J(\Lambda)$ is nilpotent and Λ/J is Artinian. Let P be a projective module such that* $\mathrm{top}(P) \cong V$ *and H be the injective envelope of W.*

(i) *If any of the following is finite, then they are all equal:*
  (a) *the multiplicity of W as a successor of V;*
  (b) *the dimension of* $\mathrm{Ext}^1_\Lambda(V,W)$ *as a vector space over* $\mathrm{End}(W)$*;*
  (c) *the multiplicity of W as a direct summand of PJ/PJ^2.*

(ii) *If any of the following is finite, then they are all equal:*
  (a) *the multiplicity of V as a predecessor of W;*
  (b) *the dimension of* $\mathrm{Ext}^1_\Lambda(V,W)$ *as a vector space over* $\mathrm{End}(V)$*;*
  (c) *the multiplicity of V as a direct summand of* $\mathrm{soc}(H/W)$.

PROOF. (i) Consider any short exact sequence $W^{(n)} = MJ \subset M \overset{f}{\twoheadrightarrow} V$. By projectivity of P, we can lift the surjection $P \twoheadrightarrow \mathrm{top}(P) \cong V$ to a map $P \to M$ which, by Nakayama's Lemma, is surjective. Therefore there is a surjection $PJ \twoheadrightarrow MJ = W^{(n)}$. Since $WJ = 0$, this last surjection induces a surjection of semisimple modules $PJ/PJ^2 \twoheadrightarrow W^{(n)}$. By semisimplicity, $W^{(n)}$ is isomorphic to a direct summand of PJ/PJ^2; and, by 50.3(i), the largest n for which this can be done is the dimension described in (b).

(ii) Let H be an injective hull of W. Then some essential extension M of W satisfies $\mathrm{soc}(M/W) \cong V^{(k)}$ if and only if $\mathrm{soc}(H/W)$ has $V^{(k)}$ as a direct summand. Therefore the result is a straightforward consequence of 50.3(ii). $\square$

50.6. REMARK. In studying Artinian serial rings, we will work in terms of multiplicities rather than the dimension of Ext. But, in order to apply later results such as 50.12 to HNP rings, we use 50.5 to rephrase them in terms of Ext.

The next few lemmas will be applied to HNP rings as well as serial rings.

50.7. LEMMA. *Let M_1, M_2 be modules of finite length over a ring Λ such that* $\mathrm{Ext}^1_\Lambda(X_1, X_2) = 0$ *for every composition factor X_i of M_i. Then* $\mathrm{Ext}^1_\Lambda(M_1, M_2) = 0$.

PROOF. We prove the lemma by induction on the sum σ of the composition lengths of M_1 and M_2. The lemma is obvious if σ equals 1 or 2. Suppose that $\sigma > 2$. Then at least one of the M_i has length > 1. We consider the former case and omit the details of the similar, latter case.

Let X_1 be any simple submodule of the non-simple M_1, and apply $\mathrm{Hom}(-, M_2)$ to the short exact sequence $X_1 \hookrightarrow M_1 \twoheadrightarrow M_1/X_1$. Part of the resulting long exact sequence is $\mathrm{Ext}^1(M_1/X_1, M_2) \to \mathrm{Ext}^1(M_1, M_2) \to \mathrm{Ext}^1(X_1, M_2)$, in which the first and last terms equal zero by induction. Hence so does the middle term. □

50.8. LEMMA. *Let A_1, A_2 be ideals of a right Noetherian ring Λ such that the rings Λ/A_i are right Artinian and $\mathrm{Ext}^1_\Lambda(X_1, X_2) = 0$ for every pair of composition factors X_i of Λ/A_i. Then $A_1 \cap A_2 = A_1A_2$.*

PROOF. Consider the short exact sequence of right Λ-modules

$$A_1/A_1A_2 \subseteq \Lambda/A_1A_2 \twoheadrightarrow \Lambda/A_1. \tag{50.8.1}$$

Since A_1 is finitely generated, A_1/A_1A_2 is a right module of finite length over the right Artinian ring Λ/A_2. Moreover, every composition factor of $(A_1/A_1A_2)_\Lambda$ is a composition factor of $(\Lambda/A_2)_\Lambda$. It follows from the hypothesis on composition factors that $\mathrm{Ext}^1_\Lambda(X_1, X_2) = 0$ for every composition factor X_1 of Λ/A_1 and X_2 of A_1/A_1A_2. Therefore $\mathrm{Ext}^1_\Lambda(\Lambda/A_1, A_1/A_1A_2) = 0$, by 50.7; and hence the exact sequence in (50.8.1) splits.

Therefore $\Lambda/A_1A_2 \cong A_1/A_1A_2 \oplus \Lambda/A_1$. Since the right-hand side is annihilated by $A_1 \cap A_2$, so is the left-hand side; and this implies that $A_1 \cap A_2 \subseteq A_1A_2$. The opposite inclusion is trivial. □

50.9. LEMMA. *Let Λ be a ring with maximal ideals $A_1 \neq A_2$ such that the rings Λ/A_1 and Λ/A_2 are Artinian. Let W_1, W_2 be simple right Λ-modules such that $W_1A_1 = 0$ and $W_2A_2 = 0$. Then:*

(i) $\mathrm{Ext}^1_\Lambda(W_1, W_2) \neq 0 \implies (A_1 \cap A_2)/A_1A_2 \neq 0$;

(ii) $\mathrm{Ext}^1_\Lambda(W_2^{\mathrm{op}}, W_1^{\mathrm{op}}) \neq 0 \implies (A_1 \cap A_2)/A_1A_2 \neq 0$;

(iii) *if Λ is right Noetherian, then* $(A_1 \cap A_2)/A_1A_2 \neq 0 \iff \mathrm{Ext}^1_\Lambda(W_1, W_2) \neq 0$.

PROOF. (i) Since $\mathrm{Ext}^1_\Lambda(W_1, W_2) \neq 0$, there is a nonsplit exact sequence $W_2 \hookrightarrow U \twoheadrightarrow W_1$, and then U is uniserial of length 2. In this situation we have $UA_1A_2 = 0$. But $U(A_1 \cap A_2) \neq 0$ (otherwise U would be a module over the semisimple Artinian ring $\Lambda/(A_1 \cap A_2) \cong \Lambda/A_1 \oplus \Lambda/A_2$ and hence semisimple). Therefore $A_1A_2 \neq A_1 \cap A_2$, as desired.

(ii) This is the left-handed version of (i).

(iii) This is immediate from (i) and 50.8. □

Some authors say that there is a *link* $A_1 \rightsquigarrow A_2$ when $(A_1 \cap A_2)/A_1A_2 \neq 0$. (Sometimes the arrow points in the opposite direction.)

We now turn towards Artinian serial rings. We recall that Morita equivalence is described in 11.3.

50.10. LEMMA. *Every ring Morita equivalent to a right Artinian right serial ring is itself a right Artinian right serial ring.*

PROOF. It is enough to show that a ring Λ is right Artinian and right serial if and only if Λ has the two Morita invariant properties: that Λ is right Artinian; and that every finitely generated indecomposable projective right Λ-module is uniserial.

For the nontrivial 'only if' part of this statement, let Λ be right Artinian and right serial. Then every free right Λ-module F of finite rank is a direct sum of uniserial modules. Therefore, by the Krull-Schmidt theorem [**Jacobson 80**, p.115], every indecomposable direct summand of any such F (that is, every indecomposable finitely generated projective module) is uniserial. □

50.11. LEMMA. *Let e be a primitive idempotent element of a ring Λ such that $J = J(\Lambda)$ is nilpotent and Λ/J is Artinian.*

(i) *A Λ-module M of finite length is uniserial if and only if MJ^{n-1}/MJ^n is a simple module whenever it is nonzero. In that case its unique composition series is $M \supset MJ \supset MJ^2 \supset \ldots \supset 0$.*
(ii) *A Λ-module M of finite length is uniserial if and only if $\mathrm{soc}_n(M)$ is a simple module whenever it is nonzero. In that case the ascending sequence of $\mathrm{soc}^n(M)$ forms the unique composition series for M.*
(iii) *If Λ is a right Artinian right serial ring, then the right Λ-module $e\Lambda$ is uniserial.*
(iv) *If Λ is a right Artinian right serial ring and M is a uniserial module of finite length n then M is determined (up to isomorphism) by $\mathrm{top}(M)$ and n. Indeed, if $\mathrm{top}(M) \cong \mathrm{top}(e\Lambda)$, then $M \cong e\Lambda/eJ^n$.*

PROOF. (i) If M is nonzero and uniserial, then it has a unique maximal submodule which must be MJ. Induction on length shows that each MJ^n/MJ^{n+1} is simple.

Conversely, suppose each nonzero MJ^n/MJ^{n+1} is simple. Then M/MJ is simple and so each proper submodule of M is contained in the unique maximal submodule MJ. By induction on length, MJ is uniserial. Evidently, the series given is the unique composition series of M.

(ii) The proof of this is a minor adaptation of the proof of (i), starting with $\mathrm{soc}(M)$ rather than MJ.

(iii) There are orthogonal idempotents $e_1, \ldots, e_n$ such that $e = e_1$ and $\sum_i e_i = 1$. Then the Krull-Schmidt theorem, applied to the decomposition $\Lambda_\Lambda = \oplus_i e_i\Lambda$ of the serial right module Λ, shows that $e\Lambda$ is uniserial.

(iv) By hypothesis, $e\Lambda$ maps onto $\mathrm{top}(M)$. Since $e\Lambda$ is projective, and using Nakayama's Lemma, this surjection can be lifted to a surjection $f\colon e\Lambda \twoheadrightarrow M$. Since $e\Lambda$ is uniserial, we have $\ker(f) = eJ^n$. □

50.12. THEOREM. *If Λ is a right Artinian ring and $J = J(\Lambda)$, then the following are equivalent.*

(i) *Λ is a right serial ring.*
(ii) *Λ/J^2 is a right serial ring.*
(iii) *The following two conditions hold for every simple right Λ-module V.*
   (a) *V has at most one successor.*
   (b) *If V has a successor W, then the multiplicity of W as successor is 1 (or, equivalently, $\mathrm{Ext}^1_\Lambda(V, W)$ has dimension 1 over $\mathrm{End}(W)$).*

PROOF. (i)⇒(ii) If $\Lambda = \oplus_{i=1}^n e_i\Lambda$ is a direct sum of uniserial right ideals of Λ then $\Lambda/J^2 = \oplus_{i=1}^n e_i\Lambda/e_iJ^2$ is a direct sum of uniserial right ideals of Λ/J^2.

(ii)⇒(iii) Let V be any simple Λ-module. Then $V \cong \mathrm{top}(e\Lambda)$ for some primitive idempotent e and so $V \cong \mathrm{top}(e\Lambda/eJ^2)$. Since $e\Lambda/eJ^2$ is a uniserial module, $W = eJ/eJ^2$ is either simple or zero. In the former case, W is the unique successor of V

and has multiplicity 1 or equivalently, by 50.5, $\mathrm{Ext}^1_\Lambda(V,W)$ has dimension 1 over $\mathrm{End}(W)$.

(iii)$\Rightarrow$(i) Since orthogonal idempotents with sum 1 can be lifted modulo nilpotent ideals [**Anderson and Fuller 92**, 27.1, 27.4], there is a decomposition $\Lambda = \oplus_{i=1}^n e_i\Lambda$, where the e_i are primitive orthogonal idempotents and $\mathrm{top}(e_i\Lambda)$ is simple. Let e equal any e_i. By 50.11(i), it is enough to show, for each $n \geq 1$, that $L_n = eJ^{n-1}/eJ^n$, if nonzero, is a simple Λ-module.

To show this, we first consider any right ideal, A say, such that A/AJ is simple. Consider the short exact sequence $AJ/AJ^2 \hookrightarrow A/AJ^2 \twoheadrightarrow A/AJ$. If $AJ^2 \neq AJ$, then A/AJ^2 is not semisimple. In that case, this sequence is nonsplit and each simple summand of AJ/AJ^2 is a successor to A/AJ. However, by (a), there cannot be two nonisomorphic simple summands and, by (b), there cannot be two isomorphic simple summands. Thus AJ/AJ^2 is a simple module or zero.

We return to L_n as above and proceed by induction. Of course L_1 is simple. Suppose $L_{n-1} = eJ^{n-2}/eJ^{n-1}$ is simple and let $A = eJ^{n-2}$; so $A/AJ = L_{n-1}$ is simple. The preceding paragraph shows that $AJ/AJ^2 = eJ^{n-1}/eJ^n = L_n$ is simple or zero, as required. □

50.13. Remarks (on condition (iii) above).

Neither of the two conditions (a) and (b) is redundant. For non-redundance of (a), see 50.21(ii). For non-redundance of (b), we can take the commutative Artinian local non-serial ring $\Lambda = k[x,y]$ where $x^2 = y^2 = xy = 0$. Its unique simple module is $V = \Lambda/J \cong k$, and the nonsplit short exact sequence $V \oplus V \cong xk \oplus yk = J \hookrightarrow \Lambda \twoheadrightarrow k \cong V$ shows that V is its own unique successor and, by 50.5(i), its multiplicity is 2.

There are two further conditions on a ring Λ closely related to (a) and (b):

(a′) each simple right Λ-module W has at most one predecessor V;

(b′) if a predecessor V exists, then its multiplicity is 1.

If Λ is a (right and left) Artinian serial ring, then 50.15 and 50.16 show that each of these is true; but if Λ is Artinian and right, but not left, serial then 50.21(ii) and 51.8 show, respectively, that (a′) and (b′) can fail.

Before the next result, we note that, over a right or left Artinian ring Λ, every simple left module X has the form V^{op} for some simple V_Λ. For $\Lambda/\mathrm{ann}(X)$ is primitive, so simple Artinian, and then V is the simple right Λ-module with $\mathrm{ann}(V) = \mathrm{ann}(X)$.

50.14. Theorem. *Let Λ be a right Artinian right serial ring. Then every simple left Λ-module has at most one predecessor.*

Proof. Choose any simple left Λ-module, say V^{op}. Suppose $\mathrm{Ext}^1(W_i^{\mathrm{op}}, V^{\mathrm{op}}) \neq 0$ $(i = 1,2)$ for two simple right modules W_i, and let the annihilators of V and each W_i be the maximal ideals M and N_i respectively. By 50.9(ii), we see that each $(M \cap N_i)/(MN_i) \neq 0$. Then 50.9(iii) shows that each $\mathrm{Ext}^1(V, W_i) \neq 0$; and hence $W_1 \cong W_2$, by 50.12. □

50.15. Corollary. *Let Λ be a (right and left) Artinian serial ring. Then every simple right Λ-module has at most one successor and at most one predecessor.*

Proof. Successors (when they exist) are unique by 50.12. Uniqueness of predecessors holds by 50.14. □

50.16. PROPOSITION. *Let Λ be an Artinian serial ring and let V be the unique predecessor of W. Then V has multiplicity 1 as a predecessor of W — equivalently,* $\mathrm{Ext}^1_\Lambda(V, W)$ *has dimension 1 over* $\mathrm{End}(V)$.

PROOF. 50.5 shows the equivalence of the two properties. We aim at the former; so we need to prove that W has no essential extension E such that $E/W \cong V_1 \oplus V_2$ with $V_1 \cong V_2 \cong V$. We suppose such an E exists and seek a contradiction. Note that $EJ^2 = 0$; so we may assume that $J^2 = 0$.

Choose elements $z_i \in E$ such that each $[z_i + W]\Lambda = V_i$, and let $X_i = z_i\Lambda$. Since E is an essential extension of W, each $X_i \supset W$. Therefore each X_i is uniserial of length 2 with the same composition factors V, W from the top. Since Λ is Artinian, 50.11(iv) implies that $X_1 \cong X_2$. In fact it implies that there is a primitive idempotent e of Λ such that $e\Lambda \cong X_1 \cong X_2$. Note that $X_1 + X_2 = E$, $X_1 \cap X_2 = W$ and $eJ \cong W$.

We claim that there are isomorphisms g_i: $e\Lambda \to X_i$ such that $g_1 = g_2$ when restricted to eJ. To show this, we first choose $d \in \Lambda$, a primitive idempotent with $\mathrm{top}(d\Lambda) \cong eJ$. Let $a = ea$ be the image of d in eJ. Since $d = d^2$ we have $a = ead$; and, of course, $a\Lambda = eJ$.

Now consider Λd. Since d is a primitive idempotent, Λd is uniserial; and, since $J^2 = 0$, Λd has length at most 2. However $\Lambda d \supset Jd \supseteq \Lambda a \neq 0$; so we see that $Jd = \Lambda a$.

As a first approximation to the desired homomorphisms, let f_i be any isomorphism f_i: $e\Lambda \to X_i$ and let $x_i = f_i(e)$. Since $e = e^2$ we have $x_i = x_i e$. Hence $f_i(r) = x_i r$ for each $r \in e\Lambda$. Now, $f_1(a) = x_1 a \in W = X_1 \cap X_2 = x_2 J$; so $x_1 a = x_2 b$ for some $b \in J$. Since $a = ad$, we may assume that $b = bd$. So $b \in Jd = \Lambda a$; say $b = ca$. Then $x_1 a = x_2 b = (x_2 c)a \neq 0$; and $x_2 c \notin x_2 J$ (otherwise $(x_2 c)a \in x_2 J^2 = 0$). Thus if we take $y_2 = x_2 c$ then $x_1 a = y_2 a$. Note that $y_2 \in X_2 e - W$. It follows that the map $g_2{:}e\Lambda \to X_2$ given by $g_2(r) = y_2 r$ is an isomorphism. Moreover, the relation $x_1 a = y_2 a$ shows that f_1 and g_2 agree on $a\Lambda = eJ$. Thus, taking $g_1 = f_1$, we see that the claim is proved.

Finally, define $h : E \to e\Lambda$ as follows: for each element $y \in E$, write $y = y_1 + y_2$ with $y_i \in X_i$ and let $h(y) = g_1^{-1}(y_1) + g_2^{-1}(y_2)$. Since $g_1^{-1} = g_2^{-1}$ when restricted to W, this is a well-defined mapping. It is a surjection, since each g_i^{-1} is. However the image $e\Lambda$ is projective. Thus E splits as a direct sum, a contradiction since E is an essential extension of a simple module. □

50.17. PROPOSITION. *Let Λ be an Artinian serial ring and E a right Λ-module of finite length such that* $\mathrm{soc}(E)$ *is a simple module. Then E is uniserial.*

PROOF. Recall, from 50.11(ii), that a module E is uniserial if and only if $\mathrm{soc}_n(E)$ is a simple module whenever it is nonzero.

Suppose that the result is false, let E be a counterexample of minimal length, and let $W = \mathrm{soc}(E)$. Note that the second socle layer $\mathrm{soc}_2(E) = \mathrm{soc}(E/W)$ is not a simple module, for otherwise E/W would be a counterexample of shorter length than E. Thus, by minimality, we have $E = \mathrm{soc}^2(E)$ and $\mathrm{soc}_2(E)$ is of composition length 2; say $\mathrm{soc}_2(E) = V_1 \oplus V_2$ with each V_i simple. However, this contradicts the facts [50.15, 50.16] that a predecessor is unique and of multiplicity 1. Hence the result is proved. □

50.18. COROLLARY. *Let e be a primitive idempotent of an Artinian serial ring Λ and suppose that the composition length of $e\Lambda_\Lambda$ equals the index of nilpotence of $J = J(\Lambda)$. Then $e\Lambda$ is an injective Λ-module.*

PROOF. Let n be the index of nilpotence of J. If $e\Lambda$ were not injective, it would have a proper essential extension E of finite length greater than n. Since $\mathrm{soc}(E) = \mathrm{soc}(e\Lambda)$, which is a simple module, E is again uniserial [50.17]. But then the unique composition series of E is $E \supset EJ \supset EJ^2 \supset \ldots \supset EJ^n \supset \ldots$ which is impossible since $J^n = 0$. □

50.19. THEOREM (Nakayama's Theorem). *Let Λ be an Artinian serial ring. Then every Λ-module (finitely generated or not) is a direct sum of uniserial modules.*

PROOF. Let M be a right Λ-module, and n the index of nilpotence of $J = J(\Lambda)$. Since Λ_Λ is a direct sum of uniserial submodules and M is a homomorphic image of a free module, we see that M is a sum (not necessarily direct) of uniserial submodules, each of length $\leq n$. We may assume that at least one of these uniserial submodules has length exactly n, for otherwise $MJ^m = 0$ for some $m < n$, and we may replace Λ by Λ/J^m.

Any uniserial submodule of length n is injective [50.18] and therefore a direct summand of M. If M is finitely generated, we are done by induction since any complementary summand has length less than that of M.

If M does not have finite length, we need to work harder. Let $\mathcal{F}$ be the set of *independent* families of uniserial submodules of M having length n. Here, 'independent' means that the sum of the submodules in the family is a direct sum. The set of independent families of submodules of M is partially ordered by inclusion of families (not of submodules!) and so, by a straightforward application of Zorn's Lemma, has a maximal element $\mathcal{F}_0$. Let S be the sum of the submodules in $\mathcal{F}_0$. Since S is a direct sum of injective modules over a Noetherian ring, S is itself injective [24.1] and therefore there is a decomposition $M = S \oplus M'$.

By maximality of the family $\mathcal{F}_0$, all uniserial submodules of M' have length $< n$. Therefore M' is a module over Λ/J^{n-1}; and therefore iterating the foregoing procedure a finite number of times completes the proof. □

The next result demonstrates that the hypothesis in Nakayama's Theorem, 50.19, that Λ is left, as well as right, serial is not redundant.

50.20. THEOREM. *Let Λ be an Artinian right but not left serial ring. Then Λ has an indecomposable right module E of length 3 which is not uniserial. In more detail: $W = \mathrm{soc}(E)$ is a simple module and E/W is the direct sum of two simple modules.*

PROOF. By replacing Λ by Λ/J^2 we may assume that $J^2 = 0$. Then J_Λ and ${}_\Lambda J$ become semisimple modules; and, by 50.11(i), every uniserial right or left Λ-module of finite length has length ≤ 2. Let $e_1, \ldots, e_n$ be orthogonal primitive idempotents with sum 1. Then $e_iJ = 0$ whenever $e_i\Lambda$ is a simple module. Therefore J_Λ is the direct sum of the socles of those $e_i\Lambda$ that have length 2.

We *claim* that some $e_i\Lambda$ of length 2, call it $e\Lambda$, is not injective. (This is the crux of the proof.) So we suppose, to the contrary, that every $e_i\Lambda$ of length 2 is injective. Since Λ is not left serial, some Λe_i, say Λe_1, is not uniserial. Since Λ is left Artinian, Je_1 is the unique maximal submodule of Λe_1, and hence Je_1 is not uniserial, by 50.11(i). We conclude that the semisimple module Je_1 contains a direct sum $\Lambda x \oplus \Lambda y$ of simple submodules, with $x, y \in Je_1$.

Now consider the right modules $x\Lambda$ and $y\Lambda$, and let $A = e_1 J \oplus (\oplus_{i>1} e_i\Lambda)$. Since A is a maximal right ideal and $xA = (xe_1)A = xe_1 J = 0$ (because $J^2 = 0$), we conclude that A is the right annihilator of x. The same is true of y; and hence $x\Lambda \cong y\Lambda$ via the map $\phi\colon x\lambda \to y\lambda$. Also, each is isomorphic to Λ/A and so is simple. Hence, as observed at the beginning of this proof, each of $x\Lambda$ and $y\Lambda$ is isomorphic to the socle of some $e_i\Lambda$ of length 2, say $e\Lambda$. Fix an inclusion $w\colon y\Lambda \hookrightarrow e\Lambda$ and consider the following maps

$$v\colon\quad x\Lambda \xrightarrow{\phi} y\Lambda \overset{w}{\hookrightarrow} e\Lambda$$

where $v = w\phi$. Since we are assuming that $e\Lambda$ is injective, v and w can be extended to maps $\Lambda \to e\Lambda$. Therefore v and w equal left multiplication by elements of Λ that we again call v, w (namely, the images of 1 under these maps). Since $\phi(x) = y$ we have $vx = w\phi(x) = wy$ which is nonzero because the map w is (1,1). But the relation $vx = wy \neq 0$ among elements of Λ contradicts directness of the sum $\Lambda x \oplus \Lambda y$. This contradiction proves our claim.

So there is a non-injective uniserial module of length 2, say $e\Lambda$. Since $e\Lambda$ is not injective, it has an essential extension E of length 3; and E is not uniserial because every uniserial Λ-module of finite length has length ≤ 2. Since E is an essential extension of $e\Lambda$, we have $\operatorname{soc}(E) = \operatorname{soc}(e\Lambda)$, a simple module, necessarily essential in E. Now $E/\operatorname{soc}(E)$ has length 2 and is not uniserial of length 2 (or else E would be uniserial, by 50.11(ii)). So $E/\operatorname{soc}(E)$ is a direct sum of two simple modules. □

The next example of a one-sided serial ring illustrates two aspects of their structure discussed in 50.13. Part (i) below shows that predecessors on the right can be non-unique in right serial rings. Thus 50.13(a′) fails to hold in this example. Part (ii) demonstrates that, although successors on the left have multiplicity 1 when they exist, they are not necessarily unique. This shows, as promised in 50.13, that condition (a) in (the left-hand version of) 50.12(iii) is not redundant.

50.21. Example. There is an Artinian right but not left serial ring Λ with three simple right modules U_i and three simple left modules ${}_iU$ $(i = 1, 2, 3)$ and the following predecessor-successor pairs.

(i) (U_2, U_1) and (U_3, U_1).
(ii) $({}_1U, {}_2U)$ and $({}_1U, {}_3U)$.

Moreover, each successor has multiplicity 1.

Proof. Let Λ be the following triangular matrix ring (obviously right and left Artinian), over any field k.

(50.21.1)
$$\Lambda = \begin{bmatrix} k & 0 & 0 \\ k & k & 0 \\ k & 0 & k \end{bmatrix}$$

Because of its triangular form, Λ has precisely three simple right modules and three simple left modules, and their additive groups are all that of k. We let $\operatorname{top}(e_{ii}\Lambda) = U_i$ and $\operatorname{top}(\Lambda e_{ii}) = {}_iU$ $(i = 1, 2, 3)$. Note that each $U_i^{\mathrm{op}} = {}_iU$.

We have $e_{11}\Lambda = [k, 0, 0]$, $e_{22}\Lambda = [k, k, 0]$ and $e_{33}\Lambda = [k, 0, k]$. Note that the only inter-column operations under which Λ is closed are: add multiples of columns 2 and 3 to column 1. Hence the three modules $e_{ii}\Lambda$ are uniserial. Therefore Λ is right serial. Moreover, since, by 50.11(iv), every finite length uniserial module is a factor module of one of the uniserial summands of Λ, the only predecessor-successor pairs among the U_i are those displayed in (i); and by 50.12, their successor multiplicities are all 1.

For (ii), let tr denote matrix transpose, and consider $\Lambda e_{11} = [k\ k\ k]^{\text{tr}}$, $\Lambda e_{22} = [0, k, 0]^{\text{tr}}$, and $\Lambda e_{33} = [0, 0, k]^{\text{tr}}$. Note that the only inter-row operations under which Λ is closed are: add multiples of row 1 to rows 2 and 3. Thus the unique maximal submodule of Λe_{11} is $[0\ k\ k]^{\text{tr}} = {}_2U \oplus {}_3U$. Hence Λ is not left uniserial, the predecessor-successor pairs are as in (ii), and the successor multiplicities are again 1. □

51. Artinian Principal Ideal Rings

The main results of this short section describe the structure of Artinian principal ideal rings and their modules. Briefly, these rings are not only serial rings, but also direct sums of full matrix rings over Artinian valuation rings [51.3], and are self-injective [51.6]. Since they are serial rings, Nakayama's Theorem 50.19 describes their modules. As in the preceding section, we will start by using one-sided hypotheses.

51.1. DEFINITION. A ring Λ is called a *right valuation ring* if Λ is uniserial as a right module over itself. In particular, if Λ is a right and left Artinian right and left valuation ring, we call Λ an *Artinian valuation ring.*

51.2. LEMMA. *Let Λ be a ring, suppose that $J = J(\Lambda)$ is nilpotent of index n, and let π be any element of $J - J^2$.*

(i) *If Λ is a right Artinian right valuation ring, then the right ideals of Λ are the ideals $J^i = \pi^i\Lambda$ $(0 \le i \le n)$.*
(ii) *If Λ is also a left Artinian left valuation ring, then every left ideal is one of the ideals listed in* (i), *and each $\pi^i\Lambda = \Lambda\pi^i$.*

PROOF. (i) The fact that the right ideals of Λ are the ideals J^i is a special case of 50.11(i). Therefore $\pi\Lambda$ must be one of these ideals; and the only possibility is clearly $\pi\Lambda = J$.

To see that $\pi^2\Lambda = J^2$, we use the fact that J is a 2-sided ideal: $\pi^2\Lambda = \pi J = \pi\Lambda J = J^2$. All cases $\pi^i\Lambda = J^i$ $(i > 2)$ are handled similarly.

(ii) The left-handed version of (i) shows that the left ideals of Λ are the ideals J^i. The same proof as in (i) shows that $J = \Lambda\pi$, and this immediately implies the rest of (ii). □

51.3. THEOREM. *The following conditions are equivalent for a right Artinian ring Λ.*

(i) *Λ is an indecomposable principal right ideal ring.*
(ii) *Λ is an indecomposable ring whose ideals are principal right ideals.*
(iii) *Λ is a right serial ring with exactly one simple right module.*
(iv) *$\Lambda \cong M_n(\Omega)$, a full matrix ring over a right Artinian right valuation ring Ω.*

If, in addition to these conditions, Λ is a left Artinian left principal ideal ring, then the ring Ω in condition (iv) is also a left Artinian left valuation ring.

PROOF. Choose a decomposition:

$$\Lambda = \oplus_i e_i\Lambda \tag{51.3.1}$$

where $e_1, e_2, \ldots$ is a finite set of orthogonal primitive idempotents whose sum is 1. Let $J = J(\Lambda)$.

(i)⇒(ii) This is trivial.

(ii)⇒(iii) We start by showing that if Λ is a right Artinian principal right ideal ring and $J^2 = 0$ then Λ is right serial. (We are not assuming, yet, that Λ is indecomposable.) Suppose, then, that Λ is not right serial. Then, for some i, it must be that the semisimple module e_iJ has length 2 or more. For ease, let's assume that $i = 1$. Let $V = \text{top}(e_1\Lambda)$. Then either:

(a) e_1J has a direct summand isomorphic to $V^{(2)}$; or

(b) e_1J has a simple direct summand $W \not\cong V$.

We show that each of these situations yields the contradiction that Λ has an ideal which is not principal as a right ideal.

Suppose that (a) occurs. Recall from 50.2 that $\text{top}(e_i\Lambda) \cong \text{top}(e_1\Lambda)$ implies $e_i\Lambda \cong e_1\Lambda$. Therefore, in this situation, the multiplicity of V in the semisimple module J is at least double the number of terms $e_i\Lambda$ in (51.3.1) such that $\text{top}(e_i\Lambda) \cong V$; i.e. at least double the multiplicity of V in Λ/J. However, since J is cyclic, there is a surjection $\Lambda/J \twoheadrightarrow J$; and so the multiplicity of V in J can be at most that of V in Λ/J. So (a) does not occur.

Now suppose that (b) occurs. Let A be the sum of all simple submodules of Λ which are not isomorphic to W. This is fully invariant, so is an ideal. In the ring Λ/A, let B' be the sum of all submodules isomorphic to W. Once again, B' is an ideal. Let B be its inverse image in Λ. We now consider the number of composition W-factors of B'. For each e_i with $e_i\Lambda \cong e_1\Lambda$, there is a contribution of at least one; and for each e_i with $\text{top}(e_i\Lambda) \cong W$ there is a contribution of at least one. So the multiplicity of W in B' is greater than that of W in Λ/J. Arguing as for (a), we see that B' cannot be cyclic and hence neither can B.

This contradiction completes the proof that Λ is right serial; and it also shows that if $e_i\Lambda$ is not simple then it has length 2 and, moreover, $\text{top}(e_i\Lambda) \cong e_iJ$. Consequently the only possible successor of a simple Λ-module X is X itself.

Now consider the general case when J^2 may not be zero, but now Λ is assumed indecomposable. We know from above that Λ/J^2 is right serial. Hence, by 50.12, the same is true of Λ. Furthermore, for each i, consecutive composition factors of the uniserial module $e_i\Lambda$ are predecessor and successor of each other when viewed as Λ/J^2-modules and so, from above, are isomorphic. Thus each composition factor of $e_i\Lambda$ is isomorphic to $\text{top}(e_i\Lambda)$. So, if we let $A = \Sigma\{e_i\Lambda \mid e_i\Lambda \cong e_1\Lambda\}$, then all composition factors of A are isomorphic to $V = \text{top}(e_1\Lambda)$. Similarly, if B is the sum of the remaining terms in (51.3.1) then $A \oplus B = \Lambda$ and no composition factors of B_Λ is isomorphic to V. Evidently A and B are fully invariant right ideals of Λ, and hence (2-sided) ideals. Indecomposability of the ring Λ shows that $B = 0$, and thus Λ has only the one simple module.

(iii)⇒(iv). Let V be the unique simple Λ-module. Then $\text{top}(e_i\Lambda) \cong V$ for every i in (51.3.1), and so, by 50.2(ii), $e_i\Lambda \cong e_1\Lambda$. Thus $e_1\Lambda$ is a progenerator and $\Lambda \cong M_n(\Omega)$ where n is the number of terms in (51.3.1) and $\Omega = \text{End}(e_1\Lambda)_\Lambda$. We see from 11.3 that the lattice of right ideals of Ω is isomorphic to the lattice of submodules of $e_1\Lambda$ and hence Ω is a right Artinian right valuation ring.

(iv)⇒(i). We prove a more general result in the next theorem.

Supplementary statement. Suppose that Λ is also a left Artinian principal left ideal ring. Then by the left module analogue of (i)$\Rightarrow$(iii), Λ is a left serial ring. Since, as was shown above, Λ and Ω are Morita equivalent, Ω is left Artinian and, by 50.10, left serial. To show that the left serial ring Ω is a left valuation ring, it suffices to show that the identity element of Ω is a primitive idempotent, and this holds since Ω is a right valuation ring. $\square$

51.4. THEOREM. *For any principal right ideal ring Ω (not necessarily Artinian) and each n, the matrix ring $M_n(\Omega)$ is a principal right ideal ring. Moreover, the generator of any right ideal of $M_n(\Omega)$ can be chosen to be a lower triangular matrix.*

PROOF. Let M be any right ideal of $\Lambda = M_n(\Omega)$, and let K be the set of first columns of matrices in M. Then K is a right Λ-module. Moreover, since matrix multiplication from the right does column operations, M is the set of matrices whose columns are all in K.

The set B_1 of (1,1)-entries of elements of K is a right ideal of Ω, and hence $B_1 = b_1\Omega$ for some $b_1 \in \Omega$. Let the first column of the matrix x displayed below be any element of K whose (1,1)-entry is b_1.

$$x = \begin{bmatrix} b_1 & 0 & 0 & \cdots & 0 \\ * & b_2 & 0 & \cdots & 0 \\ * & * & b_3 & \cdots & 0 \\ \vdots & \vdots & \vdots & \ddots & \vdots \\ * & * & * & \cdots & b_n \end{bmatrix} \tag{51.4.1}$$

Let B_2 be the set of (2,1)-entries of elements of K whose (1,1)-entry is 0. Then $B_2 = b_2\Omega$. Choosing an element of K with (1,1)-entry 0 and (2,1)-entry b_2 yields the second column of x. Continuing in this way we get the matrix x. It is routine to verify that $x\Lambda = M$. $\square$

51.5. COROLLARY. *Let Λ be a right Artinian principal right ideal ring. If any simple Λ-module X has a successor, that successor is X itself.*

PROOF. This follows directly from 51.3 since, if the ring decomposes, then X will be a module over one of its indecomposable components, as also will be any nonsplit extension of any simple Λ-module by X. $\square$

51.6. COROLLARY. *Every Artinian principal ideal ring is self-injective.*

PROOF. We may assume that the given ring Λ is indecomposable and so, by 51.3, is a full matrix ring over an Artinian valuation ring. Therefore, by Morita equivalence [11.3], we may assume that Λ is an Artinian valuation ring. Let $J = J(\Lambda) = \Lambda\pi = \pi\Lambda$ and let n be the index of nilpotence of J. Then, as noted in 51.2, the right ideals of Λ are the ideals $\Lambda\pi^i = \pi^i\Lambda$ $(0 \le i \le n)$.

To see that Λ_Λ is injective, let $f\colon \pi^i\Lambda \to \Lambda$ be a right Λ-homomorphism. Then $\operatorname{im}(f)$ has length at most i, and hence $\operatorname{im}(f) \subseteq \pi^i\Lambda = \Lambda\pi^i$. Say $f(\pi^i) = c\pi^i$. Then left multiplication by c is a homomorphism $\Lambda_\Lambda \to \Lambda_\Lambda$ that extends f. Therefore, by Baer's Criterion, Λ_Λ is injective.

[Alternatively, since the composition length of the unique uniserial summand Λ_Λ of the Artinian serial ring Λ equals the index of nilpotence of J, injectivity of Λ_Λ is a special case of 50.18.] $\square$

50.20 and the remarks following it showed that symmetric hypotheses are needed in order to obtain several of the results. The result and example which now follow concern some questions related to one-sided conditions on the ring Λ. In particular, 51.8 provides the example promised in 50.13.

51.7. COROLLARY. *Let Λ be a right and left serial Artinian ring which is also a principal right ideal ring. Then Λ is a principal left ideal ring.*

PROOF. We may, without loss, assume that Λ is an indecomposable ring. Then 51.3 tells us that Λ has only one simple right module V say. Hence it has only one simple left module, V^{op}. Therefore, by 51.3 again, Λ is a principal left ideal ring. □

51.8. EXAMPLE. Let $F = k(y)$, the field of rational functions in an indeterminate y over a ground field k. Then form the ring T of twisted polynomials over F in an indeterminate x, with $f(y)x = xf(y^2)$ for each $f(y)$ in F. Now let $\Lambda = T/x^2T$. Then:

(i) Λ is an Artinian right valuation ring.
(ii) Λ is not a principal left ideal ring and not a left serial ring.
(iii) Λ has a unique simple left module which is its own successor of multiplicity 2.
(iv) Λ has a unique simple right module which is its own predecessor of multiplicity at least 2.

PROOF. (i) One can see that T, and hence Λ, is a principal right ideal ring by using the Euclidean algorithm. Evidently $J = x\Lambda \cong F \cong \Lambda/J$; so these are simple right Λ-modules and Λ is a right Artinian right valuation ring with $J^2 = 0$.

To see Λ is left Artinian, note that $k(y)$ is 2-dimensional over $k(y^2)$, generated by 1 and y. (To see that, given any rational function $a(y)/b(y)$, write it as $(a(y)b(-y))/(b(y)b(-y))$. The denominator is then a polynomial in y^2; and the numerator can be written as $c(y^2) + yd(y^2)$ for some polynomials c and d.) Hence J is a 2-dimensional and Λ a 3-dimensional left vector space over $k(y^2)$. Thus Λ is left Artinian.

(ii) Note that $\Lambda/J \cong F$ and $J \cong (\Lambda/J)^{(2)}$ as a left module. Since $J^2 = 0$, J has the same structure as a left Λ-module as it does as a Λ/J-module. Hence ${}_{\Lambda}J$ is not principal and Λ is not a principal left ideal ring. We thus know from 51.7 that Λ is not a left serial ring (although it is easy to check this directly).

(iii) The ring Λ, as a left Λ-module, has the unique simple Λ/J on top and $J \cong (\Lambda/J)^{(2)}$. Hence, by 50.5, the multiplicity of Λ/J as its own successor is 2.

(iv) Λ has only one simple module $V = \Lambda/J$ but is not a left serial ring. Then 50.20 shows that Λ has an indecomposable right module E which is an essential extension of its simple socle and $E/\operatorname{soc}(E)$ is the direct sum of two simple modules. Since V is the unique simple module, this shows that V has multiplicity at least 2 as a predecessor of itself. □

52. Equivalent Functors

This short appendix defines the notion of equivalent functors, which is mentioned in a few places in this book.

52.1. DEFINITIONS. A *natural transformation* $\psi\colon F \to G$ of covariant functors F, G defined on some category $\mathcal{C}$ is a collection $\{\psi_M\colon F(M) \to G(M) \mid M \in \mathcal{C}\}$ of morphisms in the image category, such that the diagram

$$(52.1.1)\qquad \begin{array}{ccc} F(M) & \xrightarrow{\psi_M} & G(M) \\ \downarrow F(f) & & \downarrow G(f) \\ F(N) & \xrightarrow{\psi_N} & G(N) \end{array}$$

commutes for every $f\colon M \to N$ in $\mathcal{C}$. We call ψ an *equivalence of functors* $F \cong G$ if every ψ_M is an isomorphism in the image category. (Another common terminology is that the isomorphisms $F(M) \cong G(M)$ are *natural* for all $M \in \mathcal{C}$.) Similarly, if F, G are contravariant functors, the same definitions apply except that the vertical arrows in (52.1.1) point upward.

One example, mentioned after 1.6, is that the double dual functor $P \to P^{**}$, when applied to finitely generated projective modules over any ring, is equivalent to the identity functor.

If F, G are functors of two variables, say contravariant in the first variable and covariant in the second variable, comparable definitions apply, with diagram (52.1.1) replaced by

$$(52.1.2)\qquad \begin{array}{ccc} F(B,X) & \xrightarrow{\psi_{B,X}} & G(B,X) \\ \downarrow F(\alpha,\xi) & & \downarrow G(\alpha,\xi) \\ F(A,Y) & \xrightarrow{\psi_{A,Y}} & G(A,Y) \end{array}$$

where $\alpha\colon A \to B$ and $\xi\colon X \to Y$.

For an example, we discuss in §53 the definition of $\mathrm{Ext}^n_\Lambda(X, W)$ in terms of either a projective resolution of X or an injective resolution of W. More precisely, these two definitions yield equivalent functors, in the sense defined above.

53. Ext and Tor

This section reviews some basic facts that we need about Ext and Tor for modules over an arbitrary ring Λ, sometimes restated in a form convenient to this book. We give a few proofs, when needed to avoid an overly long chain of references to the literature, or when they follow easily from previously stated facts. Most of our detailed references are to [**Rotman 09**], [**Weibel 94**] and [**Cartan-Eilenberg 56**]. But many of the facts quoted can be found in other books on homological algebra.

53.1. Ext. We presume that the reader knows that $\mathrm{Ext}^n_\Lambda(A, X)$ is a contravariant, additive functor of A, and can be obtained using a projective resolution of A [**Rotman 09**, p. 370]. Also, $\mathrm{Ext}^n_\Lambda(A, X)$ is a covariant additive functor of X, and can be obtained using an injective resolution of X [**Rotman 09**, p. 365]. Moreover, the groups $\mathrm{Ext}^n_\Lambda(A, X)$ are independent of which resolution is used [**Weibel 94**, 2.7.6], [**Rotman 09**, 6.57, 6.40, 6.67].

53.2. Notation. Let $\alpha\colon B_\Lambda \to A_\Lambda$ be a module homomorphism. The induced homomorphism: $\mathrm{Ext}^n_\Lambda(B, X) \to \mathrm{Ext}^n_\Lambda(A, X)$ is denoted by $\mathrm{Ext}^n_\Lambda(\alpha, X)$, and abbreviated to $(\alpha, 1)$ (no subscript or superscript). Similarly, for $\phi\colon X \to Y$ we write $\mathrm{Ext}^n_\Lambda(A, \phi) = (1, \phi)$. Also, $(\alpha, 1)(1, \phi)$ is abbreviated to (α, ϕ), an unambiguous product because

$$(53.2.1)\qquad (\alpha, 1)(1, \phi) = (1, \phi)(\alpha, 1).$$

This is often called *bifunctor commutativity.*

The notation (α, ϕ) is confusing when function-composition is involved, because Ext is contravariant in α but covariant in ϕ. A better notation is to view α and ϕ themselves as acting on $\mathrm{Ext}^n_\Lambda(A, X)$ for all n via $(\alpha, 1)$ and $(1, \phi)$ respectively. Then, for $\mathcal{E} \in \mathrm{Ext}^n_\Lambda(A, X)$ define $\alpha\mathcal{E}\phi = (\alpha, \phi)(\mathcal{E})$ (when dealing with right modules).

53.3. PROPOSITION. *Given homomorphisms of right modules* $C \xrightarrow{\alpha_2} B \xrightarrow{\alpha_1} A$ *and* $X \xrightarrow{\phi_1} Y \xrightarrow{\phi_2} Z$ *and* $\mathcal{E} \in \mathrm{Ext}^n_\Lambda(A, X)$, *we have:*

$$(\alpha_2\alpha_1)\mathcal{E}(\phi_1\phi_2) = \alpha_2(\alpha_1\mathcal{E})(\phi_1\phi_2) = (\alpha_2\alpha_1)(\mathcal{E}\phi_1)\phi_2. \tag{53.3.1}$$

PROOF. By symmetry it is enough to prove the first equality, and by bifunctor commutativity it suffices to prove the case that $\phi_1 = 1 = \phi_2$. The proof is then completed by the fact that $\mathrm{Ext}^n_\Lambda(A, B)$ is contravariant in A. □

Specializing to the case $C = B = A$ and $X = Y = Z$ we get:

53.4. COROLLARY. $\mathrm{Ext}^n_\Lambda(A, X)$ *is an* $(\mathrm{End}(A), End(X))$*-bimodule for any right* Λ*-modules* A, X. □

53.5. Long exact sequences of Ext. Let $\mathcal{E}\colon C \overset{\beta}{\hookrightarrow} B \overset{\alpha}{\twoheadrightarrow} A$ be a short exact sequence of right Λ-modules, and let X be any right Λ-module. When we say *'apply* $\mathrm{Hom}_\Lambda(-, X)$ *to* $\mathcal{E}$*'* we mean to obtain the contravariant long exact sequence of Ext, which starts as follows [**Rotman 09**, 6.63]:

$$\begin{gathered} 0 \to \mathrm{Hom}_\Lambda(A, X) \xrightarrow{(\alpha,1)} \mathrm{Hom}_\Lambda(B, X) \xrightarrow{(\beta,1)} \mathrm{Hom}_\Lambda(C, X) \xrightarrow{\partial} \\ \mathrm{Ext}^1_\Lambda(A, X) \xrightarrow{(\alpha,1)} \mathrm{Ext}^1_\Lambda(B, X) \xrightarrow{(\beta,1)} \mathrm{Ext}^1_\Lambda(C, X) \xrightarrow{\partial} \mathrm{Ext}^2_\Lambda(A, X) \to \dots \end{gathered} \tag{53.5.1}$$

and continues infinitely to the right, with the *connecting homomorphisms* ∂ mapping $\mathrm{Ext}^n_\Lambda(C, X) \to \mathrm{Ext}^{n+1}_\Lambda(A, X)$ for each $n \geq 0$.

In more detail, the maps in (53.5.1) are homomorphisms of abelian groups that are *natural in its first variable for short exact sequences,* in the following sense. Suppose we are given a morphism of short exact sequences — that is, a trio of module homomorphism f, g, h such that the next diagram commutes.

$$\begin{array}{cccccccccccc} \mathcal{E}'\colon & 0 & \to & C' & \xrightarrow{\beta'} & B' & \xrightarrow{\alpha'} & A' & \to & 0 \\ & & & \downarrow{\scriptstyle h} & & \downarrow{\scriptstyle g} & & \downarrow{\scriptstyle f} & & \\ \mathcal{E}\colon & 0 & \to & C & \xrightarrow{\beta} & B & \xrightarrow{\alpha} & A & \to & 0 \end{array} \tag{53.5.2}$$

Then, for each n, the following diagram also commutes [**Rotman 09**, p. 28, 6.63].

$$\begin{array}{ccccccccc} \to \mathrm{Ext}^n_\Lambda(A, X) & \xrightarrow{(\alpha,1)} & \mathrm{Ext}^n_\Lambda(B, X) & \xrightarrow{(\beta,1)} & \mathrm{Ext}^n_\Lambda(C, X) & \xrightarrow{\partial} & \mathrm{Ext}^{n+1}_\Lambda(A, X) \to \\ \downarrow{\scriptstyle (f,1)} & & \downarrow{\scriptstyle (g,1)} & & \downarrow{\scriptstyle (h,1)} & & \downarrow{\scriptstyle (f,1)} \\ \to \mathrm{Ext}^n_\Lambda(A', X) & \xrightarrow{(\alpha',1)} & \mathrm{Ext}^n_\Lambda(B', X) & \xrightarrow{(\beta',1)} & \mathrm{Ext}^n_\Lambda(C', X) & \xrightarrow{\partial} & \mathrm{Ext}^{n+1}_\Lambda(A', X) \to \end{array} \tag{53.5.3}$$

Note that if f, g or h is an isomorphism, then so is each $(f, 1)$, $(g, 1)$, or $(h, 1)$, respectively, because functors take isomorphisms to isomorphisms.

Similarly, applying $\mathrm{Hom}(X, -)$ to $\mathcal{E}$ yields the covariant long exact sequence of Ext, which begins [**Rotman 09**, 6.46]

$$\begin{gathered} 0 \to \mathrm{Hom}(X, C) \to \mathrm{Hom}(X, B) \to \mathrm{Hom}(X, A) \xrightarrow{\partial} \\ \mathrm{Ext}^1_\Lambda(X, C) \to \mathrm{Ext}^1_\Lambda(X, B) \to \mathrm{Ext}^1_\Lambda(X, A) \xrightarrow{\partial} \mathrm{Ext}^2_\Lambda(X, C) \to \dots \end{gathered} \tag{53.5.4}$$

and is natural in its second variable for short exact sequences [**Rotman 09**, 6.47]. But since Ext is covariant in its second variable, the given vertical arrows in (53.5.2) point upwards, and (53.5.3) is replaced by:

$$
\begin{array}{ccccccccc}
\to \operatorname{Ext}^n_\Lambda(X,C) & \xrightarrow{(1,\beta)} & \operatorname{Ext}^n_\Lambda(X,B) & \xrightarrow{(1,\alpha)} & \operatorname{Ext}^n_\Lambda(X,A) & \xrightarrow{\partial} & \operatorname{Ext}^{n+1}_\Lambda(X,C) \to \\
\downarrow{\scriptstyle (1,h)} & & \downarrow{\scriptstyle (1,g)} & & \downarrow{\scriptstyle (1,f)} & & \downarrow{\scriptstyle (1,h)} \\
\to \operatorname{Ext}^n_\Lambda(X,C') & \xrightarrow{(1,\beta')} & \operatorname{Ext}^n_\Lambda(X,B') & \xrightarrow{(1,\alpha')} & \operatorname{Ext}^n_\Lambda(X,A') & \xrightarrow{\partial} & \operatorname{Ext}^{n+1}_\Lambda(X,C') \to
\end{array}
\tag{53.5.5}
$$

53.6. COROLLARY. *The homomorphisms in the long exact sequences of* Ext *are* $\operatorname{End}_\Lambda(X)$*-module homomorphisms (on the appropriate side).*

PROOF. Except for the terms involving ∂, this follows from bifunctor commutativity. For ∂, see [**Cartan-Eilenberg 56**, V:4.1]. □

53.7. LEMMA. *If B is a projective module, then $\operatorname{Ext}^n_\Lambda(B,X) = 0$ for all X and all $n \geq 1$. Conversely, if $\operatorname{Ext}^1_\Lambda(B,X) = 0$ for all X, then B is projective.*

PROOF. For the nontrivial first statement, let B be projective. Then it has a projective resolution $0 \to B \to B \to 0$. So the first statement follows from the projective resolution definition of Ext. □

53.8. PROPOSITION (Dimension shifting). *Consider any short exact sequence $C \hookrightarrow B \twoheadrightarrow A$ of right modules. Then for every X_Λ and every $n \geq 1$, the following hold.*

(i) *If B is projective, then $\operatorname{Ext}^{n+1}_\Lambda(C,X) \cong \operatorname{Ext}^n_\Lambda(A,X)$ as* $\operatorname{End}(X)$*-modules.*
(ii) *If B is injective, then $\operatorname{Ext}^n_\Lambda(X,A) \cong \operatorname{Ext}^{n+1}_\Lambda(X,C)$ as* $\operatorname{End}(X)$*-modules.*

PROOF. (i) Since B is projective we have $\operatorname{Ext}^n_\Lambda(B,X) = 0$ for all $n \geq 1$ [53.7]. Hence all of the connecting homomorphisms $\operatorname{Ext}^n_\Lambda(A,X) \to \operatorname{Ext}^{n+1}_\Lambda(C,X)$ $(n \geq 1)$ in the contravariant long exact sequence (53.5.1) for Ext are isomorphisms. These are $\operatorname{End}(X)$-isomorphisms by 53.6.

(ii) This is proved in a similar fashion, but using the covariant long exact sequence (53.5.4). □

53.9. DEFINITION. A module A_Λ has *projective dimension* at most n (written $\operatorname{pd}(A) \leq n$) if there is a projective resolution $\mathcal{P}$: $\ldots \to P_1 \to P_0 \twoheadrightarrow A$ in which $P_{n+1} = 0$. We say that $\operatorname{pd}(A) = \infty$ if no such n exists [**Rotman 09**, p. 454]. In particular, $\operatorname{pd}(A) = 0$ if and only if A is projective.

The main properties of projective dimension are [**Rotman 09**, 8.6]:

53.10. PROPOSITION. *The following statements about a Λ-module A are equivalent.*

(i) $\operatorname{pd}(A) \leq n$.
(ii) *In every projective resolution $\mathcal{P} \to A$, $\ker(P_n \to P_{n-1})$ is projective, taking $P_{-1} = A$.*
(iii) $\operatorname{Ext}^k_\Lambda(A,X) = 0$ *for every $k > n$ and every X_Λ.*

□

53.11. Ext^1_Λ in terms of extensions. The functor $\mathrm{Ext}^1_\Lambda(A, X)$ can be defined in terms of extensions of X by A, and this definition has the advantage that one can explicitly view its elements and functor actions. We denote the new version of this functor by $\mathbf{e}(A, X)$.

An *extension of* X_Λ *by* A_Λ is a short exact sequence of Λ-modules,

$$\mathcal{E}: \quad 0 \to X \to M \to A \to 0 \tag{53.11.0}$$

or, in more compact notation, $\mathcal{E}$: $X \hookrightarrow M \twoheadrightarrow A$. We describe the module M as an *extension module* (or, sometimes, simply an *extension*) of X_Λ by A_Λ. Two extensions $\mathcal{E}$ and $\mathcal{F}$ of X by A are called *equivalent* (*congruent,* in some books), with notation $[\mathcal{E}] = [\mathcal{F}]$, if there is a commutative diagram of the form

$$\begin{array}{ccccccccccc} \mathcal{E}: & 0 & \to & X & \to & M & \to & A & \to & 0 \\ & & & \downarrow 1 & & \downarrow \gamma & & \downarrow 1 & & \\ \mathcal{F}: & 0 & \to & X & \to & N & \to & A & \to & 0 \end{array} \tag{53.11.1}$$

where γ is a Λ-homomorphism and hence a Λ-isomorphism [**Rotman 09**, p. 420].

The abelian group $\mathbf{e}(A, X)$ is the collection of all equivalence classes of extensions of X by A, together with the *Baer sum,* defined as follows.

Let $\mathcal{E}$: $X \overset{g}{\hookrightarrow} M \overset{f}{\twoheadrightarrow} A$ and $\mathcal{E}'$: $X \overset{g'}{\hookrightarrow} M' \overset{f'}{\twoheadrightarrow} A$ be extensions of X by A. Let D be the denominator of the quotient module in (53.11.2), and then let $\mathcal{E} + \mathcal{E}'$ be the following extension:

$$[\mathcal{E}] + [\mathcal{E}']: \quad X \overset{\phi}{\hookrightarrow} \frac{\{(m, m') \in M \oplus M' \mid fm = f'm'\}}{\{(-gx, g'x) \mid x \in X\}} \overset{\psi}{\twoheadrightarrow} A \tag{53.11.2}$$

where $\phi x = (gx, 0) + D = (0, g'x) + D$ and $\psi(m, m') + D = fm = f'm'$.

By [**Rotman 09**, 7.35], [**Weibel 94**, 3.4.5], or [**Cartan-Eilenberg 56**, XIV-1.1, 1.2], we have:

53.12. THEOREM. *$\mathbf{e}(A, X)$ is an abelian group and $\mathbf{e}(A, X) \cong \mathrm{Ext}^1_\Lambda(A, X)$. The zero element of $\mathbf{e}(A, X)$ is the class of the split extension.* □

53.13. Functor structure of $\mathbf{e}(A, X)$. Let $\mathcal{E}$ be a given extension of right Λ-modules, as in (53.13.1) and let $\alpha : A_1 \to A$ be a right Λ-module homomorphism. Then there is an extension, called $\alpha\mathcal{E}$, and a Λ-homomorphism γ, such that the bottom two squares of the first diagram in (53.13.1) are commutative; and, moreover, $[\alpha\mathcal{E}]$ is determined by α and $[\mathcal{E}]$ [**Rotman 09**, 7.28].

(53.13.1)

$$\begin{array}{cccccccc} \beta(\alpha\mathcal{E}): & X & \hookrightarrow & M_2 & \twoheadrightarrow & A_2 \\ & \downarrow 1 & & \downarrow & & \downarrow \beta \\ \alpha\mathcal{E}: & X & \hookrightarrow & M_1 & \twoheadrightarrow & A_1 \\ & \downarrow 1 & & \downarrow \gamma & & \downarrow \alpha \\ \mathcal{E}: & X & \hookrightarrow & M & \twoheadrightarrow & A \end{array} \qquad \begin{array}{cccccccc} \mathcal{E}: & X & \hookrightarrow & M & \twoheadrightarrow & A \\ & \downarrow \phi & & \downarrow \delta & & \downarrow 1 \\ \mathcal{E}\phi: & X_3 & \hookrightarrow & M_3 & \twoheadrightarrow & A \\ & \downarrow \theta & & \downarrow & & \downarrow 1 \\ (\mathcal{E}\phi)\theta: & X_4 & \hookrightarrow & M_4 & \twoheadrightarrow & A \end{array}$$

This (and our convention that homomorphisms act on the left of right modules) implies:

$$\big[\beta(\alpha\mathcal{E})\big] = \big[(\beta\alpha)\mathcal{E}\big]. \tag{53.13.2}$$

Similar comments apply to the second diagram in (53.13.1) [**Rotman 09**, 7.29], giving:

(53.13.3) $$\big[(\mathcal{E}\phi)\theta\big] = \big[\mathcal{E}(\phi\theta)\big].$$

To summarize, so far we have proved:

53.14. LEMMA. $\mathbf{e}(A,X)$ *is a functor contravariant in* A *and covariant in* X. *In particular,* (53.13.2) *and* (53.13.3) *hold.* □

Now that we have the two functor actions of module homomorphisms α and ϕ on each of $\mathbf{e}(A,X)$ [53.14] and $\mathrm{Ext}^1_\Lambda(A,X)$ [53.2], we can sharpen the statement $\mathbf{e}(A,X) \cong \mathrm{Ext}^1_\Lambda(A,X)$ to the statement that $\mathbf{e}(A,X)$ and $\mathrm{Ext}^1_\Lambda(A,X)$ are equivalent functors.

53.15. THEOREM. *The functor* $\mathbf{e}(A_\Lambda, X_\Lambda)$ *is additive in each variable. Moreover, there is an isomorphism* $\psi_{AX} : \mathbf{e}(A,X) \to \mathrm{Ext}^1_\Lambda(A,X)$ *of abelian groups such that, for every* $\alpha\colon A' \to A$, $\phi\colon X \to X'$, $[\mathcal{E}] \in \mathbf{e}(A,X)$, *and* $\tilde{\mathcal{E}} = \psi_{AX}[\mathcal{E}]$ *we have:*

(i) $[\alpha\mathcal{E}] \overset{\psi_{A'X}}{\longmapsto} \alpha\tilde{\mathcal{E}}$ *and* $[\mathcal{E}\phi] \overset{\psi_{AX'}}{\longmapsto} \tilde{\mathcal{E}}\phi$; *and*
(ii) $[\alpha(\mathcal{E}\phi)] = [(\alpha\mathcal{E})\phi]$ *(i.e. bifunctor commutativity in* $\mathbf{e}$*).*

PROOF. This proof consists of references to proofs of its various parts, mainly from [**Rotman 09**] (often with an unpleasant mismatch of notation). Define the function $\mathbf{e}(A,X) \overset{\psi_{AX}}{\longrightarrow} \mathrm{Ext}^1_\Lambda(A,X)$ as in [**Rotman 09**, p. 422]. This is shown to be an isomorphism of abelian groups in [**Rotman 09**, 7.35].

The first relation in (i) is [**Rotman 09**, p. 430, Formula III].

The second relation in (i) is [**Rotman 09**, p. 429, Formula II].

(ii) Since ψ_{AX} is an isomorphism, (ii) is equivalent, by repeated use of (i), to bifunctor commutativity $\alpha(\tilde{\mathcal{E}}\phi) = (\alpha\tilde{\mathcal{E}})\phi$ in $\mathrm{Ext}^1_\Lambda(A,X)$. See (53.2.1). Alternatively, there is an explicit statement and proof in [**Mac Lane 63**, Chap. 3, Lemma 1.6].

Additivity. We want to prove that $[(\alpha_1+\alpha_2)\mathcal{E}] = [\alpha_1\mathcal{E}] + [\alpha_2\mathcal{E}]$ (and the analogous relation for ϕ). By means of (i) and the isomorphism ψ_{AX}, we can reduce this to the corresponding additivity in Ext^1_Λ, which we already know. □

The proof of the next result makes use of the equivalence between $\mathbf{e}$ and Ext^1_Λ.

53.16. COROLLARY. *Let* A, X *be modules over any ring* Λ *such that either:*

(i) A *is simple and* $\mathrm{Ext}^1_\Lambda(A,X)$ *has dimension 1 over* $\mathrm{End}(A)$; *or*
(ii) X *is simple and* $\mathrm{Ext}^1_\Lambda(A,X)$ *has dimension 1 over* $\mathrm{End}(X)$.

Then, up to Λ*-isomorphism, there is exactly one nonsplit extension module of* X *by* A.

PROOF. By 53.15 we may use $\mathbf{e}$ in place of Ext^1_Λ.

(i) The two rows of the following diagram display nonsplit extensions $\mathcal{D}, \mathcal{E}$ of X by A, with respective extension-modules M and N. We want to show that $M \cong N$.

(53.16.1) $$\begin{array}{ccccccc} \mathcal{D}: & X & \hookrightarrow & M & \twoheadrightarrow & A \\ & \downarrow 1 & & \downarrow \gamma & & \downarrow \alpha \\ \mathcal{E}: & X & \hookrightarrow & N & \twoheadrightarrow & A \end{array}$$

Since $\mathbf{e}(A,X)$ has dimension 1 over the division ring $\mathrm{End}(A)$ and both $\mathcal{D}$ and $\mathcal{E}$ are nonzero (i.e. nonsplit), there is an automorphism α of A such that, in the notation of (53.13.1), we have $[\mathcal{D}] = \alpha[\mathcal{E}]$. Therefore there exists $\alpha \in \operatorname{Hom}(M, N)$ such that diagram (53.16.1) commutes; and, since 1 and α are isomorphisms, so is γ.

(ii) This is almost the same as (i). □

53.17. Long exact sequence of Tor. In dealing with Tor, we assume that the reader is familiar with the definitions of the covariant functor $\operatorname{Tor}_n^\Lambda(A, X)$ in terms of a projective resolution of its first variable, and in terms of a projective resolution of its second variable [**Rotman 09**, p. 346, 355]. These two definitions define isomorphic abelian groups [**Rotman 09**, 6.32]; and the Tor analogue of bifunctor commutativity, see (53.2.1), applies here, too [**Cartan-Eilenberg 56**, III-7.1].

Let $\mathcal{E}\colon C \hookrightarrow B \twoheadrightarrow A$ be a short exact sequence of right Λ-modules and let X be any left Λ-module. When we say 'apply $- \otimes_\Lambda X$ to $\mathcal{E}$' we mean to obtain the long exact sequence of Tor which ends as follows [**Rotman 09**, 6.30]:

$$\begin{aligned} &\cdots \to \operatorname{Tor}_2^\Lambda(A, X) \xrightarrow{\partial} \operatorname{Tor}_1^\Lambda(C, X) \to \operatorname{Tor}_1^\Lambda(B, X) \to \operatorname{Tor}_1^\Lambda(A, X) \\ &\xrightarrow{\partial} C \otimes_\Lambda X \to B \otimes_\Lambda X \to A \otimes_\Lambda X \to 0\,. \end{aligned} \tag{53.17.1}$$

Symmetrically, (53.17.1) has an analogue, with X in the first variable, in which case X is a right module and A, B, C are left modules.

Our main use of Tor is the following corollary, which follows immediately from (53.17.1).

53.18. Corollary. *A left module ${}_\Lambda X$ is flat if and only if* $\operatorname{Tor}_1^\Lambda(A, X) = 0$ *for every right Λ-module A. Similarly, A_Λ is flat if and only if* $\operatorname{Tor}_1^\Lambda(A, X) = 0$ *for every left Λ-module ${}_\Lambda X$.* □

54. Open Problems

54.1. Problem. Is there a more direct argument than that given in 4.12 that if R is a basic idealizer from S then ${}_R S$ is flat?

54.2. Problem. By 5.4, every short exact sequence $X \hookrightarrow M \twoheadrightarrow Y$ is split if X, Y are semisimple modules of finite length and $M \cong X \oplus Y$. Miyata's Theorem [**Miyata 67**] shows this is true for arbitrary X, Y if M is a finitely generated module over a commutative Noetherian ring; and [**Guralnick 81**] shows this is true for any commutative ring provided that M is finitely presented. Is there a noncommutative analogue of Miyata's result?

54.3. Problem. We know that every HNP ring R is an intersection of basic chain idealizers from a Dedekind prime ring [30.8]; and if R has infinitely many nontrivial towers, this will be an infinite intersection.

When is an infinite intersection of idealizers of disjoint basic chains an HNP ring? Are there intersection versions of the directed union results used in §25, the construction of an example with infinitely many non-trivial towers?

54.4. Problem. Are all Dedekind right closures S of a given HNP ring R Morita equivalent? We have seen, in 30.6, that if some S is a finitely generated (right) R-module, the answer is 'yes' and that then, by 30.5, all such S_R are finitely generated.

54.5. PROBLEM. For any HNP ring R, 34.3 proves unique presentability by P of any homomorphic image $W^{(t)}$ where P is any finitely generated projective module with udim $P \geq 2$ and W is a simple R-module. [**Odenthal 89**], for a class of rings which includes all HNP rings with enough invertible ideals (i.e. all faithful towers trivial), proves unique presentability of any homomorphic image P/N of P provided that udim$(N) \geq 2$.

Is there a common generalization for HNP rings of these two results? It is worth noting that such a generalization would have the Cancellation Theorem 34.6 as an immediate consequence.

54.6. PROBLEM. Suppose that R is an HNP ring with only one simple right module up to isomorphism and suppose that module is faithful. We know, (see 40.4) that its ideal class group $G(R)$ is a finite cyclic group.

(a) What finite cyclic groups can occur? Indeed, can $G(R)$ be nontrivial?

(b) Over such a ring R, does stable isomorphism of finitely generated projective modules imply isomorphism?

54.7. PROBLEM. What abelian groups can occur as the ideal class group $\mathcal{G}$ of a simple Dedekind prime ring?

There are examples of simple Dedekind prime rings R where $G(R)$ is known and non-trivial. In [**McR 01**, 12.7.7], there is an example where $\mathbf{K}_0(R) \cong \mathbb{Z}^{(n)}$ for any $n \geq 2$; in which case, by 35.18, $G(R) \cong \mathbb{Z}^{(n-1)}$. Similarly, in [**McR 01**, 12.7.7], there is an example where $\mathbf{K}_0(R) \cong \mathbb{Z} \oplus \mathbb{Z}/(2)$ and so $G(R) \cong \mathbb{Z}/(2)$. We know that any abelian group can occur as the ideal class group of a commutative Dedekind domain [**Claborn 66**]. Is the same true for simple Dedekind prime rings?

54.8. PROBLEM. The examples described in §42 come from [**Klingler-Levy 95**, §3] as a consequence of their Theorem 2.13 which shows that the category $\mathcal{F}(R)$ of R-modules of finite length can have wild representation type. Is it true, for an HNP ring R, that either $\mathcal{F}(R)$ has wild representation type, or else the isomorphism classes in it can be classified? i.e. Is there a 'tame-wild theorem' for modules of finite length over HNP rings?

Bibliography

[Anderson and Fuller 92] F. W. Anderson and K. R. Fuller, *Rings and Categories of Modules, Second Edition*, Springer, New York, 1992.

[Asano 39] K. Asano, "Arithmetische Idealtheorie in nichtkommutativen Ringen", *Jap. J. Math.* **16** (1939), pp.1–36.

[Asano 50] K. Asano, "Zur Arithmetik in Schiefringen. II.", *J. Inst. Polytech. Osaka City Univ. Ser. A. Math.* **1** (1950), pp. 1–27.

[Bass 63] H. Bass, "Big projective modules are free", *Israel J. Math.* **7** (1963), pp. 24–31.

[Canning-Holland 94] R. C. Cannings and M. P. Holland, "Right ideals of rings of differential operators", *J. Algebra* 167 (1994), pp. 116–141.

[Cartan-Eilenberg 56] H. Cartan and S. Eilenberg, *Homological Algebra*, Princeton University Press, Princeton, 1956.

[Chatters 72] A. W. Chatters, "A decomposition theorem for Noetherian hereditary rings", *Bull. London Math. Soc.* **4** (1972), pp.125–126.

[Chevalley 36] C. Chevalley, *L'arithmetic dans les algébres de matrices,* Act. Sci. Ind. **323**, Hermann, Paris, 1936.

[Claborn 66] L. Claborn, "Every abelian group is a class group", *Pacific J. Math.* **18** (1966), pp. 219–222.

[Cohn 77] P. M. Cohn, *Skew field constructions,* London Mathematical Society Lecture Note Series **27** Cambridge University Press, Cambridge-New York-Melbourne, 1977.

[Cohn 91] P. M. Cohn, *Algebra. Vol. 3. Second edition.* John Wiley and Sons, Ltd., Chichester, 1991.

[Cozzens 70] J. H. Cozzens, "Homological properties of the ring of differential polynomials", *Bull. Amer. Math. Soc.* **76** (1970), pp. 75–79.

[Curtis-Reiner 81] C. W. Curtis and I. Reiner, *Methods of Representation Theory* Vol 1, Wiley-Interscience. John Wiley and Sons, Inc., New York, 1981.

[Eisenbud-Griffith 71] D. Eisenbud and P. Griffith, "Serial rings" *J. Algebra* **17** (1971), pp. 389–400.

[Eisenbud-Robson 70a] D. Eisenbud and J. C. Robson, "Modules over Dedekind prime rings", *J. Algebra* **16** (1970), pp. 67–85.

[Eisenbud-Robson 70b] D. Eisenbud and J. C. Robson, "Hereditary Noetherian prime rings", *J. Algebra* **16** (1970), pp. 86–104.

[Ely 74] R. E. Ely, "Multiple idealizers and hereditary Noetherian prime rings", *J. London. Math. Soc.* **7** (1974), pp. 673–680.

[Facchini 75] A. Facchini, *Module Theory: Endomorphism Rings and Direct Sum Decompositions in some classes of Modules,* Birkhaüser, Boston, 1975.

[Fitting 35] H. Fitting,"Primarkomponentenzerlegung in nichtkommutativen Ringen", *Math. Ann.* **111** (1935), pp. 19-41.

[Goodearl 73] K. R. Goodearl, "Idealizers and nonsingular rings", *Pac. J. Math.* **48** (1973), pp. 395–402.

[Goodearl 74] K. R. Goodearl, "Localization and splitting in hereditary Noetherian prime rings", *Pac. J. Math.* **53** (1974), pp. 137–151.

[Goodearl-Warfield 79] K. R. Goodearl and R. B. Warfield, Jr., "Simple modules over hereditary Noetherian prime rings", *J. Algebra* **57** (1979), pp. 82–100.

[Goodearl-Warfield 89] K. Goodearl and R. Warfield, Jr., *An Introduction to Noncommutative Rings,* Cambridge Univ. Press, Cambridge, 1989.

[Griffith-Robson 70] P. Griffith and J. C. Robson, "A theorem of Asano and Michler", *Proc. Amer. Math. Soc.* **24** (1970), pp. 837–838.

[Guralnick 81] R. M. Guralnick, "Roth's theorems and decomposition of modules", *Linear Algebra Appl.***39** (1981), pp.155–165.

[Harada 64] Manabu Harada, "On generalization of Asano's maximal orders in a ring", *Osaka J. Math.* **1** (1964), pp. 61–68.

[Harada 66] Manabu Harada, "Hereditary semi-primary rings and tri-angular matrix rings", *Nagoya Math. J.* **27** (1966), pp. 463–484.

[Hart-Robson 70] R. Hart and J. C. Robson, "Simple rings and rings Morita equivalent to Ore domains", *Proc. London Math. Soc.* **21** (1970), pp. 232–242.

[Hodges 87] T. J. Hodges, "K-theory and right ideal class groups for HNP rings", *Trans. Amer. Math. Soc.* **302** (1987), pp. 751–767.

[Jacobinski 71] H. Jacobinski, "Two remarks about hereditary orders", *Proc. Amer. Math. Soc.* **28** (1971), pp. 1–8.

[Jacobson 43] N. Jacobson, *The Theory of Rings*, Amer. Math. Soc. Math. Surveys 1, New York, 1943.

[Jacobson 80] N. Jacobson, *Basic algebra. II.* Second edition, W. H. Freeman and Company, New York, 1989.

[Kaplansky 52] I. Kaplansky, "Modules over Dedekind rings and valuation rings", *Trans. Amer. Math. Soc.* **72** (1952), pp. 327–340.

[Klingler-Levy 95] L. Klingler and L. S. Levy, "Wild torsion modules over Weyl algebras, and general torsion modules over HNPs", *J. Algebra* **172** (1995), pp. 273–300.

[Knebusch 67] M. Knebusch, "Elementarteilertheorie über Maximalordnungen", *J. für die Reine u. Angewandte Math.* **226** (1967), pp. 175–183.

[Koifman 70] L. A. Koifman, "Rings over which each module has a maximal submodule" (Russian), *Mat. Zametki* **7** (1970), pp. 359–367.

[Kupisch 74] H. Kupisch, "Moduln über einreihigen ringen", *Math. Z.* **137** (1974), pp. 151–153.

[Kuzmanovitch 72] J. J. Kuzmanovitch, "Localizations of HNP rings", *Trans. Amer. Math. Soc.* **173** (1972), pp. 137–157

[Lenagan 71] T. H. Lenagan, "Bounded Asano orders are hereditary" *Bull. London Math. Soc* **3** (1971), pp. 67–69.

[Lenagan 73] T. H. Lenagan, "Bounded hereditary Noetherian prime rings", *J. London Math. Soc. (2)* **6** (1973), pp. 241–246.

[Letzter-Makar-Limonov 90] G. Letzter and L. Makar-Limanov, "Rings of differential operators over rational affine curves", *Bull. Soc. Math. France* **118** (1990), pp. 193–209.

[Levy-Robson 74] L. S. Levy and J. C. Robson, "Matrices and pairs of modules", *J. Algebra* **29** (1974), pp. 427–454.

[Levy-Robson 99a] L. S. Levy and J. C. Robson, "Hereditary Noetherian prime rings 1: Integrality and simple modules", *J. Algebra* **218** (1999), pp. 307–337.

[Levy-Robson 99b] L. S. Levy and J. C. Robson, "Hereditary Noetherian prime rings 2: Finitely generated projective modules", *J. Algebra* **218** (1999), pp. 338–372.

[Levy-Robson 00] L. S. Levy and J. C. Robson, "Hereditary Noetherian prime rings 3: Infinitely generated projective modules", *J. Algebra* **225** (2000), pp. 275–298.

[Levy-Robson-Stafford 94] L. S. Levy, J.C.Robson and J. T. Stafford, "Hidden matrices", *Proc. London Math. Soc.* **69** (1994), pp. 277-308.

[Mac Lane 63] S. Mac Lane, *Homology,* Academic Press, Berlin, 1963.

[McConnell-Robson 73] J. C. McConnell and J. C. Robson, "Homomorphisms and extensions over certain differential polynomial rings", *J. Algebra* **26** (1973), pp. 319–342.

[McR 01] J. C. McConnell and J. C. Robson, *Noncommutative Noetherian Rings,* revised edition, Graduate Studies in Mathematics, **30**, American Mathematical Society, Providence, RI , 2001.

[Michler 69] G. O. Michler, "Asano orders", *Proc. London Math. Soc. (3)* **19** (1969), pp. 421–443.

[Miyata 67] Takehiko Miyata, "Note on direct summands of modules", *J. Math. Kyoto Univ.* **7** (1967) pp. 65–69.

[Nakayama 41] Tadasi Nakayama, "On Frobeniusean algebras. II." *Ann. of Math.* **(2) 42**, (1941), pp. 1–21.

[Odenthal 89] C. J. Odenthal, "Presentations over HNP rings with enough invertible ideals and torsionfree cancellation over neoclassical orders". *J. Algebra* **127** (1989), pp. 290–319.

[Ore 32] O. Ore, "Formal Theorie der linearen differential Gleichungen", *J. Reine Angew. Math.* **168** (1932), pp. 233–252.

[Puninski 01] G. Puninski, *Serial rings*, Kluwer, Dordrecht, 2001.

[Reiner 75] I. Reiner, *Maximal Orders*, Academic Press, New York, 1975.

[Robson-Small 74] J. C. Robson and L. W. Small, "Hereditary prime P.I. rings are classical hereditary orders", *J. London Math. Soc. (2)* **8** (1974), pp. 499–503.

[Robson 68] J. C. Robson, " Non-commutative Dedekind rings", *J. Algebra* **9** (1968), pp. 249-265.

[Robson 72] J. C. Robson, "Idealizers and hereditary Noetherian prime rings", *J. Algebra* **22** (1972), pp. 45–81.

[Robson 72b] J. C. Robson, "Idealizer rings", *Ring theory (Proc. Conf., Park City, Utah, 1971)*, Academic Press, New York, 1972, pp. 309–317.

[Robson 75] J. C. Robson, "The coincidence of idealizer subrings", *J. London. Math. Soc.***10** (1975), pp. 338–348.

[Rotman 09] J. Rotman, *An Introduction to Homological Algebra, Second Edition,* Springer, New York, 2009.

[Silver 67] L. Silver, "Non-commutative localizations and applications", *J. Algebra* **7** (1967), pp. 44–76.

[H. J. S. Smith 1861] H. J. S. Smith, "On systems of linear indeterminate equations and congruences", *Philos. Trans. Roy. Soc. London* **151** (1861), pp. 293–326.

[H. J. S. Smith 1873] H. J. S. Smith, "Arithmetical notes", *Proc. London Math. Soc.* **4** (1873), pp. 236–253.

[Smith 81] S. P. Smith, "An example of a ring Morita equivalent to the Weyl algebra A_1" *J. Algebra* **73** (1981), pp. 552–555.

[Stafford 81] J. T. Stafford, "Generating modules efficiently: algebraic K-theory for noncommutative Noetherian rings" *J. Algebra* **69** (1981), pp. 312-346.

[Stafford-Warfield 84] J. T. Stafford and R. B. Warfield, Jr., "Hereditary orders with infinitely many idempotent ideals", *J. Pure and Applied Algebra* **31** (1984), pp. 217–225.

[Stafford-Warfield 85] J. T. Stafford and R. B. Warfield, Jr., "Constructions of hereditary Noetherian rings and simple rings", *Proc. London Math. Soc.* **51** (1985), pp. 1–20.

[Steinitz 1911] E. Steinitz, "Rechteckige Systeme und Moduln in algebraischen Zahlkörpern" I, II, *Math. Ann.* **71** (1911), pp. 328–354, **72** (1912), pp. 297–345.

[Warfield 75] R. B. Warfield, Jr., "Serial Rings and Finitely Presented Modules", *J. Algebra* **37** (1975), pp. 187–222.

[Warfield 80] R. B. Warfield, Jr., "The number of generators of a module over a fully bounded ring", *J. Algebra* **66** (1980) pp. 425–447

[Webber 70] D. B. Webber, "Ideals and modules of simple Noetherian hereditary rings", *J. Algebra* **16** (1970), pp. 239–242.

[Weibel 94] C. A. Weibel, *An introduction to homological algebra,* Cambridge University Press, Cambridge, 1994.

[Zaks 74] A. Zaks, Hereditary Noetherian rings, *J. Algebra* **29** (1974), pp. 513-527.

Index of Symbols

Index of Terminology